Magnetic Quantum Dots for Bioimaging

Bioimaging is a sophisticated, non-invasive, and non-destructive technique for the direct visualization of biological processes. Highly luminescent quantum dots combined with magnetic nanoparticles or ions form an exciting class of new materials for bioimaging. These materials can be prepared in cost-effective ways and show unique optical behaviors. *Magnetic Quantum Dots for Bioimaging* explores leading research in the fabrication, characterization, properties, and application of magnetic quantum dots in bioimaging.

- Covers synthesis, properties, and bioimaging techniques
- Discusses modern manufacturing technologies and purification of magnetic quantum dots
- Explores thoroughly the properties and extent of magnetization to various imaging techniques
- Describes the biocompatibility, suitability, and toxic effects of magnetic quantum dots
- Reviews recent innovations, applications, opportunities, and future directions in magnetic quantum dots and their surface-decorated nanomaterials

This comprehensive reference offers a road map of the use of these innovative materials for researchers, academics, technologists, and advanced students working in materials engineering and sensor technology.

Magnetic Quantum Dots for Bioimaging

Edited by
Amin Reza Rajabzadeh
Seshasai Srinivasan
Poushali Das
Sayan Ganguly

CRC Press is an imprint of the
Taylor & Francis Group, an **informa** business

First edition published 2023
by CRC Press
6000 Broken Sound Parkway NW, Suite 300, Boca Raton, FL 33487-2742

and by CRC Press
4 Park Square, Milton Park, Abingdon, Oxon, OX14 4RN

CRC Press is an imprint of Taylor & Francis Group, LLC

ISBN: 978-1-032-32146-2 (hbk)
ISBN: 978-1-032-33491-2 (pbk)
ISBN: 978-1-003-31987-0 (ebk)

DOI: 10.1201/9781003319870

Typeset in Times
by Newgen Publishing UK

Contents

Editors

Amin Reza Rajabzadeh, PhD, is Associate Professor and former Chair of the Biotechnology Program at the School of Engineering Practice and Technology, McMaster University, Canada. Dr. Rajabzadeh specializes in the field of biochemical engineering with a focus on biosensors, bioseparation and purification, bioprocess monitoring and control, bioreactor design, and environmental engineering. Dr. Rajabzadeh is a Professional Engineer of Ontario and is a member of the Canadian Engineering Education Association and the American Society for Engineering Education. Dr. Rajabzadeh was a MacPherson Leadership in Teaching and Learning (LTL) Fellow from 2017 to 2019.

Seshasai Srinivasan, PhD, is Chair of the Software Engineering Technology program at McMaster University's Faculty of Engineering. Prior to this, he held a Research Scientist and a part-time Instructor position in the Department of Mechanical and Industrial Engineering of Ryerson University, a postdoctoral position at the Laboratory of Food Process Engineering of the Swiss Federal Institute of Technology (ETH-Zurich) in Switzerland and a Research Associate position in the Engine Research Center of the University of Wisconsin–Madison. He maintains a vibrant interdisciplinary research program in the areas of biosensors, quantum dots, Li-ion batteries, food rheology, and thermodiffusive flows.

Poushali Das, PhD, is currently a senior postdoctoral research fellow at McMaster University, Canada. Before that she was at Bar-Ilan University, Israel. She earned a PhD in 2019 at the Indian Institute of Technology, Kharagpur, India. Her research interests include multifunctional luminescent quantum dots and applications in sensors, antioxidant properties and the biomedical field, polymer/quantum dot nanocomposites, and MXene/polymer nanocomposites sonochemical synthesis of graphene-based nanocomposites.

Sayan Ganguly, PhD, is currently a senior postdoctoral researcher at University of Waterloo, Canada. Before that he was at Bar-Ilan University, Israel. He earned a PhD at the Indian Institute of Technology, Kharagpur, India. His primary research interests include superabsorbent hydrogels, composite hydrogels, polymer-graphene nanocomposites, MXene-polymer systems, polymer composites for EMI shielding, and conducting polymer composites.

Contributors

Kapil Agrawal
R. C. Patel College of Pharmacy, Shirpur
Shirpur, Maharashtra, India

Sarfaraz Ahmed
Department of Pharmaceutical Chemistry
Global Institute of Pharmaceutical Education and Research
Baksaura, Uttarakhand, India

Himani Bajaj
Adarsh Vijendra Institute of Pharmaceutical Sciences (AVIPS)
Shobhit University
Gangoh, Saharanpur, Uttar Pradesh, India

Varsha Brahmkhatri
Centre for Nano and Material Sciences
Jain University
Bengaluru Karnataka, India

Nurhan Onar Çamlıbel
Textile Engineering Department
Pamukkale University
Denizli, Turkey

Neha Chauhan
Laxminarayan Dev College of Pharmacy
Bharuch, Gujarat, India

Disha Dutta
Devsthali Vidyapeeth College of Pharmacy
Rudrapur, Uttarakhand, India

Mohd Vaseem Fateh
Department of Pharmacy
Sam Higginbottom University of Agriculture, Technology and Sciences (SHUATS), Naini
Naini, Allahabad, Uttar Pradesh, India

Sayan Ganguly
Department of Chemistry
and
Bar-Ilan Institute for Nanotechnology and Advanced Materials
Bar-Ilan University
Ramat Gan, Israel

Bina Gidwani
Columbia Institute of Pharmacy
Raipur, India

Sanjay Kumar Gupta
Shri Rawatpura Sarkar Institute of Pharmacy
Kumhari, Durg, Chhattisgarh, India

Fahadul Islam
Department of Pharmacy
Faculty of Allied Health Sciences
Daffodil International University
Dhaka, Bangladesh

Akhlesh K Jain
SLT Institute of Pharmaceutical Sciences
Guru Ghasidas Vishwavidyalaya, Bilaspur, Chhattisgarh, India

Vishal Jain
University Institute of Pharmacy
Pt. Ravishankar Shukla University
Raipur, Chhattisgarh, India

Varsha Lisa John
Department of Chemistry
CHRIST (Deemed to be University)
Bangalore, India

Veenu Joshi
Center for Basic Science
Pt. Ravishankar Shukla University
Raipur, India

Sharuk L. Khan
Department of Pharmaceutical Chemistry
N.B.S. Institute of Pharmacy
Ausa, Maharashtra, India

Parveen Kumar
Exigo Recycling Pvt Ltd
Noida, Uttar Pradesh, India

Preeti Kush
Adarsh Vijendra Institute of Pharmaceutical Sciences
Shobhit University Gangoh
Saharanpur, Uttar Pradesh, India

Vivek Mewada
Department of Pharmaceutics
Institute of Pharmacy
Nirma University
Ahmedabad, India

Gopi Krishna Moku
Department of Physical Sciences
Kakatiya Institute of Technology and Science
Warangal, Telangana, India

Prashant Subhash Palghadmal
SND College of Pharmacy Yeola Babhulgaon
Nashik, Maharashtra, India

Ravindra Kumar Pandey
Columbia Institute of Pharmacy
Raipur, India

Navin Chandra Pant
Six Sigma Institute of Technology and Science
Rudrapur, Uttarakhand, India

Vimal Patel
Department of Pharmaceutics
Institute of Pharmacy
Nirma University
Ahmedabad, India

Nilesh S. Patil
RD & SH National College & SWA Science College
Bandra, Maharashtra, India

Pranita Rananaware
Centre for Nano and Material Sciences
Jain University
Bengaluru, Karnataka, India

Dipti Rawat
Nanotechnology Laboratory
Department of Physics and Materials Science
Jaypee University of Information Technology
Waknaghat, Solan, India

VJ Reddy
Department of Physical Sciences
Kakatiya Institute of Technology and Science
Yerragattu Gutta, Warangal, Telangana, India

Varsha Sahu
Department of Pharmaceutical Sciences
Utkal University
Bhubaneshwar, Odhisa, India

Hiral Shah
Department of Pharmaceutics
Arihant School of Pharmacy & BRI
Gandhinagar, India

Jigar Shah
Department of Pharmaceutics
Institute of Pharmacy
Nirma University
Ahmedabad, India

Shiv Shankar Shukla
Columbia Institute of Pharmacy
Raipur, India

Falak A. Siddiqui
Department of Pharmaceutical Chemistry
N.B.S. Institute of Pharmacy
Ausa, Maharashtra, India

Ragini Raj Singh
Nanotechnology Laboratory
Department of Physics and Materials Science
Jaypee University of Information Technology
Waknaghat, Solan, India

Ranjit Singh
Adarsh Vijendra Institute of Pharmaceutical Sciences
Shobhit University Gangoh
Saharanpur, Uttar Pradesh, India

Rokeya Sultana
Yenepoya Pharmacy College and Research Centre
Yenepoya (Deemed to be University)
Deralakatte, Mangalore, India

Vinod T. P.
Department of Chemistry
CHRIST (Deemed to be University)
Bangalore, India

Poonam Talwan
Himachal Institute of Pharmaceutical Education and Research
Nadaun, Himachal Pradesh, India

Md. Rageeb Md. Usman
Department of Pharmacognosy
Smt. Sharadchandrika Suresh Patil College of Pharmacy
Chopda, Maharashtra, India

Swathi Vangala
Telangana Social Welfare Residential Degree College for Women
Bhupalapally, Telangana, India

Astha Verma
Shri Rawatpura Sarkar Institute of Pharmacy
Kumhari, Durg, Chhattisgarh, India

Amber Vyas
University Institute of Pharmacy
Pt. Ravishankar Shukla University
Raipur, Chhattisgarh, India

Mayank Yadav
AVIPS
Shobhit University
Gangoh, Saharanpur, Uttar Pradesh, India

Venu Yakati
Department of Chemical and Biological Engineering
University of Alabama
Tuscaloosa, Alabama, USA

1 Introduction to Magnetic Quantum Dots

Bina Gidwani[1], Varsha Sahu[2], Veenu Joshi[3], Shiv Shankar Shukla[1], Ravindra Kumar Pandey[1], Akhlesh K Jain[5], and Amber Vyas[4]*
[1]Columbia Institute of Pharmacy, Raipur, India
[2]Department of Pharmaceutical Sciences, Utkal University, Bhubaneshwar, Odhisa, India
[3]Center for Basic Science, Pt. Ravishankar Shukla University, Raipur, India
[4]University Institute of Pharmacy, Pt. Ravishankar Shukla University, Raipur, India
[5]SLT Institute of Pharmaceutical Sciences, Guru Ghasidas Vishwavidyalaya, Bilaspur, Chhattisgarh, India

CONTENTS

1.1 INTRODUCTION

One of the most significant fields of science is nanotechnology, which operates at the cellular and molecular level and contributes to significant advancements in engineering, biology, and medicine [1]. In nanotechnology different types of nanoscale materials are available, which possess specific optical, electrical, catalytic, and

DOI: 10.1201/9781003319870-1

magnetic properties, which are different from bulk material and single molecules. In the nanotechnology field, some materials, such as organic, inorganic or semiconductor materials, are used in the form of rods, wire, and cube, and particles are available, which are worked as nanoscale tools. These wide ranges of nano tools are used in electronic devices, biomedicine, food industries, and forensic science, etc. In the biomedical field, nanoscale device or nanoparticles are already used for diagnosis, treatment, drug delivery, genetic, molecular imaging, and gene therapy [2, 3].

Recently, biomedical imaging has gain vast areas of research, from diagnosis to treatment and various clinical applications. This imaging involves molecular imaging, in-vitro and in-vivo imaging, etc. This imaging technique includes optical imaging, spectroscopic imaging, and functional imaging [4]. For diagnosis and treatment of diseases, it is essential to detect the specific target and track the targets in living cells, such as at the sub-cellular level, which play an important role in biological detection and investigation. In recent eras, various imaging probes or modalities are available and used for clinical diagnosis, treatment, animal imaging, preclinical imaging, and molecular imaging, such as magnetic resonance imaging (MRI), positron emission tomography (PET), computed tomography (CT), ultrasound (US), optical imaging, etc. [5, 6, 7]. These imaging probes are also known as biomarkers, which help to gets the image of the particular targeting pathway. For the molecular imaging purposes, it is essential to have the biomarker of imaging agent interact with surrounding environment and cause a molecular change, which gives the imaging of that particular area [8]. Some of the imaging agents are referred to as endogenous molecules or exogenous probes, which are used to visualize, characterize, and quantify the biological system. Imaging techniques or imaging agents were classified according to sensitivity, complexity, and resolution [3]. The different types of imaging probes are shown in Figure 1.1.

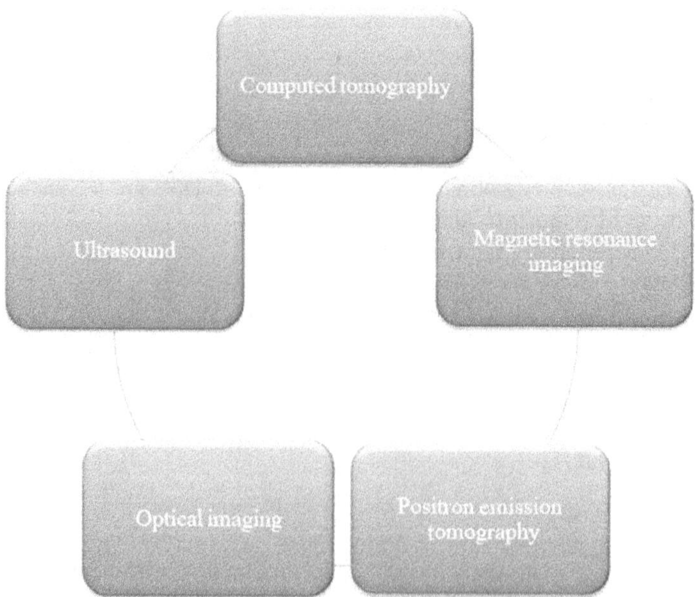

FIGURE 1.1

1.2 TYPES OF IMAGING PROBES

Magnetic resonance imaging (MRI): It is an imaging method that employs radio waves and a strong magnetic field to image different human and animal organs [30]. It is the noninvasive diagnostic method that is most frequently employed. High spatial resolution, sensitivity, and specificity are all features of MRI images. Some exogenous contrast agents were necessary for the MRI approach [13]. The common contrast agents utilized in MRI technology include super paramagnetic iron oxide nanoparticles like Fe_3O_4, FeCO, and $MnFe_2O_4$; paramagnetic complexes like Ga^{3+} or Mn^{2+} based chelates; and paramagnetic ion containing nanoparticles like Gd_2O_{3-}MnO. Tissue contrast, molecular imaging, and morphological and functional estimation are all possible with MRI imaging techniques [7, 10].

Ultrasound (US): The most popular imaging agents are both highly safe and inexpensive. Perfusion imaging and lesion or region characterization are also possible uses for ultrasound. Contrast chemicals can interact with the ultrasonic beam in ultrasonography [7, 8]. Common contrast materials for ultrasonography include colloidal suspension, perfluorocarbon emulsion, and gas-filled micro bubbles. Contrast compounds typically have features that make them suitable for vascular circulation, biodegradability, and imaging [10].

Optical imaging (OI): This is the most widely used, least expensive, and non-invasive method. This method makes use of light as a radiation source. Fluorescence and bioluminescence imaging are two optical imaging techniques that are used to examine the transmission of non-ionizing radiation or light photons from tissue [8]. Several proteic fluorophores, including organic or inorganic dyes and green or red fluorescent proteins, are used as imaging agents. In addition to this, iron oxide nanoparticles with a fluorescent characteristic and quantum dots were employed for in-vitro and in-vivo imaging [10].

Computed tomography (CT): It delivers non-invasive medical imaging using X-ray scanning, which uses tissue reflectance to create a computed image that may be used to gain information about the body in the form of vascular and morphological images [7, 8]. In the body, diverse compositions like fat, water, and tissue exist, which have different X-ray absorption rates. Therefore, this class of imaging agents can image the anatomical structure of the lungs, bones, and tumor cells. Iodine, gastrografin, and barium salt are a few imaging agents employed in computed tomography for clinical purposes [7, 10].

Positron emission tomography (PET): It is a subset of radionuclide imaging techniques that uses radioactive materials as radiotracers to detect changes in metabolic events and build images. PET makes use of radio nuclides such as 16F, 64Cu, and 68Ga. For instance, 2-[18F]fluoro-2-deoxy-D-glucose, sometimes known as (18F) FDG, has been a popular PET imaging probe for the detection and treatment of cancer [8]. The use of PET permits the localization and analysis of activity measurements in a certain organ or tissue. In PET, radioactive material is combined with a medication and administered intravenously; the gamma rays that are released are then detected, creating a three-dimensional image. The major drawback of this technique is that it has poor spatial resolution and no limitation of tissue penetration, which limits its use in clinical application [7].

Every imaging probe or technique has its own advantages and limitations. Among all the imaging techniques, there is no one perfect technique for imaging that fulfills all the requirements for medical imaging [3]. To overcome this problem, hybrid or conjugated imaging probes are used, such as PET-CT, PET-MRI, ultrasound –MRI, and optical-CT, etc. [5, 7, 8]. This conjugation provides synergistic effects of each other's techniques. But the integration of two technique are a little difficult. To overcome this problem, nanoparticles are used for bioimaging [8, 11]. Nanoparticles have high surface area and provide various surface functionalization or modifications, which produce multifunctional nanoparticles for bioimaging. Recently various functionalized nanoparticles are produced that are widely applicable for imaging [11].

1.3 NANOPARTICLES AND THEIR TYPES

Nanoparticles have great potential for biomedical science because of their unique size and physicochemical properties. It is reported that the sizes of nanoparticles range from 1-100nm, but sometimes they exceed 1-1000nm. They have a large effective surface area and inherent properties such as size and structure. Changes in size, shape, structure, and surface modification change their pharmacokinetic properties; increase blood circulation time; provide controlled, sustained, or targeted release; and enhance biodistribution and bioavailability. They also enhance the permeability of drugs through carriers and cause the accumulation of drugs at target sites [8]. In the pharmaceutical field, various types of nanocarriers or nanoparticles are available, such as polymeric nanoparticles, inorganic nanoparticles, and lipid nanoparticles. In all types of nanoparticles, there are certain advantages or disadvantages. The selection of nanoparticles for the treatment and diagnosis of diseases are done according to their application, type of disease, route of administration, advantages, and limitations [1]. The conjugation of biomaterial, such as protein, peptide, nucleic acid, and antibodies with nanoparticles, creates great advances in bioengineering, medicine, molecular biology, diagnostics, and therapy. Recent progress of nanotechnology in the field of biomedicine include the electronic, optical, and magnetic properties' modification of functional nanoparticles [2, 12]. The types of nanoparticles are shown in Figure 1.2.

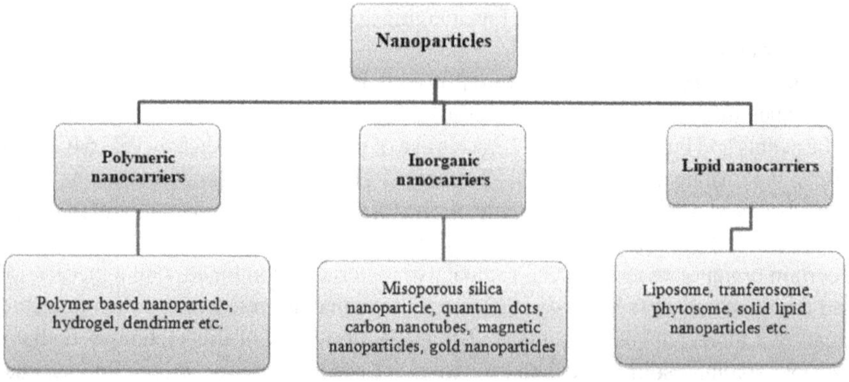

FIGURE 1.2

Recently in all these type of nanoparticles, inorganic nanoparticles are widely used because they are the only nanoparticles which are mostly used for diagnosis, molecular imaging, as well as treatment. They provide great advantages over other nanoparticles for bioimaging and diagnosis. Inorganic nanoparticles have great potential towards optical and magnetic properties, which have gained much attention in the past few years [13]. In inorganic nanoparticles, there are various nanoparticles present, such as gold nanoparticles, silver nanoparticles, quantum dots, metallic nanoparticles, magnetic nanoparticles, and zinc oxide nanoparticles etc. The combination of various nanoparticles, such as gold NP, silver NP, quantum dots, and liposomes, shows a new class of nanocomposition, which is widely applicable for diagnosis as well as treatment [8].

1.4 QUANTUM DOTS

Over the past few years, quantum dots and magnetic nanoparticles separately have been highly recommended for imaging, sensing, drug delivery, and diagnosis probes. Quantum dots (QDs) have special optical properties, such as sharp emission and broad absorption spectra, which are governed by the quantum confinement effect and have great importance in lasers, transistors, QD glasses, bioimaging, and biolabelling, etc. [14, 15]. Magnetic nanoparticles have great magnetism properties which are widely applicable for drug targeting, MRI contrast enhancement, magnetic separation, and magnetic immobilization [14].

Some basic nanoparticles often show undesired properties in biological systems. These are either hydrophobic or hydrophilic or show agglomeration and oxidation in particular environments. For increasing their stability and biocompatibility, plain nanoparticles are modified. In the case of magnetic nanoparticles, the main concern is their super para-magnetism property. When they conjugate with biological systems, their properties might change. This occurs because of their high reactivity and surface energy [16]. And for quantum dots, single QDs are insufficient for some in-vivo bioanalysis and biolocalization studies. Nanoparticles have small sizes and large surface areas, which give various advantages for modification. This modification can be done by either the addition of polymer, surfactant, or functional groups or conjugation with other nanoparticles which improves their physicochemical property and gives an advantage of synergistic effects or dual properties in one single nanoparticle [16].

1.5 MAGNETIC QUANTUM DOTS

Magnetic quantum dots (MQDs) are a new class of nano material that is a combination of both semiconductor quantum dots and magnetic nanoparticles in a single composite provide optical and magnetic property in a single material which is advantageous to bioimaging, bioseparation, and multimodal imaging [18]. In previous various years magnetic quantum dots used under the name of "magnetic fluorescence nanoparticles." In this composition magnetic compound and organic fluorophore dyes are used. But fluorophore dyes have some limitations, such as photodegradation, when subjected to fluorescence studies; therefore quantum dots are used for the preparation of magnetic fluorescence composites. The simultaneous delivery of magnetic agents and fluorescent particles to individual cell offers the optical imaging and MRI [19]. Magnetic quantum dots provide the multifunctional drug-loaded magnetic nanoparticles, which offer an increased drug transport rate, mucus penetration, cellular imaging, and antibiotic efficiency [20].

Quantum dots (QDs): In older days fluorescent organic dyes such as rhodamine, fluorescence, isothiocynate, and pyrene were used for diagnosis, imaging, detection, and sensing, but fluorescent molecules have some disadvantages, such as a lack of photostability, broad and asymmetric emission spectra, photo bleaching, and pH dependency, which hide their multiple abilities [21, 27]. Among all the tracing agents or nanoparticles, QDs are fastest growing and widely used nanoparticles in diagnostic, biomolecular imaging and cellular imaging, drug delivery, and theranostic approaches etc. [22]. QDs provide great offerings over traditional organic dyes, such as unique optical properties, broad excitation spectra, narrow emission bandwidth, and great photostability [23].

It is reported that the first biomedical application of QDs was as a fluorescent biological label in 1990s, then continuous research occurred on the fluorescent property of QDs in biological systems. Numerous studies were performed, and continuous evolution of QDs occurred, such as in-vivo imaging, cancer targeting, and imaging, as biosensors, etc. [25].

Properties of quantum dots: They have unique properties which make them suitable for bioimaging such as:

- Broad absorption
- High quantum yield
- Symmetric photoluminescence spectra
- High resistance to photo bleaching [25]

QDs are semiconductor nanoparticles, with confined three-dimension. The size ranges from 1-10nm in diameter which shows the quantum confinement during light excitation [15, 20]. It consists of a few hundred to a few thousand atoms. The properties of QDs depend on the size and fluorescence of the QDs; therefore the fluorescent signal can be accurately tuned. Quantum dots are available in different sizes: QDs of less than 1nm can contain approximately fewer than 100 atoms, or the size ranges from more than 200nm containing 10,000 atoms. The small size allows the luminescence quantum efficiencies that are much higher than for bulk semiconductor. Most of the QDs are composed of atoms from group II and IV, or group III and V, or group IV and VI of the periodic table [27]. The most used QDs were of a CdSe core with a shell of ZnS [5, 2]. The fluorescence emission wavelength properties of QDs also depend on the size because of their quantum confinement in which the electronic state of bulk material changes to QDs [16]. In bulk semiconductor material, an exciton is bound with precise length which is called exciton Bohr radius. The properties of QDs are changed according to their size when the excitons are constrained due to their size. Visible quantum dots emit the fluorescence in the range of 400-700nm, while NIR quantum dots emit the fluorescence in the range of 700-900nm. From both types of quantum dots, NIR quantum dots have gain much interest in the biomedical field because at this range, the tissue absorbance and auto fluorescence are minimal which causes a low background noise of quantum dots [5, 2, 16].

Different quantum dot types, such as the core-shell type and alloyed type, which are categorized according to their makeup, are utilized for biological imaging. Quantum dots that have a core and shell made of inorganic materials like PbS, CdTe, or CdSe exhibit greater band gap energies surrounding the core. They are brighter and more intense. Different band gap energies exist in alloyed quantum dots [16].

High photostability, high emission quantum yield, narrow emission peak, size dependent wavelength, water solubility, particle size dependent band gape, fluorescence, and tunability are only a few of the characteristics of quantum dots. Quantum dots are useful for studying dynamic cell motility, protein trafficking, and DNA detection due to their unique features [2].

As compared to organic dyes, the inorganic nature of quantum dots has reduced photo bleaching property under the light which allows the real-time monitoring of cell and biological events over a longer extended period of time. It also gives real-time imaging and monitoring of the cell. But for the complex application in bioanalysis and bioimaging, such as in-vivo localization studies, using only optical tracing is insufficient.

The surface chemistry of QDs provides the various functionalizations which improves the pharmacokinetic and biological applicability [22, 24]. The optical and tracking properties of QDs also depend on their surface structure. To provide high optical properties, there are various surface functionalizations available which can increase the photoluminescence and quantum yield and reduce the toxicity of QDs [24]. As an example, ZnS coating over Cd QDs passivates the surface of the quantum dots and lessens the toxicity of Cd-QDs by stopping the leaching of Cd from the QD core. Depending on the need, this core particle can be further customized with additional layer using a variety of chemicals. To achieve a hydrophobic or hydrophilic surface for biomedical applications, core structures can be stabilized by adding an organic layer around the core [12].

1.6 MAGNETIC NANOPARTICLES (MNPs)

Magnetic nanoparticles composed of iron, iron oxide, maghemite, platinum, cobalt, nickel, gadolinium chelates, and some nanomagnetic nanocrystals with paramagnetic metallic ion [16]. Because they are used in biomedical applications such as magnetic resonance imaging (MR), bioseparation, cancer therapy employing NIR lasers, and targeted drug administration, magnetic nanoparticles have previously attracted a lot of attention [5]. Such nanoparticles have a large constant magnetic moment and function as paramagnetic atoms with quick magnetic responses when the size of magnetic nanoparticles is below a crucial value at room temperature [7]. The possibility of coagulation is quite unlikely at room temperature due to the attractive properties of the magnetic nanoparticles, which range in size from 10 to 20 nm. With the help of an external magnetic field, magnetic nanoparticles can target the desired location because of their reduced particle size, high coupling capacity, and high specific surface area [3].

Properties of magnetic nanoparticles: Magnetic nanoparticles has some unique property which make them suitable for biomedical application and imaging:

- They have a large surface area.
- They rapidly separate the target molecule from mixture.
- They can be easily manageable through external magnetic field [34].

Since magnetic nanoparticles have a large surface volume ratio compared to bulk materials, their local environment affects their surface area, which in turn affects their surface magnetic properties. This causes magnetic nanoparticles to behave differently from bulk materials in terms of their magnetism properties. One of the size-dependent characteristics of magnetic nanoparticles is superparamagnetism [7, 23]. However, in magnetic nanoparticles, magnetic anisotropy is less than thermal energy. In bulk materials, magnetic anisotropy is greater than thermal energy. Although this thermal energy is sufficient to reverse the magnetic spin direction, it is insufficient to outweigh the energy of the spin-spin coupling process. Superparamagnetism is a property of nanoparticles that results in zero net magnetization [27].

The blocking temperature is the temperature at which ferromagnetic particles transit to superparamagnetism. Single ferromagnetic and superparamagnetic particles align in the presence of an external magnetic field. While superparamagnetic nanoparticles lose their net magnetization in the absence of an external magnetic field, ferromagnetic nanoparticles do not.

Iron oxide magnetic nanoparticles have low toxicity in magnetic nanoparticles and have been employed in clinical settings without any problems. The magnetic iron oxide nanoparticles known as "Feridex" were given the green light by the US Food and Drug Administration (FDA) for use in clinical settings [23, 27]. Magnetite (Fe3O4), maghemite (-Fe2O3), and hematite (-Fe2O3) are examples of superparamagnetic iron oxide nanoparticles (SPIONs) that are primarily employed as contrast agents for imaging, diagnostics, targeting, and treatment. Fe2+ and other ions in SPION make them appropriate for biological applications [7].

For biomedical purposes the magnetic nanoparticles are preferred which show superparamagnetic behavior at room temperature, and for diagnosis it should be stable in water and physiological environment at pH 7. The colloidal property of magnetic nanoparticles depends on their charges and their surface chemistry. The biocompatibility and toxicity of magnetic nanoparticles are dependent on the nature of material which have magnetic properties (examples: cobalt, nickel, iron, and magnetite) in their core and coating materials [1]. Magnetic nanoparticles, e.g., Fe_3O_4, have great importance in nanomedicine due to their property of being manipulated and detected with external magnetic fields. They can basically be preferred as cell labelling material, but their poor signal intensity limits their use in the biomedical field. They can be widely applicable in MRI, bioseparation, biosensing, cancer therapy, and targeted delivery [5, 16]. The modification of magnetic nanoparticles provides good stability, biocompatibility, and low toxicity which make them suitable for biomedical applications, such as noninvasive magnetic resonance imaging agents, as separating agent for various biological materials and detection of cancerous cells.

1.7 LIMITATIONS OF MNPs AND QUANTUM DOTS IN BIOIMAGING

Magnetic nanoparticles has some major limitations which limit their uses in biomedical and pharmaceutical fields, such as they do not provide long term tracking because their magnetic resonance signal is lost due to cell proliferation, especially in rapidly dividing cells. Therefore to overcome this, cellular internalization methods are widely used [7]. For imaging purposes through magnetic nanoparticles, a simultaneous visualization of multimodal with two or more imaging probes with different spectra is used. Therefore it gives poor spatial resolution, and uses of multiple probes are problematic [20]. In the biomedical imaging field, quantum dots have limits to their uses because of toxicity and their colloidal nature, because they show the hindrance effect which affects their physical and chemical stability. However many surface functionalizations and conjugations are available to address this issue, but they also cause quenching effects [27]. Other limitations of QDs are that some complex applications in bioanalysis and bioimaging, such as in-vivo localization studies through quantum dots or optical tracing, are insufficient.

The combinations of these two nanoparticles with dual natures like optical and magnetic properties on nanoscales give modifications as well as bring new advancements in biomedical science, theranostics, nanoprobes for imaging, and molecular imaging for detection and treatment [5]. On the basis of unique properties of QDs and magnetic nanoparticles, these nanoparticles have potential applications in tumor detection, biomarker enrichment, tumor cell separation, biolabelling, dynamic tracking, MRI imaging, drug delivery, etc. This improves the understanding of cancer which facilitate the diagnosis and treatment of diseases [5, 17].

Magnetic quantum dots (MQDs): Single QDs are insufficient for some in-vivo bioanalysis and biolocalization studies; therefore QDs are conjugate with magnetic ions, which provide a facility by combining the magnetic and optical properties in a single material. This unique conjugation work as a multimodal contrast agent and photostable label which are used in nano medicine and biomedical applications [25]. The combination of two techniques into a single agent can overcome the limitation of conventional techniques. The manipulations of nanoparticles with electromagnetic fields or magnetism work as a power to locate them in particular cell or track them easily in numerous biological molecules [20]. The diagrammatic representation of Functional Magnetic Quantum dots is shown in Figure 1.3.

MQDs show the superparamagnetism and luminescence emission property in a single hybrid composite. Fabrication of MQDs uses the magnetic resonance and fluorescence imaging which give the separation of cancer cell and allow the fluorescence identification of cancer cells that ultimate kill the cancer cell through combination of photodynamic and a photothermal effect. These properties of MQDs allow them to target, separate, and visualize biomolecules and work as contrast agents for MRI [22]. The advantage of MQDs involves an extremely smaller size, size tunable fluorescence, resistance to photodegradation, photostability, and high quantum yield. It also generates the auto fluorescence and minimal absorbance which provide great help in deep tissue non-invasive in-vivo imaging such as tumor targeting, lymph node mapping, biodistribution, etc. MQDs emit the fluorescence in the near infrared ranges

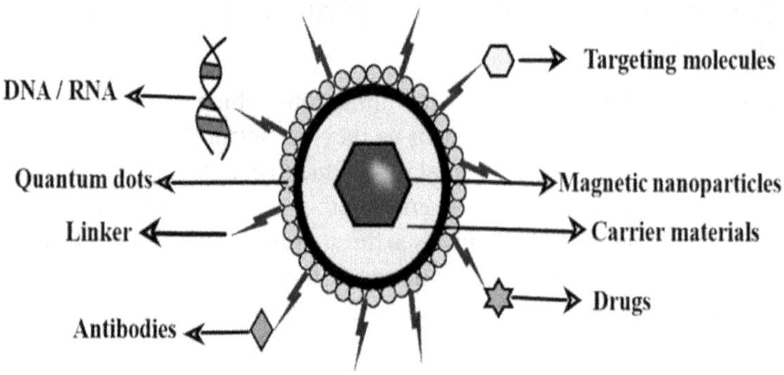

FIGURE 1.3

650-900nm [26]. MQDs work as a multimodal imaging probe and tracking agent for organ tissue by magnetic resonance imaging (MRI) on the cellular and molecular levels by fluorescence imaging.

For the preparation of magnetic quantum dots, magnetic material such as gadolinium in the form of ion and superparamagnetic iron oxide nanoparticles or SPIONs in the form of nanoparticles are used to combine with quantum dots. These SPIONs are composed of iron oxides, either magnetite (Fe_3O_4) or maghemite (Fe_2O_3), whose size is smaller than magnetic crystal domain which shows the paramagnetic property in solution and reduce their potential for aggregation, but it also induced magnetic behavior in the presence of a magnetic field [23]. SPIONs have been widely used in biomedical applications such as cell separation, single molecule manipulation, and magnetic resonance imaging [24]. Therefore a single agent comprising magnetic quantum dots provides multimodal imaging which gives great value in diagnostic information and dynamics of a disease.

MQDs has two attractive features, fluorescence and superparamagnetism, which allow the controlled movement of cell through magnetic force and monitoring using fluorescent microscope. The synergistic potential of magnetic quantum dots provides multi-modality imaging and multi-level targeting which make them suitable for both diagnostic as well as treatment [11, 17]. The magnetic quantum dots conjugate with targeting groups such as antibodies to target the cancer cell, followed by magnetic separation and detection mode, such as magnetic resonance imaging (MRI) and fluorescence analysis. For magnetolytic therapy, magnetic quantum dots conjugate with the therapeutic agents. In magnetolytic therapy, cancer cells were killed by heat generation through magnetism and released drugs which targeted the specific cancerous cell.

Most of the data show that for the preparation of magnetic quantum dots, iron oxide and other ferrites are widely used as compared to other nanoparticles. Some ferromagnetic particles such as cobalt are not used for biomedical purposes because of their toxicity. Nickel elements are also not used because of their toxic effect and low magnetism properties. However various magnetic nanoparticles are fabricated in the form of core shell structure which is widely used in the biomedicine field [2]. This composition is stable in physiological conditions, and it can be easily modified with biomolecules, such as antibodies and drugs.

1.8 TYPES OF MQDs

The magnetic quantum dots are classified according to their integration of magnetic and fluorescent property in one single molecule, that is the formation of magnetic quantum dots. There are four types of magnetic quantum dots according to their architecture mechanism and method of preparation. The types are listed in Figure 1.4.

Type I: Core shell and heterostructure: Type I magnetic quantum dots (shown in Figure 1.5) deal with the hetero structure of core or shell. In this, quantum dots and magnetic nanoparticles are simply fused together to form either core/shell or hetero structures [25, 5]. Their attachment mechanism is not fully discovered yet. The core/shell and hetero structure type of magnetic quantum dots is relatively small, approximately 50nm. This type of magnetic quantum dots are easily varied and prepared, and they have high payloads. They are widely used as contrast agents [19]. For example – through one-pot synthesis, the super paramagnetic cobalt cores were firstly precipitated and then re-dispersed in coordinating solvent before adding the CdSe precursor shell material. These magnetic quantum dots obtained a small core size of 8nm and CdSe shell of 2nm; its fluorescence emission peak is larger than CdSe QDs that is 580nm, and stroke shift is 40-50nm. These magnetic quantum dots modified the spin structure of the lowest excitation state [25]. Other examples of this type of magnetic quantum dot, in these super paramagnetic Fe2O3 magnetic nanoparticles, are attached with CdSe QDs, and this hetero structure present in the size 10nm. It shows both superparamagnetic and fluorescent properties. It is reported that the CdSe or CdS QDs combined with FePt magnetic nanoparticles form magnetic quantum dots. The FePt core has 3nm, and the CdSe shell has 3-5nm diameter,

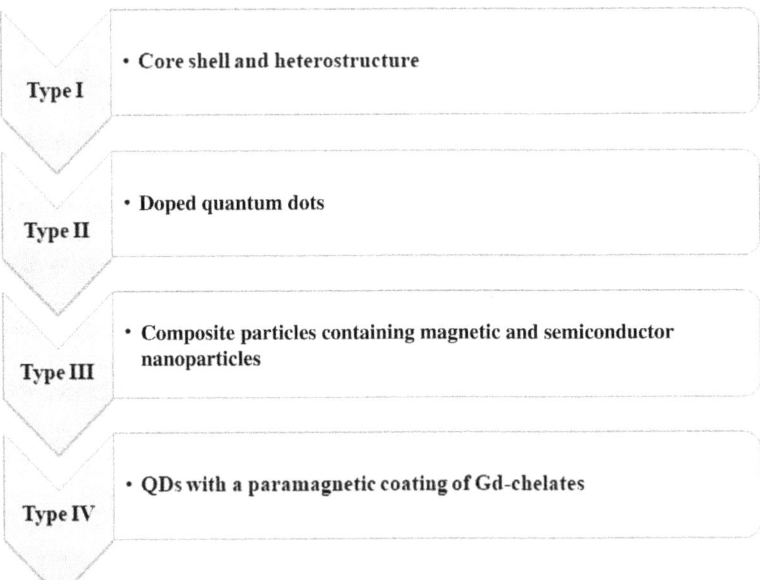

FIGURE 1.4

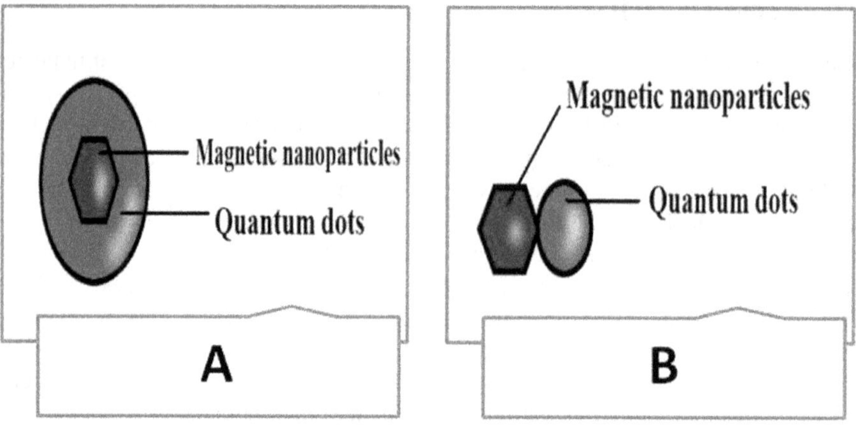

FIGURE 1.5

and their fluorescence emission was found at 465nm [6]. Another example shows the formation of heterodimer type of magnetic quantum dots, which contain CdSe QDs and Fe_2O_3 MNPs, in which Fe_2O_3 is 8-10nm in diameter and QDs size should be vary from 2-5nm. The emission color of QDs varied between 550-650nm. It also exhibited superparamagnetism properties. For the biological purpose, these MQDs are coated with silica shell through a reverse micelle technique. It also allows the attachment of functional group through amide bonds [13].

Type II: Doped quantum dots: These types of MQDs (shown in Figure 1.6) are prepared through the direct incorporation of fluorescence and magnetic properties into single nanoparticles. MQDs are created through the doping of paramagnetic ions into QDs. Early studies show that doping with transition metal ions (i.e., paramagnetic ions) increased after 1904 with high photoluminescence quantum yield [5, 25]. It is an actual incorporation of dopant ion into semiconductor core. This is the older or long traditional method of doping of paramagnetic ions into semiconductor nanoparticles. Type II magnetic quantum dots have size ranges less than 5nm because of this property. This type of magnetic quantum dot has improved the clinical applications. Paramagnetic ions have the same energy state within the QDs' band gap work as trap states, which determine the wavelength and lifetime of the luminescence of doped QDs [13]. The luminescence may originate from the paramagnetic ion itself or be involved in the band gap state which related with paramagnetic dopant. Some spectroscopic techniques can be used for the verification of location of dopant ions through X-ray spectroscopy, luminescence spectroscopy and electron paramagnetic resonance (EPR) etc. Previously this type of MQDs worked as light emitting device and in the field of spintronics, but recently it is used as multimodal imaging probes. It is reported that CdS: MN/ZnS QDs worked as bimodal imaging agents [25]. These MQDs are prepared through a reverse micelle method. It shows yellow Mn^{2+} emission and show magnetic response. It is also conjugate with HIV-1 TAT peptide which is worked as labelling agents for the imaging of brain tumor [27]. Another example shows that preparation of $CdSe/Zn_{1-X}Mn_XS$ for biomodal imaging; in which

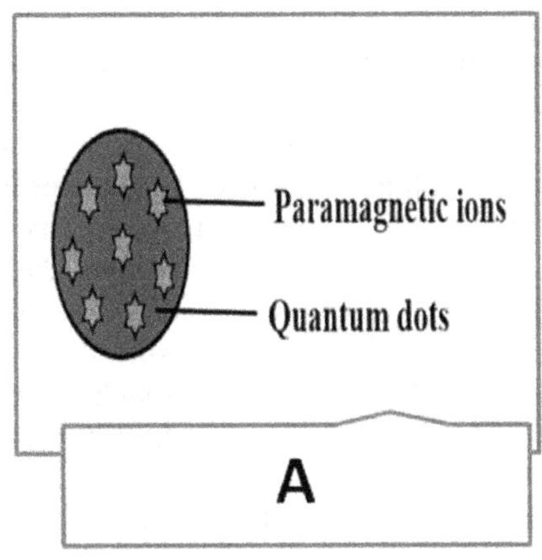

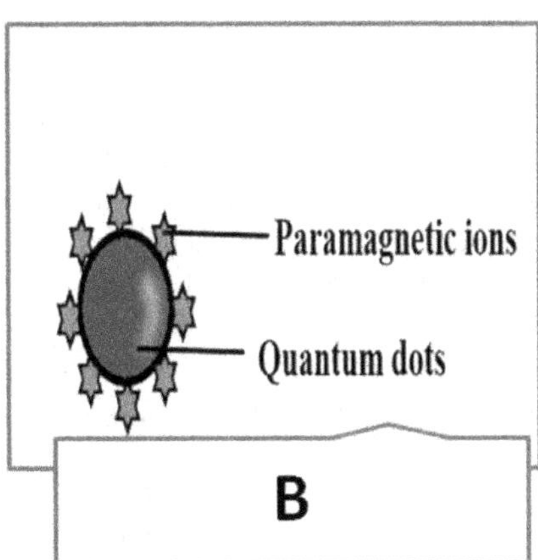

FIGURE 1.6

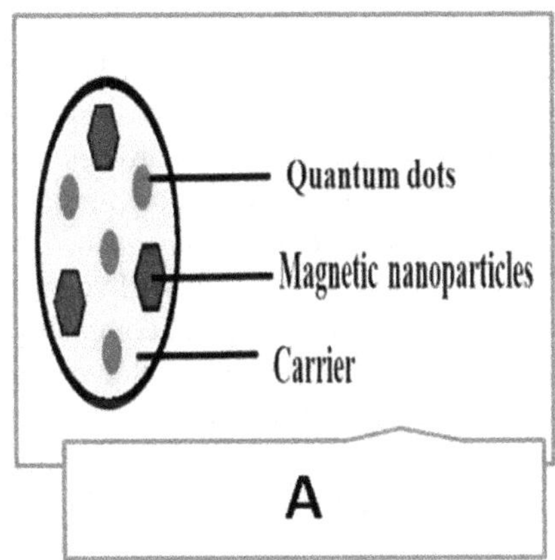

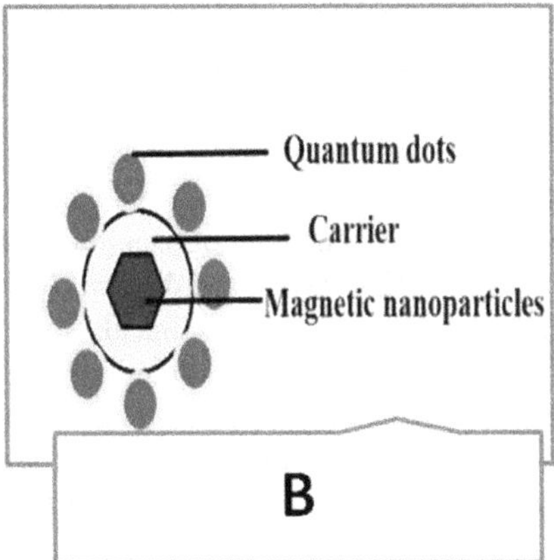

FIGURE 1.7

luminescent CdSe core was grown with ZnS shell which are already doped with Mn^{2+} and it shows the paramagnetic properties.

Type III: Composite particles containing magnetic and semiconductor nanoparticles: These types of magnetic QDs (shown in Figure 1.7) are prepared by combining the magnetic and semiconductor nanoparticles by either using the carrier material to create a new composite in which both magnetic and semiconductor nanoparticles can be integrated through attachment to the outside of carrier or combination of both. Type III are generally large size more than 50nm, but this type of magnetic quantum dots are more versatile. The QDs is incorporated in carrier particles and magnetic nanoparticles are attached to the outside of the carrier. In this as carrier particles both silica and polymers are used. These MQDs size are generally larger than the previous two types [5, 25]. For example, it is reported that in silica particles size ranges 50nm both Fe_2O_3 MNPs and CdSe/ZnS QDs are incorporated through standard micro-emulsion system, which result in the formation of mono dispersed silica nanoparticles with multiple magnetic nanoparticles and quantum dots. Incorporation of nanoparticles in silica particle reduced the QE of the CdSe/ZnS QDs. Another method reported that MQDs was prepared by the incorporation of iron oxide MNPs (diameter 30nm) into silica sphere (diameter 70nm) using a Stober approach [13]. Then silica sphere coated with the positive charged polymer which helps for the deposition of CdTe QDs and again coated with silica particle with 20nm size ranges. Finally the composition of MQDs was prepared with diameter 220nm [27]. Research shows MNPs and QDs both are conjugate outside the silica sphere with 100nm. In which CdSe/ZnS QDs attached with silica surface through amine group of silica particle and final composition of MQDs was obtained with 100nm in size which loaded with Fe_3O_4 MNPs with 14nm and CdSe/ZnS QDs with 4.5 nm size.

Type IV: Quantum dots with a paramagnetic coating of Gd-chelates: These types of magnetic quantum dots (shown in Figure 1.8) are prepared through paramagnetic coating of Gd-chelates with quantum dots. Its size ranges from >10nm. In this type of magnetic quantum dots, Quantum dots are coated with organic complexes containing paramagnetic ions which is called chelates. The lanthanide ion Gd^{3+} has a high magnetic moment and symmetric electronic ground state because of this property it is most widely used paramagnetic ions for MRI contrast agent [5, 25]. Type 4 has high QE because of there are no interaction between magnetic coating and quantum dots, but their paramagnetic payload is limited for the small particles. For reducing the toxicity and enhance the stability of Gd^{3+}, these ions are complexed in organic chelates and through an ionic interaction coordinate to the paramagnetic ion. Diethylenetriaminepentaacidic acid (Gd-DTPA) is the most important MRI contrast agent for experiment and clinical use, but there are some other different compounds are available which are used in biomedical field [19]. For the addition of fluorescent properties to the paramagnetic chelates, the organic dye is attached through covalently and non-covalently to the complexes. Another method is to attachment the paramagnetic chelate to the QDs through covalent and non-covalent formation, this technique

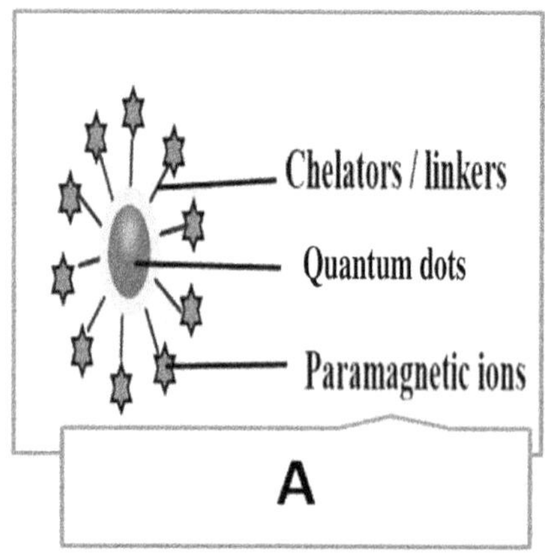

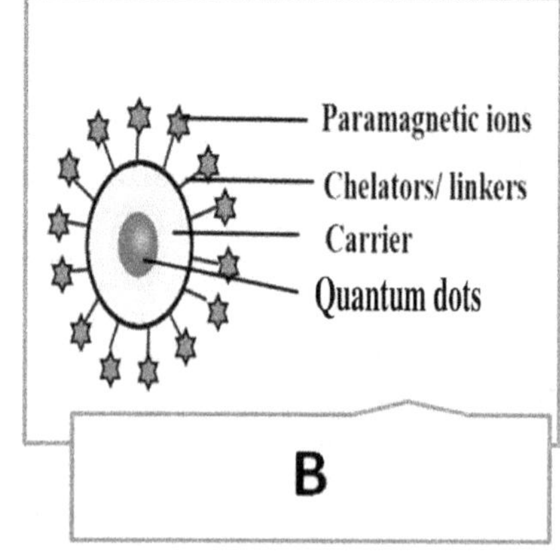

FIGURE 1.8

is used for the formation of MRI and fluorescence imaging containing biomodality [13]. For example research shows the first-time combination of paramagnetic chelates with QDs was prepared in which used lipid micelles surrounds the QDs and it is non-covalently integrated with Gd- DTPA complexes.

1.9 FUNDAMENTAL PROPERTIES OF MQDS

Some fundamental properties of magnetic quantum dots are broadly categorized into magneto-optical or magneto-electrical properties. The sp-d exchange interaction of dopant ion with semiconductor host molecules creates magnetic property of MQDs. Some of the MQDs are prepared from iron oxide MNPs and QDs which shows the excellent superparamagnetism and fluorescence properties [28]. These MQDs have some basic properties which make them suitable for multifunctional nanoprobe, such as-

- It has strong magnetic strength as well as tunable functionality, e.g., rapid and simple magnetic separation.
- It has intense, stable and size tunable fluorescence because of QDs.
- It has an extremely small size.
- It is resistance to photodegradation and photo-bleaching effects.
- It has high photostability and quantum yield.
- It produces auto-fluorescence and minimal absorbance.
- It gives imaging and visualization through simple exposure of ultraviolet light.
- Mostly MQDs are derived from iron based MNPs because of their super para-magnetic properties, abundancy, low cast and easy to synthesize [26, 28].

1.10 CRITERIA FOR THE PREPARATION OF MAGNETIC QUANTUM DOTS

The fabrication of magnetic quantum dots required fulfilling both desirable property such as magnetism and fluorescence in one single component.

- The selection of magnetic nanoparticles is done according to their magnetism and toxicity profile. For example, cobalt and nickel containing magnetic nanoparticles shows high toxicity and low magnetism property which are not suitable for bioimaging purposes.
- The selection of quantum dots is done according to their desired fluorescence and their toxicity profile.
- Surface functionalization (e.g., targeting ligand, imaging probe) plays an important role in preparation of magnetic quantum dots.
- The selection of conjugation technique is also important for the preparation of magnetic quantum dots.
- Selection of carrier or matrix materials is done according to their physico-chemical properties.

1.11 SYNTHESIS OF MAGNETIC QUANTUM DOTS

Various techniques are available for the preparation of magnetic quantum dots. This composite have different composition, architectural structure and morphology. For the synthesis of magnetic quantum dots it is necessary to integration of quantum dots and magnetic nanoparticles are done without losing its fluorescence and magnetism property [5]. There are different types of synthesis processes available which are used for preparation of magnetic quantum dots.

- **High temperature decomposition technique:** In this method organic metallic compound or precursor are heated at high temperature in the presence of organic phase, which results the decomposition of compound through particles nucleation and growth. In this method two materials are synthesized. The first one is "core" that is synthesized with core materials, e.g., Fe_3O_4, FePt, Fe_2O_3, and Co which work as template for growth of second material that is "shell" for example – CdS, CdSe, CdTe [5, 14]. In this technique two types of magnetic quantum dots were synthesized first one is core shell and second one is hetero structure. It was the first method which was used for the preparation of magnetic quantum dots; however the problem is, there are no optimization for the properties or functionalization of magnetic quantum dots, which was essential for their activity. For example FeP-CdTe magnetic quantum dots nanoparticles synthesize using high temperature decomposition [14]. Report shows that Magnetic quantum dots was prepared through CdSe quantum dots (QDs) around g-Fe2O3 magnetic cores with high temperature which is used for the labelling of live cells of mouse breast cancer cells (4T1) and human liver cells (hepG2) [44].
- **Doping technique:** In the doping method for the synthesis of magnetic quantum dots there is addition of transition element as dopant into nanocrystal lattice. In this method both fluorescence and magnetic properties are produced by dopant ion which creates the energy state. One of the examples of this method is the formation of manganese doped nanoparticles which shows photoluminescence and magnetic property [5]. Some dopant ions or paramagnetic ions which are used in this method are rare earth materials such as lanthanides Tb^{3+}, Er^{3+}, Yb^{3}. This technique is limited to dopant ions.
- **Encapsulation technique:** In this method already prepared magnetic and fluorescent nanoparticles are incorporated into matrix or carrier materials. This method reduced the various inherent problems which are associated with doping or high temperature decomposition because in this technique magnetic and fluorescent nanoparticles are prepared separately and reduce the fluorescence quenching. The formation of matrix can be done either in top-down or bottom-up approach. In this technique micelle, liposome, silica and polymers (e.g., poly lactic-co-glycolic acid) are used as carrier material or matrix [5, 23]. Silica coated CdSe-ZnS and magnetic nanoparticles such as Fe_2O3 or Fe_3O4 for bioimaging application [13].
- **Cross-linking technique:** Another method for the preparation of magnetic quantum dots is cross-linking techniques. Primarily it required SPIONs as

TABLE 1.1
Characterization Techniques with Their Uses

Sr. No.	Characterization Techniques	Uses
1.	Transmission electron microscopy (TEM)	For structure and size of the MQDs
2.	X-ray photoelectron spectroscopy (XPS)	Magnetic quantum dots are used to quantitatively analyze atomic composition and chemistry.
3.	X-ray diffraction (XRD)	For phase analysis, crystalline variants of magnetic quantum dots
4.	Inductively coupled plasma optical emission spectrometer (ICP)	elemental analysis of the MQDs
5.	UV–vis absorption	For measuring the optical property
6.	Photoluminescence (PL),	For measuring the purity and optical property
7.	superconducting quantum interference device (SQUID)	For measuring the magnetic property of magnetic quantum dots.

magnetic nanoparticles and quantum dots as fluorescent materials. Coupling ligand should have functional groups such as thiol, siloxane or carboxyl etc. Cross-linking is done either in chemical linker or charge attraction. Cross-linking reduce the various limitations of other techniques [5, 23].

1.12 CHARACTERIZATION OF MAGNETIC QUANTUM DOTS

There are various methods which are used for the characterization of magnetic quantum dots, such as TEM, XPS, and XRD. These are used for the characterization of size, structure, and chemical composition, and SQUID and PL spectroscopy are for the detection of magnetism and optical properties [26, 29]. The characterization techniques are listed in Table 1.1.

1.13 APPLICATION OF MAGNETIC QUANTUM DOTS

In previous years QDs were demonstrated for in-vitro imaging, and these nanoparticles has been widely used as bioimaging agent. However continuous development leads to enhancing their imaging capability and other properties. Different integrations occur with quantum dots such as quantum dots-liposome/ gold nanoparticles etc. [23]. Recently various research has shown that integration of QDs with magnetic materials gains much attention because of fluorescence and magnetic property present in single particles. The diagrammatic representation of applications of magnetic quantum dots is shown in Figure 1.9.

- **Multimodal imaging:** A variety of multimodalities are available which combine optical and magnetism imaging property in one single compound. It is

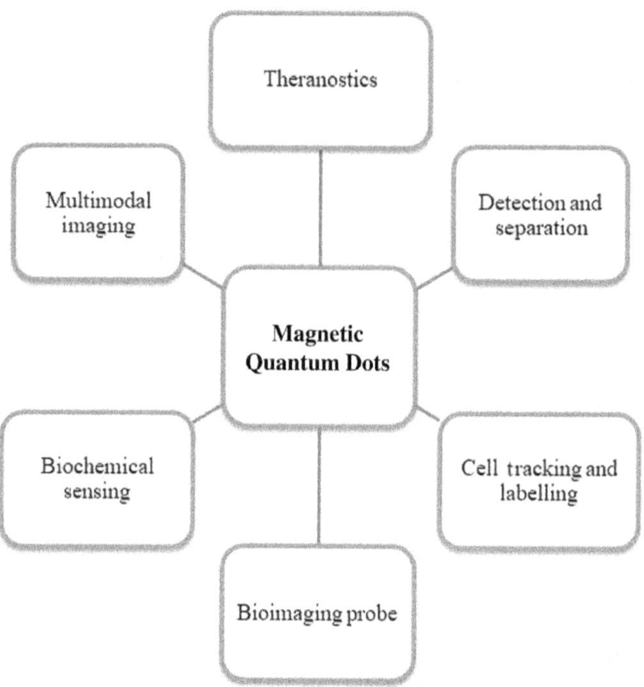

FIGURE 1.9

reported that firstly combining manganese ion with QDs through doping technique which are used for arterial labelling and shows the multimodalities QDs which can cross the blood brain barrier which indicate the QDs conjugate therapeutic drug delivery into brain [23]. Another report shows that some paramagnetic ions such as gadolinium-diethylene triamine penta-acetic acid, Gd-DTPA combined with QDs which gives the In-vivo imaging of integrins and fluorescence detection of apoptosis. Another example of magnetic quantum dots in multimodal imaging is QDs combined with SPIONs in the form of magnetic elements. Some composition like CdSe–Fe_2O_3 coated with silica shell and used for in-vitro imaging [21]. Research shows the formation of a multilayered nanoprobe was prepared based on magnetic ferric oxide particles and quantum dots for multimodality imaging of breast cancer tumors [42].

- **Theranostics:** Nanoparticles have potential for drug delivery as carrier and when it conjugate with imaging agents it work for treatment as well as diagnosis that is called theranostics. A theranostic should be small to freely circulate in blood, it should be non-immunogenic, it can easily identify or penetrate the target site, it should deliver the drug to particular site, it should be biodegradable, and can easily remove from body after performing their work [23]. Some fluorescence agents, such as quantum dots, work as theranostic agents and conjugation of various nanoparticles play an important role in theranostic approach. Magnetic quantum dots also work as theranostic agents. They have both fluorescence and magnetic properties in one single compound. Research shows the formation of magnetic-carbon-quantum dots {Gd-CDs/AFn (DOX)/ FA} which are labelled with apoferritin used for bioimaging and targeted therapies [41].

- **Detection and separation:** Magnetic quantum dots have another great application of molecular imaging and separation. Some other imaging techniques such as bulk measurements cannot detect the signal from low concentration of solution, even small changes in molecular structure can affect their activity therefore and accurate detection is needed which are rare for the diagnosis purpose [23]. Magnetic quantum dots not only isolate or separate the cells but also different biological molecules [9]. Nanoparticle-based detection techniques have some unique property with similar to size. This property can lead to the detection of various bio molecules with more accuracy. The separation and detection of DNA, RNA, and protein can be done by magnetic quantum dots. It also is used in diagnostics, detection of food contamination and microbial contamination, etc. Research shows the formation of maghemite super paramagnetic core and CdSe/ZnS coated with polymer and anticylcline-E antibodies which separates the MCF-7 breast cancer cells from serum solution [9].
- **Bioimaging probe: –** Some imaging probes have great impact on bioimaging such as fluorescence microscopy and MRI, these techniques allow the imaging of live in vitro and in vivo, which provide more accurate result and clear diagnosis. For this purpose fluorescent magnetic nanoparticles work as dual nature contrast agent. It gives optical tracking of cellular processes and biological entities with magnetic manipulation. These conjugated nanoparticles work as imaging probe and labelling agents [9].
- **Biochemical sensing:** Magnetic quantum dots have great potential for biochemical sensing. Their dual nature not only detects biological and chemical molecules but also allows the manipulation of sensing agent. This property is important for chemical and biological element sensing in environments. The functionalized magnetic quantum dots is used for the sensing if viruses and DNA [9]. Research show that the magnetic quantum dots are used for the sensing of metal ions, food borne pathogens, toxins, pesticides and antibiotics etc. [19].
- **Cell tracking and cell labelling:** Bioimaging techniques are also used in the cell tracking process. According to one research, selected, aligned cells that have been immune magnetically labelled can be tracked and detected optically. Research shows the formation of tri-functionalized magnetic quantum dots nanospheres from co embedding of QDs and magnetic nanoparticles into nanospheres which functionalized with hydrazide copolymer and then coupled with IgG, avidin and biotin. These nanospheres selectively linked with apoptotic cells allow the visualization and isolation [9]. Cell tracking: Magnetic graphene QDs are prepared through hydrothermal cutting techniques. In this graphene oxide and iron oxide sheet subjected to hydrothermal cutting technique. MMOCT (magneto motive optical coherence tomography) and CFM (confocal fluorescence microscopy) both employ this as a contrast agent [30]. According to a report, high-temperature CdSe quantum dots (QDs)

were employed to generate magnetic quantum dots, which were then used to label live mouse breast cancer cell (4T1) and human liver cell (hepG2) specimens [44].

1.14 TOXICITY CONCERN FOR MAGNETIC QUANTUM DOTS

Toxicity of the magnetic quantum dots is one of the major concerns. The toxicity of nanoparticles depends on many factors such as chemical composition, size, surface property, concentration etc. For toxicology study of magnetic quantum dots still needed research there are very few reports about their degradation, bio distribution, excretion and immune response. For individual compound quantum dots as well as magnetic nanoparticles both have potential toxicity. For quantum dots its precursor molecule is metal components which lead to toxicity due to (reactive oxygen species) ROS causing cell death [5, 25]. Some research shows surface modification of quantum dots reduce the toxicity, this surface modification can be done by either coating polymers, modify their surface chemistry or changes their composition materials such as carbon, InP, ZnO, and graphene, etc. These modifications give the opportunities to quantum dots for biomedical application [23]. For magnetic nanoparticles or ions, some toxicological tests of iron oxide give the biosafety result. It is assumed that the iron is highly regulated with in body and stored in the ferritin protein. It is reported that some paramagnetic ions such as Mn 2+, Gd 3+ has nephrogenic risk with kidney impaired patients, and also cause toxicity. For reducing the risk of toxicity various chelating agent or carrier materials are used. So overall the toxicity of magnetic quantum dots is related with their magnetic materials and quantum dots. Screening of acute and chronic effect of magnetic quantum dots still needs to investigate, and this is the most challenging part for used of MQDS in biomedical field and clinical applications [5, 23].

1.15 CONCLUSION AND FUTURE PROSPECTS

In recent years multi-functional nanoparticles have gained much attention in biological and biomedical field. It includes various ranges of fields such as personalized medicine, pharmacogenomics, molecular imaging, theranostics, diagnosis, and treatment. Hence there has been various research conducted for the technique which is more accurate, more sensitive, and less time consuming. The formation of multifunctional magnetic quantum dots through combination of magnetic nanoparticles and quantum dots to target the specific cell has great advantages such as magnetic as well as optical properties. This multifunctional probe has the advantage of the magnetic field, and their fluorescent part has the capability of tagging of several target molecules which made them attractive tool for diagnostic and imaging. This chapter deals with the fundamental, classification and recent progress of the magnetic quantum dots. Magnetic quantum dots are utilized for bioseparation, detection, multi-modality imaging, diagnosis and treatment. These newer generation molecules with dual characters have huge potential in diagnosis and treatment of some disease but also require some improvement. This multimodal tool is still in the early stage of research. The major concern with magnetic quantum dots are biostability because of longtime circulation in blood, their toxicity, because their mechanism is still unknown, their accumulation in tissue and biocompatibility. This major area of magnetic quantum dots requires further research studies in the future.

REFERENCES

1) Abolfazl Akbarzadeh, Mohammad Samiei and Soodabeh Davaran "Magnetic nanoparticles: preparation, physical properties, and applications in biomedicine" Nanoscale Research Letters 2012, 7:144.
2) Anca Armăşelu "Quantum dots and fluorescent and magnetic nanocomposites: Recent investigations and applications in biology and medicine" in Nonmagnetic and Magnetic Quantum Dots, 2018. London: IntechOpen. http://dx.doi.org/10.5772/intechopen.70614
3) Amitabha Acharya "Luminescent magnetic quantum dots for in vitro/in vivo imaging and applications in therapeutics" Journal of Nanoscience and Nanotechnology Vol. 13, 3753–3768, 2013. doi:10.1166/jnn.2013.7460
4) Sun Z, Ng KH and Ramli N "Biomedical imaging research: a fast-emerging area for interdisciplinary collaboration" Biomed Imaging Interv J 2011; 7(3):e21. doi: 10.2349/biij.7.3.e21
5) Lihong Jing, Ke Ding, Stephen V. Kershaw, Ivan M. Kempson, Andrey L. Rogach and Mingyuan Gao "Magnetically engineered semiconductor quantum dots as multimodal imaging probes" Adv. Mater. 2014, 26, 6367–6386 doi: 10.1002/adma.201402296
6) Ke Ding, Lihong Jing, Chunyan Liu, Yi Hou, Mingyuan Gao "Magnetically engineered Cd-free quantum dots as dual-modality probes for fluorescence/magnetic resonance imaging of tumors" Biomaterials 35 (2014) 1608–1617 http://dx.doi.org/10.1016/j.biomaterials.2013.10.078
7) Raghuraj Singh Chouhan, Milena Horvat, Jahangeer Ahmed, Norah Alhokbany, Saad M. Alshehri and Sonu Gandhi "Magnetic nanoparticles—a multifunctional potential agent for diagnosis and therapy" Cancers 2021, 13, 2213. https://doi.org/10.3390/cancers13092213
8) Zhe Liu, Fabian Kiessling and Jessica Gatjens "Advanced nanomaterials in multimodal imaging: Design, functionalization, and biomedical applications" Journal of Nanomaterials Volume 2010, Article ID 894303, 15 pages. doi:10.1155/2010/894303
9) Serena A. Corr, Yury P. Rakovich, Yurii K. Gunko "Multifunctional magnetic-fluorescent nanocomposites for biomedical applications" Nanoscale Res Lett (2008) 3:87–104. doi: 10.1007/s11671-008-9122-8
10) Karunanithi Rajamanickam "Multimodal molecular imaging strategies using functionalized nano probes" J Nanotechnol Res 2019; 1 (4): 119–135. doi: 10.26502/jnr.2688-85210010
11) DipaDutta, Jagriti Gupta, Dinbandhu Thakur, Dhirendra Bahadur "Magnetically engineered SnO2 quantum dots as a bimodal agent for optical and magnetic resonance imaging" Materials Research Express, 2017. https://doi.org/10.1088/2053-1591/aa9ac3
12) Sarka Nevolova and Petr Skladal "Nanomaterials for biomedical imaging and targeting" Microchimica Acta (2022) 189: 163. https://doi.org/10.1007/s00604-022-05215-7
13) Subramanian Tamil Selvan, "Silica-coated quantum dots and magnetic nanoparticles for bioimaging applications" (Mini-Review), FA110 Biointerphases volume5, no.3, September 2010, American Vacuum Society. doi: 10.1116/1.3516492

14) Shuang Deng, Gang Ruan, Ning Han and Jessica O Winter "Interactions in fluorescent-magnetic heterodimer nanocomposites" Nanotechnology 21 (2010) 145605 (6pp) http://dx.doi.org/10.1088/0957-4484/21/14/145605
15) F. Henneberger and J. Puls "Diluted magnetic quantum dots" Introduction to the Physics of Diluted Magnetic Semiconductors" Chapter 5, 2010, Springer Series in Materials Science 144, DOI 10.1007/978-3-642-15856-8_5
16) Erandi Munasinghe, Maheshi Aththapaththu and Lakmal Jayarathne "Magnetic and quantum dot nanoparticles for drug delivery and diagnostic systems" Colloid Science in Pharmaceutical Nanotechnology. doi: http://dx.doi.org/10.5772/intechopen.88611
17) S. Tamil Selvan, Pranab K. Patra, Chung Yen Ang and Jackie Y. Ying "Synthesis of silica-coated semiconductor and magnetic quantum dots and their use in the imaging of live cells" Angew. Chem. 2007, 119, 2500–2504 doi: 10.1002/ange.200604245
18) Jackie Y. Ying, Yuangang Zheng, and S. Tamil Selvan, "Synthesis and applications of quantum dots and magnetic quantum dots" Proc. of SPIE Vol. 6866, 686602, (2008), Colloidal Quantum Dots for Biomedical Applications III, doi: 10.1117/12.784053.
19) Jincheng Xiong, Huixia Zhang, Linqian Qin, Shuai Zhang, Jiyue Cao and Haiyang Jiang "Magnetic fluorescent quantum dots nanocomposites in food contaminants analysis: Current challenges and opportunities" Int. J. Mol. Sci. 2022, 23, 4088. https://doi.org/10.3390/ijms23084088
20) Syed Rahin Ahmed, Jinhua Dong, Megumi Yui, Tatsuya Kato, Jaebeom Lee and Enoch Y Park "Quantum dots incorporated magnetic nanoparticles for imaging colon carcinoma cells" Journal of Nanobiotechnology 2013, 11:28
21) Chariya Kaewsaneha, Pramuan Tangboriboonrat, Duangporn Polpanich, and Abdelhamid Elaissari "Multifunctional fluorescent-magnetic polymeric colloidal particles: Preparations and bioanalytical applications" ACS Applied Materials & Interfaces, 2015. doi: 10.1021/acsami.5b07515
22) Florian Part, Christoph Zaba, Oliver Bixner, Tilman A Grünewald, Herwig Michor, Seta Küpcü, Monika Debreczeny, Elisabetta De Vito Francesco, Andrea Lassenberger, Stefan Schrittwieser, Stephan Hann, Helga Lichtenegger, and Eva-Kathrin Ehmoser "The doping method determines para- or superparamagnetic properties of photostable and surface modifiable quantum dots for multimodal bioimaging" Chemistry of Materials, 2018. doi: 10.1021/acs.chemmater.8b00431
23) Kalpesh D. Mahajan, Qirui Fan, Jenny Dorcéna, Gang Ruan and Jessica O. Winter "Magnetic quantum dots in biotechnology – synthesis and applications" Biotechnol. J. 2013, 8 doi: 10.1002/biot.201300038
24) Satyapriya Bhandari, Rumi Khandelia, Uday Narayan Pan and Arun Chattopadhyay "Surface complexation based biocompatible magnetofluorescentNanoprobe for targeted cellular imaging" ACS Applied Materials & Interfaces 7, no. 32 (2015): 17552–17557.
25) Rolf Koole, Willem J. M. Mulder, Matti M. van Schooneveld, Gustav J. Strijkers, Andries Meijerink and Klaas Nicolay "Magnetic quantum dots for multimodal imaging" Volume 1, Advanced Review, 2009. doi: 10.1002/wnan.014
26) Ajoy K. Saha, Parvesh Sharma, Han-Byul Sohn, Siddhartha Ghosh, Ritesh. K. Das, Arthur F. Hebard, Huadong Zeng, Celine Baligand, Glenn A. Walter and Brij M. Moudgil "Fe doped CdTeS magnetic quantum dots for bioimaging" J. Mater. Chem. B, 2013, doi: 10.1039/C3TB20859A.
27) Ali Tufani, Anjum Qureshi, Javed H. Niazi "Iron oxide nanoparticles based magnetic luminescent quantum dots (MQDs) synthesis and biomedical/biological applications: a review" Materials Science & Engineering C, 2020. https://doi.org/10.1016/j.msec.2020.111545

28) Mahima Makkar and Ranjani Viswanatha "Recent advances in magnetic ion-doped semiconductor quantum dots" Current Science, Vol. 112, No. 7, 10 April 2017.
29) Gui Huan Du, Zu Li Liu, Qiang Hua Lu, Xing Xia, Li Hui Jia, Kai Lun Yao, Qian Chu and Su Ming Zhang "Fe3O4/CdSe/ZnS magnetic fluorescent bifunctional nanocomposites" Nanotechnology 17 (2006) 2850–2854, http://dx.doi.org/10.1088/0957-4484/17/12/004
30) Wei Li, Stephen J. Matcher, "Novel magnetic graphene quantum dot as dual modality fluorescence/MMOCT contrast agent for tracking epithelial stem cells" Proc. of SPIE Vol. 10079, 100790X, 2017, doi: 10.1117/12.2252761.
31) Jing Ruan, Kan Wang, Hua Song, Xin Xu, Jiajia Ji and Daxiang Cui "Biocompatibility of hydrophilic silica-coated CdTe quantum dots and magnetic nanoparticles" Nanoscale Research Letters 2011, 6:299.
32) YuankuiLeng, Weijie Wu, Li Li, Kun Lin, Kang Sun, Xiaoyuan Chen, and Wanwan Li "Magnetic/fluorescent barcodes based on cadmium-free near-infrared-emitting quantum dots for multiplexed detection" Adv. Funct. Mater. 2016, doi: 10.1002/adfm.201602900
33) Heesun Yang, Swadeshmukul Santra, Glenn A. Walter and Paul H. Holloway "GdIII-functionalized fluorescent quantum dots as multimodal imaging probes" Adv. Mater. 2006, 18, 2890–2894 doi: 10.1002/adma.200502665
34) Yoshiaki Maeda, Tomoko Yoshino and Tadashi Matsunaga "Novel nanocomposites consisting of in vivo-biotinylated bacterial magnetic particles and quantum dots for magnetic separation and fluorescent labeling of cancer cells" J. Mater. Chem., 2009, 19, 6361–6366 doi: 10.1039/b900693a
35) Shuai Zhou, Qianwang Chen, Xianyi Hua and Tianyun Zhao "Bifunctional luminescent superparamagneticnanocomposites of CdSe/CdS-Fe3O4 synthesized via a facile method" J. Mater. Chem., 2012, 22, 8263. doi: 10.1039/c2jm16783b
36) Gary Sitbon, Sophie Bouccara, Mariana Tasso, Aurelie Francois, Lina Bezdetnaya, Frederic Marchal, Marine Beaumont and Thomas Pons "Multimodal Mn-doped I–III–VI quantum dots for near infrared fluorescence and magnetic resonance imaging: from synthesis to in vivo application" Nanoscale, 2014, 6, 9264–9272 doi: 10.1039/c4nr02239d
37) Xiaowan Li, Chenyu Li and Ligang Chen "Preparation of multifunctional magnetic–fluorescent nanocomposites for analysis of tetracycline hydrochloride" New J. Chem, 2015 doi: 10.1039/c5nj01365h
38) San Kyeong, Cheolhwan Jeong, Han Young Kim, Do Won Hwang, Homan Kang, Jin-Kyoung Yang, Dong Soo Lee, Bong-Hyun Jun and Yoon-Sik Lee "Fabrication of mono-dispersed silica-coated quantum dot-assembled magnetic nanoparticles" RSC Adv., 2015. doi: 10.1039/C5RA03139G
39) Ling Wang, Guangzhen Wang, Yitong Wang, Huizhong Liu, Shuli Dong and Jingcheng Hao "Fluorescent hybrid nanospheres induced by single-stranded dna and magnetic carbon quantum dots" New J. Chem., 2019, doi: 10.1039/C8NJ06157B
40) Guifen Jie, Jinxin Yuan, Tingyu Huanga and Yanbin Zhao "Electrochemiluminescence of dendritic magnetic quantum dots nanostructure and its quenching by gold nanoparticles for cancer cells assay" Electroanalysis 2012, 24, No. 5, 1220–1225. https://doi.org/10.1002/elan.201200062
41) Hanchun Yao, Li Su, Man Zeng, Li Cao, Weiwei Zhao, Chengqun Chen, Bin Du and Jie Zhou "Construction of magnetic-carbon-quantumdots-probe-labeled apoferritinnanocages for bioimaging and targeted therapy" International Journal of Nanomedicine 2016:11 4423–4438 http://dx.doi.org/10.2147/IJN.S108039
42) Qiang Ma, Yuko Nakane, Yuki Mori, Miyuki Hasegawa, Yoshichika Yoshioka, Tomonobu M. Watanabe, Kohsuke Gonda, Noriaki Ohuchi and Takashi Jin

"Multilayered, core/shell nanoprobes based on magnetic ferric oxide particles and quantum dots for multimodality imaging of breast cancer tumors" Biomaterials 33 (2012) 8486e8494 http://dx.doi.org/10.1016/j.biomaterials.2012.07.051

43) Fei Ye, Åsa Barrefelt, Heba Asem, Manuchehr Abedi-Valugerdi, Ibrahim El-Serafi, Maryam Saghafian, Khalid Abu-Salah, Salman Alrokayan, Mamoun Muhammed, Moustapha Hassan "Biodegradable polymeric vesicles containing magnetic nanoparticles, quantum dots and anticancer drugs for drug delivery and imaging" Biomaterials 35 (2014) http://dx.doi.org/10.1016/j.biomaterials.2014.01.041

44) Alex W.H. Lin, Chung Yen Ang, Pranab K. Patra, Yu Han, Hongwei Gu, Jean-Marie Le Breton, Jean Juraszek, Hubert Chiron, Georgia C. Papaefthymiou, Subramanian Tamil Selvan, Jackie Y. Ying "Seed-mediated synthesis, properties and application of g-Fe2O3–CdSe magnetic quantum dots" Journal of Solid State Chemistry 184 (2011) 2150–2158 doi:10.1016/j.jssc.2011.05.043

45) Bodhisatwa Das, Agnishwar Girigoswami, Pallabi Pal, Santanu Dhara "Manganese oxide-carbon quantum dots nano-composites for fluorescence/magnetic resonance (T1) dual mode bioimaging, long term cell tracking, and ROS scavenging" Materials Science & Engineering C, 2019 https://doi.org/10.1016/j.msec.2019.04.077

46) Dana A. Schwartz, Nick S. Norberg, Quyen P. Nguyen, Jason M. Parker, and Daniel R. Gamelin "Magnetic quantum dots: Synthesis, spectroscopy, and magnetism of Co2+- and Ni2+-doped ZnONanocrystals" J. Am. Chem. Soc. 2003, 125, 13205–13218. https://doi.org/10.1021/ja036811v

47) Dong Kee Yi, S. Tamil Selvan, Su Seong Lee, Georgia C. Papaefthymiou, Darshan Kundaliya and Jackie Y. Ying "Silica-coated nanocomposites of magnetic nanoparticles and quantum dots" J. Am. Chem. Soc. 2005, 127, 4990–4991.

48) Hongwei Gu, Rongkun Zheng, XiXiang Zhang and Bing Xu "Facile one-pot synthesis of bifunctional heterodimers of nanoparticles: A conjugate of quantum dot and magnetic nanoparticles" J. Am. Chem. Soc. 2004, 126 , 5664–5665.

49) Jaeyun Kim, Ji Eun Lee, Jinwoo Lee, Jung Ho Yu, Byoung Chan Kim, Kwangjin An, Yosun Hwang, Chae-Ho Shin, Je-Geun Park, Jungbae Kim and Taeghwan Hyeon "Magnetic fluorescent delivery vehicle using uniform mesoporous silica spheres embedded with monodisperse magnetic and semiconductor nanocrystals" J. Am. Chem. Soc. 2006, 128, 688–689.

50) Jinhao Gao, Wei Zhang, Pingbo Huang, Bei Zhang, Xixiang Zhang and Bing Xu "Intracellular spatial control of fluorescent magnetic nanoparticles" J. Am. Chem. Soc. 2008, 130, 3710–3711.

51) Su Xi, Xu Yi, CheYulan, Liao Xin, Jiang Yan "A type of novel fluorescent magnetic carbon quantum dots for cells imaging and detection" Journal of Biomedical Materials Research: Part A, doi: 10.1002/jbm.a.35468 .

52) Surinder P. Singh "Multifunctional magnetic quantum dots for cancer theranostics" Journal of Biomedical Nanotechnology Vol. 7, 95–97, 2011. Journal of Biomedical Nanotechnology Vol. 7, 95–97, 2011. doi:10.1166/jbn.2011.1219

53) Linlin Li, Hongbo Li, Dong Chen, Huiyu Liu, Fangqiong Tang, Yanqi Zhang, Jun Ren, and Yi Li "Preparation and characterization of quantum dots coated magnetic hollow spheres for magnetic fluorescent multimodal imaging and drug delivery" Journal of Nanoscience and Nanotechnology Vol.9, 2540–2545, 2009 doi:10.1166/jnn.2009.dk04

54) Yi Wu, Hui Zou, Ying Zhang, Mingyao Mou, Qianqian Niu, Zhengyu Yan and Shenghua Liao, "The loading of luminescent magnetic nanocomposites Fe3O4@polyaniline/carbon dots for methotrexate and its release behavior in vitro" Journal of Nanoscience and Nanotechnology Vol. 20, 701–708, 2020. doi:10.1166/jnn.2020.16900

55) Sumera Khizar, Nasir M. Ahmad, Nadia Zine, Nicole Jaffrezic-Renault, Abdelhamid Errachid-el-salhi and Abdelhamid Elaissari "Magnetic nanoparticles: From synthesis to theranostic applications" ACS Appl. Nano Mater. 2021, 4, 4284–4306. https://doi.org/10.1021/acsanm.1c00852
56) Maria I. N. da Silva, Alexandra A. P. Mansur, Vanessa Schatkoski, Klaus W. H. Krambrock, Juan Gonzalez and Herman S. Mansur "Fluorescent-magnetic nanostructures based on polymer-quantum dots conjugates" Macromol. Symp. 2012, 319, 114–120. DOI: 10.1002/masy.201100178
57) Fangchao Cui, Jian Ji, Jiadi Sun, Jun Wang, Haiming Wang, Yinzhi Zhang, Hong Ding, Yong Lu, Dan Xu and Xiulan Sun "A novel magnetic fluorescent biosensor based on graphene quantum dots for rapid, efficient, and sensitive separation and detection of circulating tumor cells" Analytical and Bioanalytical Chemistry, 2019. https://doi.org/10.1007/s00216-018-1501-0
58) Jian Cao, Haifeng Niu, Jiang Du, Lili Yang, Maobin Wei, Xiaoyan Liu, Qianyu Liu and Jinghai Yang "Fabrication of P(NIPAAm-co-AAm) coated optical-magnetic quantum dots/silica core-shell nanocomposites for temperature triggered drug release, bioimaging and in vivo tumor inhibition" Journal of Materials Science: Materials in Medicine (2018) 29:169 https://doi.org/10.1007/s10856-018-6179-5
59) Hui Xia, Ruijie Tong, Yanling Song, Fang Xiong, Jiman Li, Shichao Wang, Huihui Fu, Jirui Wen, Dongze Li, Ye Zeng, Zhiwei Zhao and Jiang Wu "Synthesis and bioapplications of targeted magnetic-fluorescent composite nanoparticles" J Nanopart Res (2017) 19:149. doi: 10.1007/s11051-017-3833-7
60) Qi Xiao and Chong Xiao "Preparation and characterization of silica-coated magnetic–fluorescent bifunctional microspheres" Nanoscale Res Lett (2009) 4:1078–1084. doi: 10.1007/s11671-009-9356-0
61) Pengfei Zhang, Mohamed S. Draz, Anwen Xiong, Wannian Yan, Huanxing Han and Wansheng Chen "Immunoengineered magnetic-quantum dot nanobead system for the isolation and detection of circulating tumor cells" Zhang et al. J Nanobiotechnol (2021) 19:116 https://doi.org/10.1186/s12951-021-00860-1
62) Bo Wei, Congyu Zhou, Zhengjun Yao, Linling Xu, Zhejia Li, Li Wan, Jinsen Hou, Jintang Zhou "Lightweight and high-efficiency microwave absorption of reduced graphene oxide loaded with irregular magnetic quantum dots" Journal of Alloys and Compounds 886 (2021) 161330. https://doi.org/10.1016/j.jallcom.2021.161330
63) Shanmugapriya V., Bharathi S., Esakkinaveen D., Arunpandiyan S., Selvakumar B., Sasikala G., Jayavel R. and Arivarasan A. "Structural, optical, and magnetic properties of gd doped cdte quantum dots for magnetic imaging applications" ECS Journal of Solid State Science and Technology, 2022, 11 013010. doi: 10.1149/2162-8777/ ac4bad
64) Anup Kale, Sonia Kale, Prasad Yadav, Haribhau Gholap, Renu Pasricha, J P Jog, Benoit Lefez, Beatrice Hannoyer, Padma Shastry and Satishchandra Ogale "Magnetite/CdTe magnetic–fluorescent composite nanosystem for magnetic separation and bio-imaging" Nanotechnology 22 (2011) 225101. http://dx.doi.org/10.1088/0957-4484/22/22/225101
65) C.-H. Yang, K.-S. Huang, Y.-S. Lin, K. Lu, C.-C. Tzeng, E.-C. Wang, C.-H. Lin, W.-Y. Hsu and J.-Y. Chang "Microfluidic assisted synthesis of multi-functional polycaprolactone microcapsules: incorporation of CdTe quantum dots, Fe3O4 superparamagnetic nanoparticles and tamoxifen anticancer drugs" Lab Chip, 2009, 9, 961–965.
66) Hyungrak Kim, Marc Achermann, Laurent P. Balet, Jennifer A. Hollingsworth, and Victor I. Klimov "Synthesis and characterization of co/cdse core/shell nanocomposites: Bifunctional magnetic-optical nanocrystals" J. Am. Chem. Soc. 2005, 127, 544–546.

67) Swadeshmukul Santra, Heesun Yang, Paul H. Holloway, Jessie T. Stanley and Robert A. Mericle "Synthesis of water-dispersible fluorescent, radio-opaque, and paramagnetic CdS:Mn/ZnS quantum dots: A multifunctional probe for bioimaging" J. Am. Chem. Soc. 2005, 127, 1656–1657.
68) Jinhao Gao, Bei Zhang, Yuan Gao, Yue Pan, Xixiang Zhang and Bing Xu, "Fluorescent magnetic nanocrystals by sequential addition of reagents in a one-pot reaction: A simple preparation for multifunctional nanostructures" 9 J. Am. Chem. Soc. 2007, 129 , 11928–11935.
69) Daniele Gerion, Julie Herberg, Robert Bok, Erica Gjersing, Erick Ramon, Robert Maxwell, John Kurhanewicz, Thomas F. Budinger, Joe W. Gray, Marc A. Shuman and Fanqing Frank Chen "Paramagnetic silica-coated nanocrystals as an advanced MRI contrast agent" J. Phys. Chem. C 2007, 111, 12542–12551.
70) Willem J. M. Mulder, Rolf Koole, Ricardo J. Brandwijk, Gert Storm, Patrick T. K. Chin, Gustav J. Strijkers, Celso de Mello Donegá, Klaas Nicolay and Arjan W. Griffioen "Quantum dots with a paramagnetic coating as a bimodal molecular imaging probe" Nano Lett., Vol. 6, No. 1, 2006.
71) Er-Qun Song, Jun Hu, Cong-Ying Wen, Zhi-Quan Tian, Xu Yu, Zhi-Ling Zhang, Yun-Bo Shi and Dai-Wen Pang "Fluorescent-magnetic-biotargeting multifunctional nanobioprobes for detecting and isolating multiple types of tumor cells" ACS nano 5, No. 2, 761–770, 2011.

2 Synthesis Approaches of Magnetic Quantum Dots

Nurhan Onar Çamlıbel
Textile Engineering Department, Pamukkale University, Denizli, Turkey

CONTENTS

2.1 INTRODUCTION

QDs are unique inorganic fluorescent nanoparticles (2–10 nm) formed from atoms from groups II or groups IV and VI (e.g., ZnS, ZnSe CdSe, CdTe, CdS, PbSe) or groups III and V (e.g., GaAs, GaN, InP, InN and InAs) by combining metallic elements and non-metallic elements called as metal chalcogenide and perovskite quantum dots and exhibit luminescence and glow with size dependence when excited with a specific wavelength light (Tufani et al. 2021; Kumar 2018; Brett 2022; Armaselu 2017). Recently, carbon quantum dots (CQDs) and graphene quantum dots (GQDs) instead of semiconductor QDs attracted much interest because of its much lower toxicity, biocompatibility and photostability, great optical property, good solubility and photoluminescence properties chemical stability and easy synthesis (Scaria 2020; Dincer 2022).

In comparison with the fluorescent proteins and organic dyes, QDs have outstanding tunable electrical and optical properties including high quantum yield, large effective Stokes shifts, broad absorption, broad excitation spectra, narrow emission bands for multicolor imaging, size-tunable emission wavelength, brightness, photostability, good chemical stability and high photobleaching threshold (Armaselu 2017; Su 2014; Sun 2010; Hsu 2011).

DOI: 10.1201/9781003319870-2

QDs are synthesized with several techniques such as the microwave-assisted aqueous "green" route, photolithography, hot injection method, wet chemical synthesis, surfactant micelles, coprecipitation or organic solvent synthesis at high temperature. Optical and electrical properties of QDs highly depend on the extent of their surface defects. These can be decreased by surface modification with inorganic or organic capping agents while ensuing an increase on quantum yields of QDs by passivation or other means, targeted delivery, circulation and colloidal stability. Various approaches have been investigated to promote their optical performance, chemical stability, photostability and environmental stability and to hinder the leakage of toxic metals and to control dispersibility, including passivation of QDs as core (e.g., CdS, CdSe, PbS, PbSe, InP, CuInS2, Si) with cross-liked inorganic thin shells (ZnS, ZnSe) by successive ionic layer adsorption and reaction (SILAR), sonochemical, heat-up methods, cation exchange and ligand or polymer bridging such as thiol-based coating, silica coating for stearic stabilization on QD surface and polymers or oxides encapsulation (PMMA, PVA, PVP, liposomes, glyconanospheres, dendrimers, SiO2, TiO2, ZrO2, etc.) (Zhang 2022; Selvan 2010; Brett 2022; Armaselu 2017; Part 2018; Bhandari 2015; Evans 2010).

QDs are great contrast and tracing agents for biomedical labelling and bioimaging (Mahato 2017). Several methods for conjugation of QDs with polymers, antibodies, proteins, drugs, ligands, and subcellular organelles were researched enabling bioimaging, phototherapy and quantitatively detection of target molecules. (Tufani 2021; Ishikawa 2011). However optical tracing with single QDs nanocrystals is inadequate for complex in-vivo applications (Part et al. 2018). For multimodality imaging, various nanoparticles such as QDs, magnetic nanoparticles, lantanides, SW CNT, Au nanoparticles could be combined enabling emission at NIR region well known as "optical transmission window" to in-vivo imaging and visible region (400-700 nm) (Ma 2012). The combination of QDs and magnetic nanoparticles (MNPs) with multimodal fluorescence and magnetic properties offer a great solution.

Magnetic nanoparticles such as monodisperse iron oxides (magnetite (Fe3O4), maghemite (γ-Fe2O3) and hematite (α-Fe2O3)), Co, Mn, Fe, CoFe, FePt and SmCo5 nanoparticles with excellent magnetic, structural and biochemical properties have high potential for biomedical applications (Wu 2015; Frey 2009).

There has been increasing interest in MQDs composed of fluorescent materials (fluorescent proteins, organic dyes and QDs) and magnetic nanoparticles (MNPs) as multifunctional nanosystems with multimodal luminescence/fluorescence and magnetic functionality enabling combination of fluorescence imaging together with computed tomography (CT), magnetic resonance imaging (MRI), positron emission tomography (PET) etc. and diagnostics (Point-of care (PoC) disease diagnosis), treatment, magnetic separation, labelling, therapy, cell harvesting, tracking and drug release applications (Koole et al. 2009; Tufani et al. 2021; Oluwafemi et al. 2021; Majahan et al. 2013; Selvan 2010; Armaselu 2017). However, MQDs still have some limitations and challenges such as the reduction of the PLQY, a high hydrodynamic diameter, difficult synthesis methods, their dispersibility in aqueous medium, cytotoxicity of (especially heavy metal based) QDs, stability in blood, circulation life time, ease of functionalization for targeted delivery (Galiyeva 2021; Bhandari 2015). The preparation of nanosized (<100 nm) multifunctional and multimodal MQDs and MQDs composites modified with some polymers, micelles or silica have great importance because these particles could be employed in more advanced cell separation (< 30 nm to cross the nuclear membrane) and labelling applications (<15–20 nm for labelling minute subcellular features) (Evans 2010).

Most of MQDs are originated from paramagnetic and superparamagnetic agents such as ions (e.g., gadolinium Gd(III)) or iron based MNPs (e.g., iron oxide [Fe_2O_3] nanoparticles, SPIONs) because of their abundancy, easy to synthesize and low cost (Tufani 2021). MQDs have the potential to be employed as contrast agents in MRI. Novel multicomponent MQDs suggested long-term stability, spectral flexibility and enabling excitation in NIR-I, NIR-II and NIR-III as contrast agent for merging magnetic resonance and fluorescence imaging techniques (Koole et al. 2009; Selvan 2010; Armaselu 2017).

Four different techniques to fabricate MQDs were described in the following ways:

- Heterocrystalline growth in which QD is overgrown with a layer of a magnetic material and linked to an MNP
- Encapsulation or assembly of separately QDs and MNPs or single MQDs in silica gel micelles/liposomes and polymers (silica or polymer coated MQDs)
- Doping of paramagnetic transition metal ions into QDs
- Miscellaneous methods (CNT based iron oxide and QD conjugation with subsequent attachment of biomolecules) (Armaselu 2017; Tufani et al. 2021; Selvan 2010; Koole et al. 2009; Acharya 2013).

2.2 SYNTHESIS METHODS OF MQDs

2.2.1 Heterocrystalline Growth

Magnetic quantum dots could be synthesized at heterostructures with core-shell configurations or two distinct asymmetric nanoparticles by heterodimer architecture (Das 2021; Tufani 2021). Figure 2.1 showed MQDs synthesis approach with heterocrystalline growth method. The composition, synthesis route, analysis techniques, purposes and optical and magnetic performance of the MQDs fabricated via heterocrystalline growth method were given in Table 2.1.

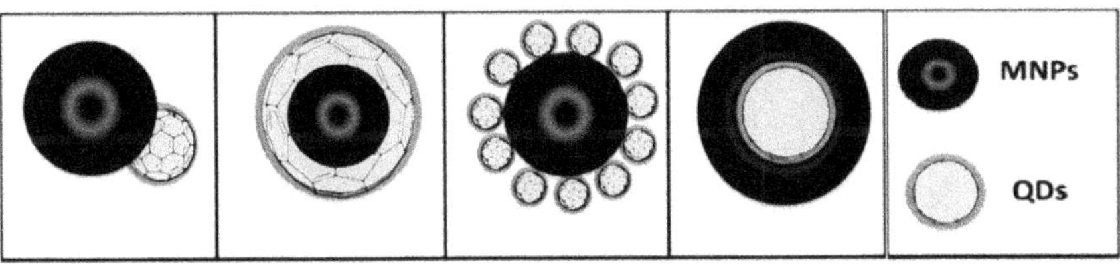

FIGURE 2.1 MQDs synthesis approach with heterocrystalline growth method. Reproduced with permission from (Das 2021); published by The Royal Society of Chemistry, 2022. Licensed by Creative Common CC BY 3.0.

TABLE 2.1
The Composition, Synthesis Route, Analysis Techniques, Purposes and Optical and Magnetic Performance of MQDs Synthesized by Heterocrystalline Growth Route

MNP Composition	QD Composition	Modifying Agent	Method	Purpose	Emission/Size-Zeta Potential	QY	Magnetism	Ref.
FePt	PbS, PbSe	OA, OLA, ODE	Coprecipitation method	Electronics, optoelectronics.	13.6 nm for FePt-PbS		0.1 emu/g at 300 K for FePt-PbS	Lee 2010
FeP	CdTe	TOP, TOPO	one-pot chemical synthesis process	Biosensing, bioimaging, biolabelling, magneto-electronics.	460 nm/~ 6 nm for FePt	~25%	No saturation at 300 K	Deng .2010
γ-Fe2O3	CdSe	OA, SA, TOP, TOPO, HAD, APS, TMAH, Igepal-CO520, BAM	Seed-mediated growth	Biolabelling,	550 to 600 nm/ 13 nm	13% to 18% for MQDs, 8% to 13% for SiO2-MQDs	~0.40emu/g, from 5 K to 15K	Lin 2011
Fe3O4	ZnS	HQ, cysteine, OLA, OA, TMAH	Seed-mediated growth	Nanoprobe for targeted cellular imaging	500 nm/>10nm (-25.5 mV)	5.9%	3.88 and 3.81 emu/g for MQDs and HQ treated MQDs	Bhandari 2015
Fe3O4	CdTe, CdS	GSH, MPA, TGA	simple aqueous self-assembly route for MQDs, microwave technique for QDs	Subcellular labelling	550-600 nm/15-20 nm	7%	9.2 emu/g	Evans 2010
Fe3O4	CdTe	TGA, sodium citrate, CTAB, hCC49 antibodies Fab region	LbL self-assembly	fluorescentmagneto nanoprobes, biolabelling	540 nm/50 nm (DLS)		65 emu/g	Ahmed 2013

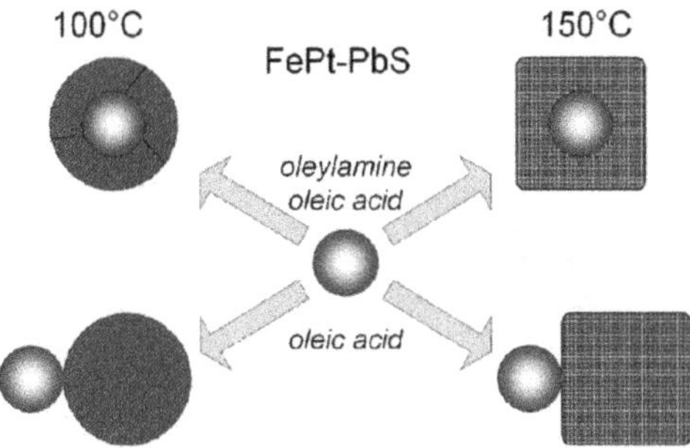

FIGURE 2.2 Different structures of FePt-PbS under different synthesis conditions. Reproduced with permission from Lee (2010). Copyright (2010) American Chemical Society.

Lee et al. (2010) reported the synthesize of magnetic core-shell FePt-PbS and FePt –PbSe MQDs for electronics and optoelectronics (Figure 2.2). The results with zero-field-cooled (ZFC) scans at H=100 Oe for FePt-PbS cubic core (3.5 nm) -shells (~4 nm) demonstrated that FePt-PbS cubic core-shells have two peaks at 14 K and ~70 K which demonstrate the presence of two magnetic phases in FePt-PbS MQDs fabricated at 150 °C. FePt-PbS and FePt-PbSe nanostructures exhibited a softening of magnetic response at the low temperature (5 K) with reducing approximately 20 and 80 Oe of Hc, while FePt nanocrystals showed ferromagnetic behavior with the coercive field (Hc) ≈ 1.9 kOe at 5K and superparamagnetic behavior with Hc=0 at room temperature (300 K).

Deng et al. synthesized iron phosphide-CdTe (FeP-CdTe) MQDs with various FeP:CdTe ratios by the one-pot method. The absorbance peaks originated from FeP (~215 nm) and CdTe QDs (640 nm) displayed slightly blue-shift and red shift respectively for FeP–CdTe (1.5:1) nanocomposites. The quantum yield of the FeP–CdTe NCs was ~25%. FeP NPs exhibited ferromagnetic (2.5 kOe of coercivity and 20 emu/g of magnetic saturation at 5 K) while MQDs were displaying superparamagnetic behavior at 5 K and lost their superparamagnetic behavior at 300K. Meanwhile the absence of FeP peaks in MQDs in XRD spectra showed possible changing of the FeP nanostructure throughout the synthesis of CdTe.

Lin et al. (2011) synthesized heterodimer γ-Fe_2O_3-CdSe MQD at different growth times with size-tunable emission properties by seed mediated growth method and coated MQDs with silica using amino propyl trimethoxysilane (APS) as precursor and then conjugated SiO_2-MQDs to bioanchored membrane (BAM, oleyl-O-poly(ethyleneglycol)-succinyl-N-hydroxysuccinimidyl ester) for biolabelling (Figure 2.3). The MQDs exhibited superparamagnetic behavior with a coercivity of 138Oe and a saturation magnetization of 0.40emu/g at 5K.

Bhandari et al. fabricated superparamagnetic iron oxide nanoparticle Fe3O4 (SPION)–ZnS MQDs multimodal nanocomposites with low photobleaching rate, high QY, good stability especially in blood serum, proper lifetime, superparamagnetism, and no apparent cytotoxicity. Figure 2.4 demonstrated fluorescence and magnetic performance of the MQDs and their synthesis route. Cysteine-capped ZnS were grown on the SPION surface and then MQDs were treated with 8-hydroxyquinoline (HQ) as ligands, called MQD composites (HQ treated MQDs). HQ treated MQDs with

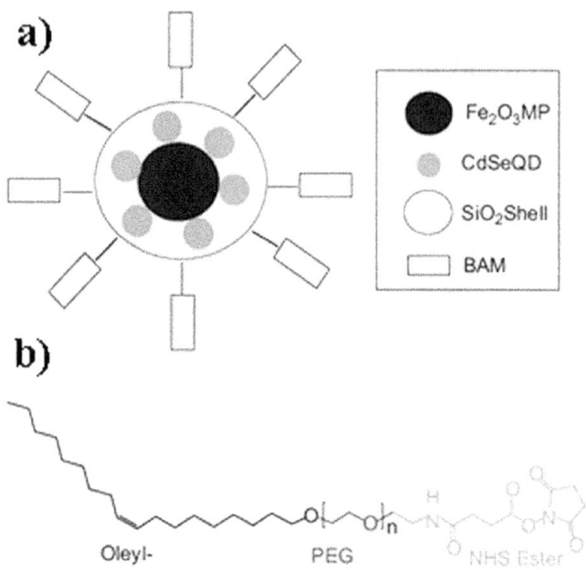

FIGURE 2.3 a) Schematic representation of BAM/SiO_2/MQDs, and b) chemical structure of BAM (Lin 2011).

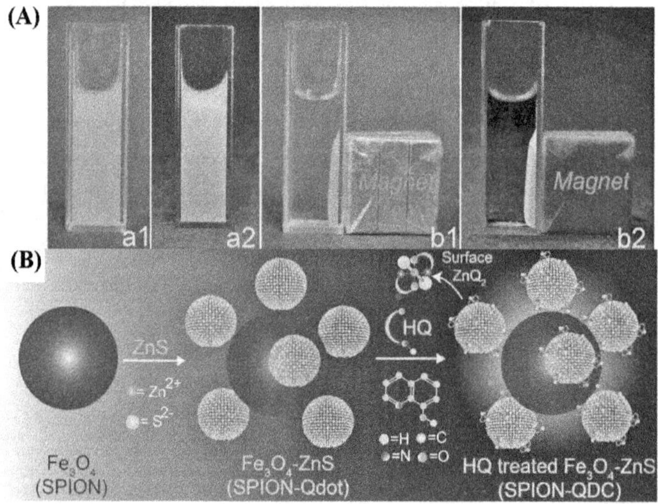

FIGURE 2.4 (A) Images of SPION-QDC in water with absence (a) and presence (b) of an external magnetic field under white (1) and UV light (365 nm) (2). (B) Synthesis of Fe_3O_4–ZnS MQDs. Reproduced with permission from (Bhandari 2015). Copyright (2015) American Chemical Society.

red-shifted emission maximum exhibited great optical property without the loss of magnetic property in comparison with the MQDs.

Evans (2010) prepared small size MQDs (CdTe–Fe_3O_4 nanocomposites) via a simple aqueous self-assembly method for binding of QDs to MNPs without solubility issues and needing thick polymer or silica coating, where QDs and MNPs also were produced by aqueous processes. For good attachment of L-glutathione (GSH)-functionalized QDs to Fe_3O_4, 3-mercaptopropanoic acid (MPA) or thioglycolic acid (TGA) was added to QDs solution and then Fe_3O_4. GSH as a stabilizer and sulfide ions source was used for the passivation of CdTe NPs with CdS shell. The MQDs exhibited superparamagnetic properties with 9.2 emu/g of magnetic saturation, strong fluorescent emission and size-tunable luminescence depending on S:Te ratio and using or not MPA or TGA.

Ahmed et al. (2013) coated citrate-modified negatively charged Fe_3O_4 NPs with positively charged CTAB and then electrostatically linked negatively charged TGA-capped CdTe QDs with CTAB to synthesize MQDs for cancer cell imaging, biosensing and magnetic separation. The MQDs were conjugated with hCC49 antibodies Fab region to evaluate its binding activity with LS174T cancer cells as fluorescent nanoprobes. MQDs indicated excellent colloidal stability, low cyctotoxicity after 64 fold dilutions, superparamagnetic properties (65 emu/g) and strong fluorescence intensity.

2.2.2 Encapsulation and Assembly Method

QDs and MNPs were encapsulated in the silica matrix, polymers or micelles/liposomes (Tufani 2021). The synthesis of colloidal NCs with monodispersability and a particle size less than 200 nm was essential to manage good distributions in the targeted cancer cells (Cho 2010). Figure 2.5 showed MQDs synthesis approaches with encapsulation or assembly of MNPs and/or QDs. In Table 2.2, the composition,

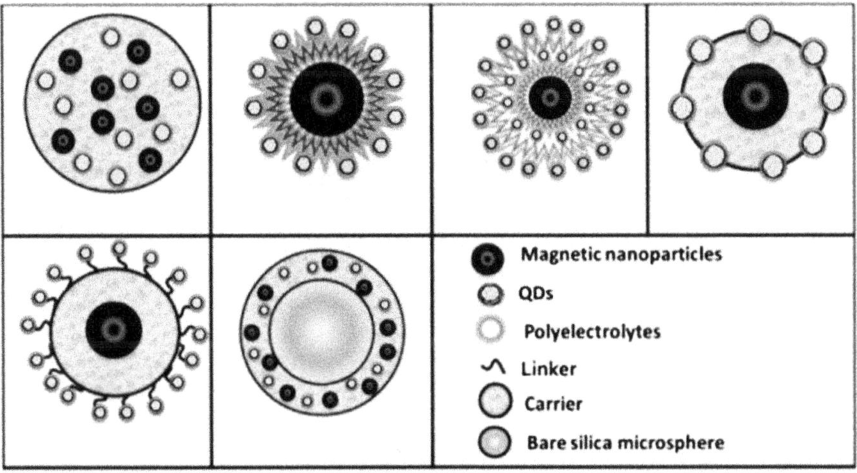

FIGURE 2.5 MQD synthesis approaches with encapsulation or assembly of MNPs and/or QDs. Reproduced with permission from Das (2021); published by The Royal Society of Chemistry, 2022. Licensed by Creative Common CC BY 3.0.

TABLE 2.2
The Composition, Synthesis Route, Analysis Techniques, Purposes, and Optical and Magnetic Performance of MQDs Synthesized by Encapsulation and Assembly Methods in Silica Matrix

MNP	QD	Modifying Agent	Silica Shell Precursor	Method
	ZnS:Mn2^{+}, Fe3O4	OA, sodium oleate, Igepal CO-520, NIPAAM, AAM, BIS, PEG-4000, DOX (drug)	TEOS, MPTS	reverse microemulsion method/ free radical polymerization method
Fe3O4	CdSe/CdS	DCC, FA, OA, ODA	TEOS, APTES	oil-in-water microemulsion for silica encapsulation
Fe3O4	CdSe/ZnS	MPA, sodium citrate	TEOS, MPS	Sol-gel process for silica coating
Fe3O4	CdTe	MPA, CA	TEOS, APTES	reverse micro-emulsion method for silica encapsulation
Fe3O4	CdSe/ZnS, CdSeTe/CdS	TOP, TBP, TOPO, $(TMS)_2S$, HDA, THPMP, CA, EDC	TEOS, APS	microemulsion method
Fe3O4	CdTe	OA, TGA	APTES	reverse microemulsion method for silica encapsulation
Fe3O4	CdTe	GSH, EDC/NHS	TEOS, APTES	reverse micro-emulsion method for silica encapsulation
Fe3O4	ZnS		MPS, TEOS	solution growth, coprecipitation method
Fe3O4	CdTe	OA, MPA, CTAB	TEOS, APS	LbL assembling, Stöber method
Fe3O4	CdTe	PEI, MPA, TGA, MPS, NIPAM, MBA,	TEOS	LbL assembling, Stöber method
Fe3O4	CuInS2/ZnS	ODE, OM, Igepal CO-520, OLA, EDC/NHS	TEOS, APTES, MPEGS	one-pot two-step reverse microemulsion method
γFe2O3	CdSe	TOPO, HDA, TOP, Igepal-CO520, TMAH, BAM	APS	reverse microemulsion method

Purpose	Emission/Size	QY	Magnetism	Ref.
Tumor therapy (TGI of 80%), biolabelling, bioimaging, drug release	589 nm for /29 nm for MQD-silica, 200 nm MQD-silica-polymer	22.4% for MQD-silica-polymer	~3 emu/g at room temp.	Cao 2018
tumor cell imaging and cell labelling	617 nm/600 nm size	42% of QDs	~0.03 emu/g at 300K (superparamagnetic)	Sun 2010
in-vivo nanohyperthermia and cancer treatment, biolabelling	600 nm for QD/ 10 nm (Fe_3O_4), 5 nm (QD), 4 nm silica shell	30 % for MPA-capped QD	15.4 emu/gr	Xu 2010
near-simultaneous multicomponent separation and analysis	550 nm/60 nm±5		15.28 emu/g with high Fe_3O_4 concentrations	Song 2014
Multimodality bioimaging	650 nm and 750 nm/150 nm			Ma 2012
Simultaneous in-vivo targeted magnetofluorescent imaging and targeting therapy	640 nm/50 nm		4 emu/g	Wang 2011
detection of latent fingerprints.	498nm/50 nm	4.25 %	3.1 emu/g	Wang 2019
targeted and tracing drug delivery carrier	330 nm/100 nm		30 emu/g	Koc 2017
Targeted drug delivery, fluorescent tracing	~675 nm/ 70-90 nm		7 emu/g	Yin 2016
thermo/pH-sensitive drug carriers for in-vivo therapy.	530 nm/190 ± 15 nm		1.4 emu/gr	Gui 2014
imaging probes for cancer diagnosis and chemotherapy.	580 nm/< 30 nm		25 emu/g at 300K	Hsu 2011
biolabelling, bioimaging, bioseparation	12-14 nm	8–10%		Selvan 2007

(continued)

TABLE 2.2 (Continued)
The Composition, Synthesis Route, Analysis Techniques, Purposes, and Optical and Magnetic Performance of MQDs Synthesized by Encapsulation and Assembly Methods in Silica Matrix

MNP	QD	Modifying Agent	Silica Shell Precursor	Method
Fe3O4/ γFe2O3	CdTe	TGA, TMAH, PDADMAC/PSS/ PDADMAC	TEOS, APS	LbL assembly process
	Fe3O4@SiO2–G2:5–carbon dot	FA, DOX (drug)	TEOS, Dendrimer	Coprecipitation, Stöber method, hydrothermal method,
	CsPbBr3/ Fe3O4	OA, OLA, ODE, PEF-4000, mesoporous polystyrene microspheres	TMOS, APS	heat injection method, encapsulation
Fe3O4	CdTeS	OA, FA, CTAB	TEOS, APS	modified co-precipitation, modified Stober method, encapsulation, hydrothermal synthesis,

synthesis route, analysis techniques, purposes, and optical and magnetic performance of MQDs synthesized by encapsulation and assembly methods in silica matrix were presented.

2.2.2.1 MQDs Embedded in Silica Matrix

Essential reasons for coating silica on QDs are its biocompatibility, great chemical stability, easily furthered conjugation with different functional groups and inhibiting toxicity that arises from QDs containing toxic heavy metals (Cao 2013; Tufani 2021). The NPs could be encapsulated with SiO_2 shells by a sol–gel process (the "Stober method") and the reverse microemulsion synthesis (Hsu 2011).

Cao et al. (2018) synthesized ZnS:Mn^{2+} QDs Fe_3O_4 QDs/SiO_2 core-shell NCs by the reverse microemulsion method and subsequently functionalized their surface by the 3-(trimethoxysilyl) propyl methacrylate (MPTS). Fluorescent, superparamagnetic, thermosensitive and water soluble ZnS:Mn^{2+} QDs Fe_3O_4 QDs/SiO_2/P(NIPAAm-co-AAm) core-shell-shell NCs were synthesized by free radical polymerization process employing N-isopropylacrylamide (NIPAAM), acrylamide (AAM), N,N-methylenebisacrylamide (BIS) for drug (Doxorubicin (DOX)) release and tumor inhibition.

Yi et al. (2005) synthesized the QDs and MPs separately and then incorporated MQDs within silica by a facile reverse microemulsion method to fabricate MQDs with the emission and magnetic properties. However, they displayed low quantum yield and needed a long synthesis time.

Purpose	Emission/Size	QY	Magnetism	Ref.
luminescent markers, magnetic separation	~600 nm/Size 220 ± 10 nm		1.34 emu/g at room temp.	Salgueirino 2006
Drug delivery, fluorescent probe	396 nm/400 nm size		31.6 emu/g	Karimi 2021
capturing circulating tumor cells (CTC)	500 nm/3-10 μm		1 emu/g	Ma 2021
fluorescence labelling and photothermal therapy	556 nm/>80 nm		~40 emu/g	Yin 2021

Sun et al. (2010) stabilized oleate-Fe_3O_4 MNPs and octadecylamine capped-CdSeS QDs with cetyltrimethylammonium bromide (CTAB) and then water soluble MNPs and QDs were encapsulated with hollow silica shell and finally MQDCs were conjugated with folic acid (FA) for cell imaging and labelling in Hela cells and NIH 3T3 fibroblast cells (Figure 2.6). Multifunctional and biocompatible silica nanocomposites showed luminescent and superparamagnetic properties with no hysteresis. It was also reported that uniform silica nanocomposites with a 600 nm size (above 100 nm) possess low cyctotoxicity (Sun 2010; Yu 2009).

Xu et al. (2010) coated MNPs with a silica film using TEOS and 3-mercatopropyltrimethoxysilane (MPS) precursor to fabricate Fe_3O_4/SiO_2-SH by sol-gel process and subsequently were combined with water-soluble CdSe/ZnS capped with 3-mercaptopropionic acid (MPA) (Figure 2.7). Multifunctional superparamagnetic, photoluminescent and highly efficient RF-absorber properties of MQDC with almost no cytotoxicity toward the Panc-1 cells and high biocompatibility were demonstrated. It was reported that stable ZnS shell and the MPA capping layer of CdSe QDs considerably reduced the cytotoxicity.

Song et al. (2014) encapsulated CA modified Fe_3O_4 MNPs with various concentrations and water-soluble CdTe QDs stabilized with MPA in silica shell ((CdTe/Fe_3O_4)@SiO2 MQDs-FMNC-fluorescence magnetic nanospheres) by reverse micro-emulsion method employing TEOS as precursor in basic media and furthered modified APTES (FMNC-NH_2) for multicomponent separation and analysis (Figure 2.8). FMNC exhibited superparamagnetic, fluorescence properties and

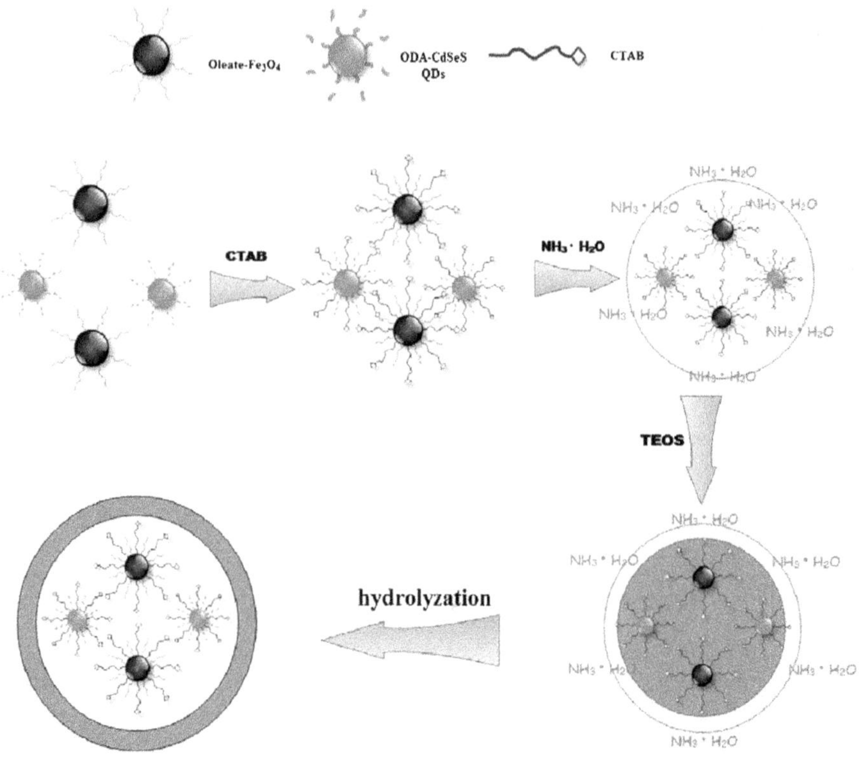

FIGURE 2.6 The synthesis of MQDs with silica encapsulation (Sun 2010).

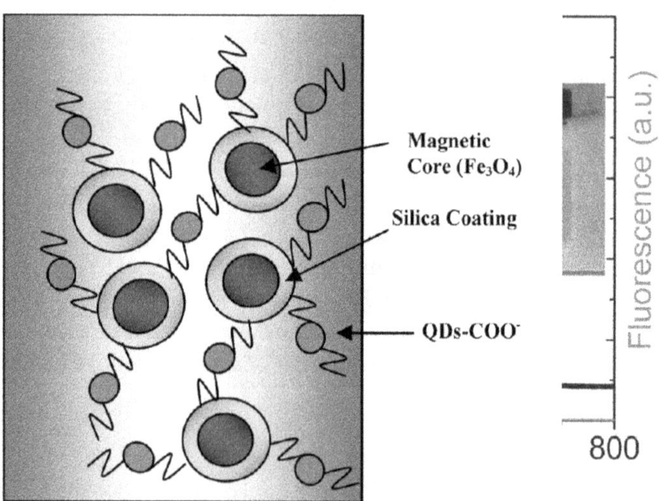

FIGURE 2.7 Schematic illustration of MQDs. Reproduced with permission from Xu (2010). Copyright (2010) American Chemical Society.

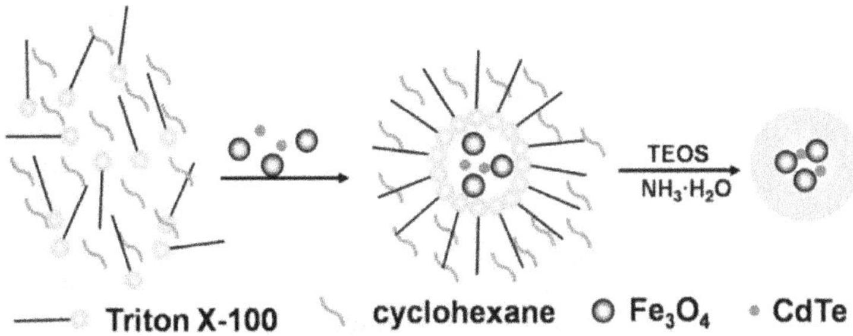

FIGURE 2.8 Schematic presentation the synthesis of (CdTe/Fe_3O_4)@SiO2 nanocomposites. Reproduced with permission from Song (2014). Copyright (2014) American Chemical Society.

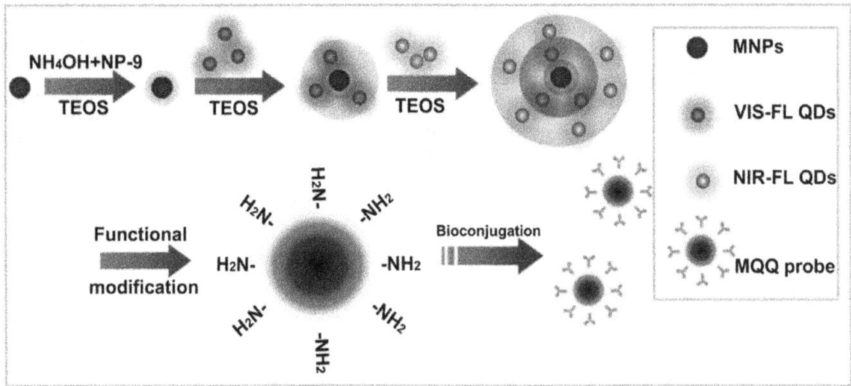

FIGURE 2.9 The synthesis route of a MQDs containing Fe_3O_4 MNPs, VIS-FL QDs and NIR-FL QDs (Ma 2012). NP-9: nonionic surfactant.

good stability and their magnetic saturation values increased while luminescence intensity decreased with increasing Fe_3O_4 concentration. It also easily demonstrated the conjugation of FMNC with immunoglobulin G (IgG), protein A, and antibody (Ab), exhibiting easy biofunctionalization.

Ma et al. (2012) synthesized multi-layered MQD nanocomposite (MQQ) probes containing MNPs (Fe_3O_4), VIS fluorescent CdSe/ZnS QDs and NIR-fluorescent CdSeTe/CdS QDs with encapsulation in silica matrix by microemulsion method for multimodality imaging in vivo and in vitro. The MQDs was modified employing APS and 3-(trihydroxysilyl)-propylmethyl-phosphonate (THPMP) and conjugated anti-HER2 antibody. 150 nm diameter of MQQ with narrow size distribution, 1914 mg $L^{-1}s^{-1}$ of r^2 relaxivity (as an effective T2 contrast agent) and low cytotoxicity of HER2-MQQ probe were demonstrated (Figure 2.9).

Wang et al. (2011) prepared fluorescent magnetic nanoprobes (FMNPs) with CdTe and Fe_3O_4 MNPs embedded in silica matrix with good biocompatibility and stability and then functionalized FMNPs with amino groups and carboxyl groups. It was reported that carboxyl functionalized FMNPs (-30.50 mV of zeta potential value) were easier conjugated with BRCAA1 protein than amino functionalized FMNPs (zeta-potential value of 24.80 mV). FMNPs-COOH was conjugated with BRCAA1 protein for specific targeted imaging of cancer cells in vivo. BRCAA1-conjugated FMNPs showed very low toxicity, lower fluorescent intensity with blue-shift and lower magnetic intensity from 4 emu/g to 3.7 emu/g than pure FMNPs. It was suggested simultaneous employing FMNPs for hyperthermia therapy of cancer cells irradiated with an altering magnetic field and imaging.

Wang et al. (2019) performed covalently binding of COOH-modified CdTe QDs with amino functionalized $Fe_3O_4@SiO_2$ NPs ($Fe_3O_4@SiO_2$- NH-CO-CdTe- MQDs). Hydrophilic MQDs with carboxyls on the surface and spherical core-shell structure at uniform size of 50 nm showed desired magnetic properties with magnetic saturation of 3.1 emu/g and broad band emission peak at 498 nm with quantum yield of 4.25%, suggested for their application in detection of latent fingerprints.

Koc et al. (2017) assembled $Fe_3O_4@SiO_2$ nanoparticles and (3-mercaptopropyl) trimethoxysilane (MPMS) capped ZnS QDs via Si-O-Si bonds. $ZnS@Fe_3O_4$ nanospheres with uniform size distribution (100 nm) exhibited maximum fluorescence intensity at 330 nm for assembling time of 18 h, which slightly higher than fluorescence intensity of ZnS and superparamagnetic properties with magnetic saturation of 30 emu/g. It was revealed that fluorescence intensity of MQDs at 330 nm increased while assembling process time was increasing (Figure 2.10).

Chen et al. (2022) synthesized hollow carbon/Fe_3O_4 magnetic quantum dots (C/MQDs) using silicon dioxide as the template by in-situ polymerization-solvothermal-calcination process.

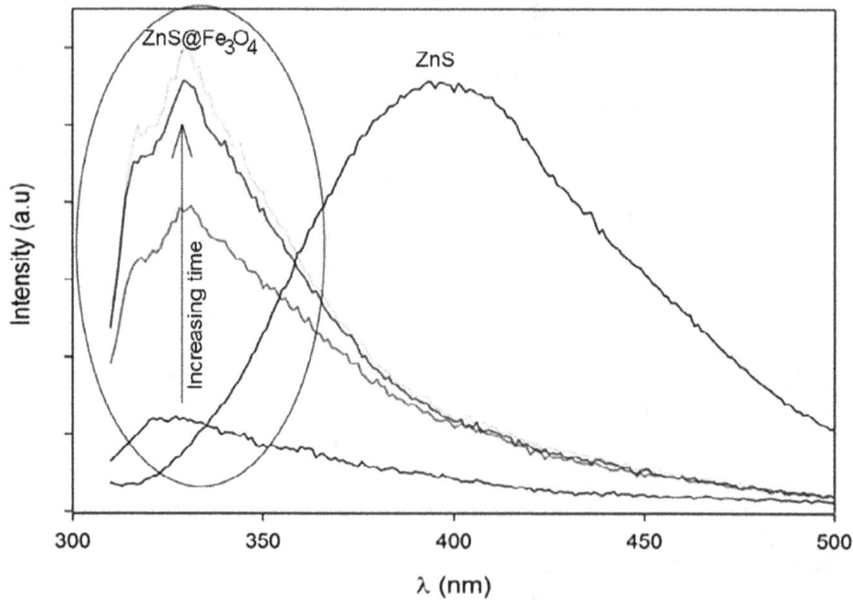

FIGURE 2.10 Fluorescence spectrum of $ZnS@Fe_3O_4$ nanospheres and ZnS QDs depending on assembling time. (t= 0, 6, 12,18 h) (Koc 2017).

Yin et al. (2016) synthesized oleic acid stabilized-Fe_3O_4 MNPs (zeta potential value of +42 mV) by coprecipitation process and surface modification process and mesoporous silica shell was assembled to MNPs. SiO_2/Fe_3O_4 NPs were modified with 3-aminopropyltrimethoxysilane. MPA-capped CdTe QDs was attached to NH-SiO_2/Fe_3O_4 by electrostatic interaction. FMNC have a uniform size distribution between 70-90 nm. The emission peak of FMNC slightly redshifted and the photoluminescence (PL) intensity was slightly reduced in comparison to the CdTe QDs while magnetic saturation of Fe_3O_4 (66 emu/g) were significantly decreasing to 7 emu/g for FMNC. It was reported that FMNC could be connected with antibody by glutaraldehyde method for labelling process. It was also suggested that drugs attached to silica shell and encapsulated with pH-sensitive or thermosensitive polymer matrix could be transported into the targeted cells under an external magnetic field and released altering pH or temperature.

Hsu et al. fabricated SiO2 nanocomposites containing Fe_3O_4 MNPs and $CuInS_2$/ZnS QDs and functionalized by APS, 2-(Methoxypoly(ethyleneoxy) propyl) trimethoxysilane (MPEGS) to attach both poly(ethyleneglycol) (PEG) and FMNC by a one-pot two-step reverse microemulsion route. Dual modality FMNCs showed superparamagnetic properties with 25 emu/g magnetic saturation at 300 K and fluorescence performance. Maximum emission wavelength of FMNCs with bright orange emission under UV light source slightly redshifted in comparison with MQDs. Pt(IV)-anticancer drugs were conjugated to FMNCs. Pt(IV)-conjugated FMNCs exhibited higher cytotoxicity to MCF-7 (human breast cancer) cells than the free Pt(IV) anticancer drug. It was also demonstrated that FMNCs are good T2 MRI contrast agents with high efficiency.

Gui et al. (2014) stabilized Fe_3O_4NPs with thiodiglycolic acid (TGA), and then modified them with polyethylenimine (PEI) while their surface was positively charging. 3-mercaptopropionic acid (MPA)-capped CdTe QDs electrostatically interacted with PEI-modified Fe_3O_4 NPs and subsequently MQDs were encapsulated in mesoporous silica NPs and modified with MPS. FMNCs were embedded by free radical polymerization of N-isopropylacrylamide (NIPAM) in the presence of chitosan (CS) and FMNCs to produce dual-embedded mesoporous FMNC/PNIPAM-g-CS which shown excellent magnetism, fluorescence, thermosensitivity originated from PNIPAM polymer, pH-sensitivity stem from CS polymer and biocompatibility properties. Encapsulation with PNIPAM-g-CS polymer did not significantly affect the PL spectrum (maximum emission peak ~530 nm) of silica embedded FMNCs and dual-embedded FMNCs have highly stable relative absorbance. The magnetic saturation values of superparamagnetic dual-embedded FMNCs was lower than only silica embedded FMNCs and Fe_3O_4 NPs due to diamagnetic nature of silica shells. It was also reported that dual embedded FMNCs incubated with HepG2 cancer cells displayed bright green PL under irradiation at 480 nm of excitation wavelength. Adriamycin (ADM) (at >0.8 mg/ml) loaded dual embedded FMNCs with high loading efficiency possessed higher toxicity than free ADM while dual embedded FMNCs without loading ADM did not exhibit toxicity to HepG2 cells. Figure 2.11 illustrated fabrication of hybrid microspheres and loading and release of ADM from hybrid microspheres. Figure 2.12 displayed the in-vitro cytotoxicity of free ADM, microspheres with and without ADM loading at various concentrations.

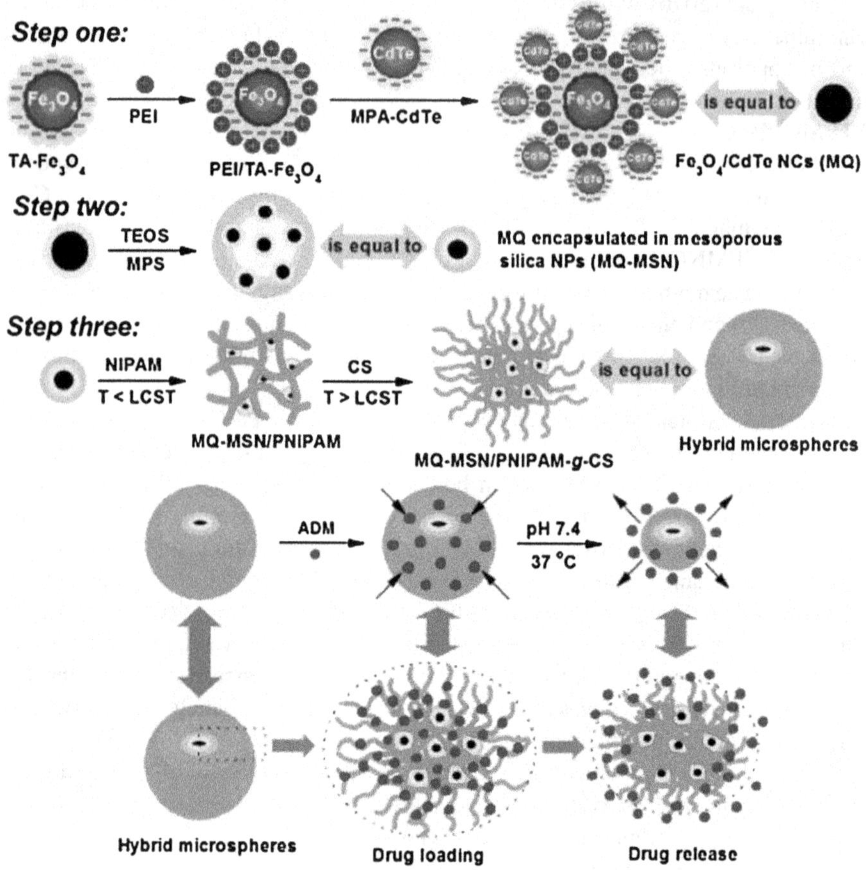

FIGURE 2.11 Schematic showing of the fabrication of hybrid microspheres and loading and release of ADM from hybrid microspheres (Gui 2014).

Selvan (2007) fabricated γ-Fe_2O_3-CdSe-NH_2 passivated with TOPO/HDA at heterodimer structure by seed-mediated nucleation and growth of QDs on the core Fe_2O_3 MPs with various growth times (1–8 min) resulting to red, yellow, green and orange emissions and high quantum yield. MQDs were salinized with using APS precursor by reverse microemulsion route and subsequently conjugated with oleyl-O-poly(ethylene glycol)succinyl-N-hydroxysuccinimidyl ester, called the bioanchored membrane (BAM). The labelling and imaging of SiO_2/MQDs treated with HepG2 human liver cancer cells, 4T1 mouse breast cancer cells and NIH-3T3 mouse fibroblast cells displayed the successful conjugation of SiO_2/MQDs with BAM.

Salgueirino-Maceira 2006 coated polyelectrolytes (alternate adsorption of PDADMAC (poly(diallyldimethyl ammonium chloride) and PSS (poly(sodium 4-styrenesulfonate) and PDADMAC) and subsequently thioglycolic acid-stabilized CdTe QDs on the surface of silica-embedded Fe_3O_4/c-Fe_2O_3 nanoparticles by layer-by-layer (LbL) assembly technique. QD patterned SiO_2/MNPs were finally

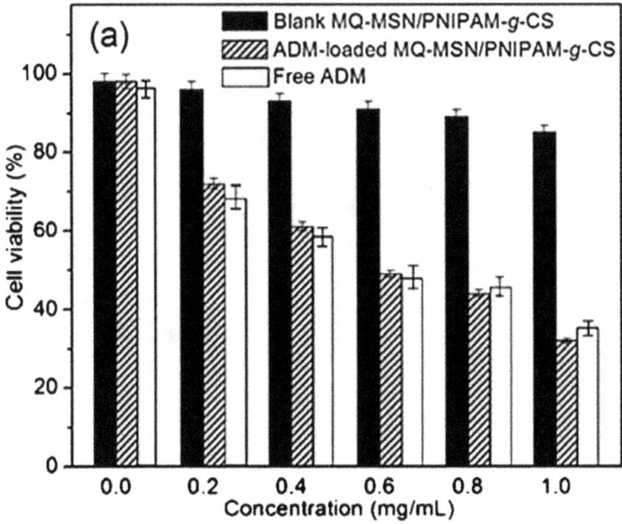

FIGURE 2.12 In-vitro cytotoxicity of free ADM, microspheres with and without ADM loading at various concentrations (Gui 2014).

encapsulated with silica shell. The FMNC exhibited superparamagnetic properties with 1.34 emu/g of magnetic saturation, size-tunable emission, strong excitonic photoluminescence. Final silica coating (20 nm thickness) caused blue-shift (10 nm) of maximum emission wavelength.

Karimi 2021 fabricated Fe_3O_4 by coprecipitation method and coated Fe_3O_4 with silica using TEOS as precursor by a modified Stober method to obtain Fe3O4@SiO2 and subsequently synthesized dendrimer-modified Fe_3O_4@SiO_2 (Fe_3O_4@SiO2–G1) nanoparticles treated with 4-dimethylaminopyridine (DMAP) as a catalyst and N,Ndicyclohexylcarbodiimide (DCC) as an activating agent at first step and dimethyl sulfoxide (DMSO), DCC and DMAP and then DMSO and melamine at second step. Fe_3O_4@SiO_2–G1 was crosslinked with N-doped carbon dots (CDs) hydrothermal method employing folic acid (FA) to obtain Fe3O4@SiO2–G2:5–CD. Fe3O4@SiO2–G2:5–CD was loaded with DOX. The carrier exhibited pH-dependent releasing, biocompatibility, good colloidal stability, and outstanding photoluminescent activity.

Ma 2021 synthesized CsPbBr3 NCs as perovskite nanocrystals by heat injection method and Fe3O4 QDs using FeCl3·6H2O, FeCl2·4H2O and PEG-4000. To prepare (CsPbBr3/Fe3O4)@MPSs@SiO2 magneto-optical microsphere, CsPbBr3 NCs and Fe3O4 QDs were encapsulated in mesoporous polystyrene microspheres (MPSs) in the presence of tetramethyl orthosilicate (TMOS), denoted as CFMS-MOMs microspheres. The microspheres were functionalized with N-maleimidobutyryloxy succinimide ester (GMBS) and MPMS and then grafted streptavidin (SA). Subsequently the microspheres were exposed with biotinylated anti-EpCAM antibody for specific cell targeting (Figure 2.13).

Yin (2021) synthesized Fe3O4 as core and mesoporous SiO_2 containing CdTeS QD as shell by LbL method for fluorescence labelling, magnetic separation and

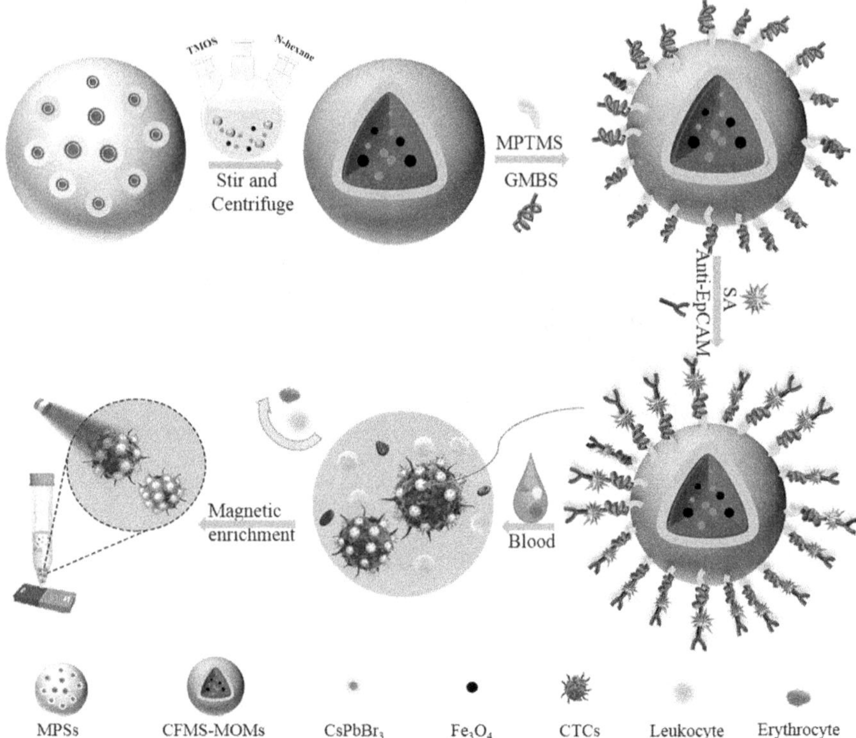

FIGURE 2.13 Schematic illustration of fabrication, functionalization and CTCs capture route of CFMS-MOMs (Ma 2021). CTCs: circulating tumor cells.

photothermal therapy for cancer cells (Figure 2.14). The nanoprobes exhibited great fluorescence and magnetic performance and specific targeted labelling behavior by modification with FA.

2.2.2.2 MQDs Embedded in Polymer Beads or Micelles

MQDs could be designed by encapsulation of QDs and MNPs into polymer materials and bounding of MNPs and QDs by covalent bond (Tufani 2021). Table 2.3 presented composition, synthesis route, analysis techniques, purposes, and optical and magnetic performance of MQDs synthesized by encapsulation and assembly methods in polymer beads, micelles or liposomes.

Li et al. carried out DNA-capped CdTe and CdTe/CdS core/shell QD synthesis in presence of L-glutathione (GSH) and thiourea as stabilizer and sulfide ions source and DNA-capped Fe_3O_4 NPs synthesis by conjugation NH2-modified DNA using N-hydroxysuccinimide (NHS) to carboxyl-modified Fe3O4 nanoparticles using 1-(3-dimethylaminopropyl)-3-ethylcarbodiimide hydrochloride (EDC). DNA-Fe3O4 NPs, DNA-QDs and two different aptamers self-assembled to produce MQAP and MQAM for cancer diagnosis, prognosis and magnetic isolation of rare cancer cells. MQAPs shown high magnetic response, high capture efficiency (CE), capture purity,

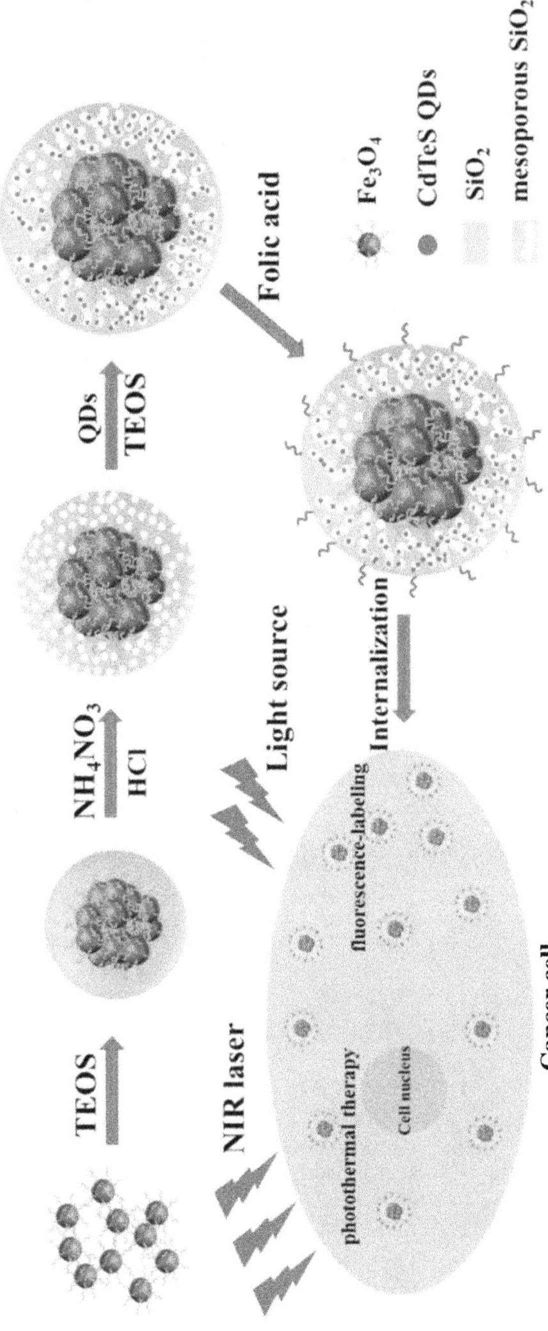

FIGURE 2.14 Schematic illustration of the synthesis of Fe_3O_4@ CdTeS QDs@SiO_2 nanocomposites (Yin 2021).

TABLE 2.3
The Composition, Synthesis Route, Analysis Techniques, Purposes, and Optical and Magnetic Performance of MQDs Synthesized by Encapsulation and Assembly Methods in Polymer Beads, Micelles or Liposomes

MNP Composition	QD Composition	Modifying Agent	Polymer	Method
Commercial Fe_3O_4 (15 nm)	CdTe/CdS	GSH, thiourea, DNA, NHS/EDC	Aptamer SC2, Aptamer SH2	hybridization chain reaction for MQAP and MQAM
Fe_3O_4	CdSe/ZnS	EDC/NHS, OA, OM, TOP, OA, ODE, MPA	PSEMBs	combining conventional swelling method with high-temperature swelling method
Iron oxide (SPIO)	Commercial organic QDs	OA, OM	PLA-TPGS	modified nanoprecipitation method
Iron oxide (IONPs)	Commercial PbS	sodium oleate, OA, ODE	PLGA	double-emulsion technique, thermal decomposition method
Fe_3O_4	CdSe/ZnS	TOPO, TOP, $(TMS)_2S$	PEI	Coprecipitation, hydrothermal method
Iron oxide	AgInS2/ZnS, CdTe/CdS, CdHgTe/CdS	OLA, GSH, TAA, TMAH,	dextran	Assembly approach based on covalent linking
Fe_3O_4 (~10 nm)	CdSeTe/ZnS	EDC-NHS, PTX (drug)	Polystyrene, PLGA	oil-in-water emulsion and coating methods
Iron oxide (5-10 nm), MWCNT	CdTe (2-5 nm)	MPA, EDC/NHS, COOH-PEG-COOH, DEG	PAH	LBL assembly
Fe_3O_4	ZnS:Mn		Chitosan	encapsulation
Iron oxide (IONPs)	CdS	GSH, TMAH, NHS/EDC, TPP, PD (drug)	Chitosan	encapsulation
Fe_3O_4	CdTe@ZnS	Glutaraldehyde, GSH, EDC/NHS, TGA, PEG-2000, DOX (drug)	CMCS, chitosan	encapsulation

Purpose	Emission/Size	QY	Magnetism	Ref.
ultrasensitive cell capture and tumor cell detection	639 nm/ 263 nm for MQAPs (hydrodynamic size)	19.4% for DNA-CdTe/CdS		Li 2018
fast separation and multiplexed assays	551 nm /12 μm for beads	40% reduction of fluorescence intensity with addition of 12.5 μg/mg MNPs to beads	~0.5 emu/g of beads with 12.5 μg/mg MNPs	Li 2011
medical diagnosis and treatment	652 nm/ 325.8 ± 5.2 nm			Tan 2011
bioimaging, hyperthermia treatments	1100nm/100 nm to 1 μm		55 A·m2/kg at 300K	Ortgies 2016
cells detection, biolabelling, in-vivo deep-tissue tracking	592 nm/57.8 nm	40 %	55 emu/g at 300 K	Lou 2011
Dualmode bioimaging.	607 nm/44 nm			Koktysh2011
cancer diagnosis and chemotherapy	790 nm/150 nm for micropsheres			Cho 2010
Cancer diagnosis and therapy	720 nm	14%	65 emu/g, at room temp.	Chen 2010
simultaneous biolabelling, imaging, cell sorting and separation	595 nm/ 7 nm		10.45 emu/g at room temp.	Liu 2012
healthcare diagnostic system, dual-mode imaging	505 nm/200-300 nm			Walia 2017
biolabelling, drug delivery, bimodal imaging	586nm/185 nm	71.2% PLQY	36.8 emu/g at 298 K	Ding 2017

(continued)

TABLE 2.3 (Continued)
The Composition, Synthesis Route, Analysis Techniques, Purposes, and Optical and Magnetic Performance of MQDs Synthesized by Encapsulation and Assembly Methods in Polymer Beads, Micelles or Liposomes

MNP Composition	QD Composition	Modifying Agent	Polymer	Method
Fe_3O_4	Commercial Qdot 655	EDC, PSS, Au NRs, Taxol (drug)	PLGA	seedless growth method, nanoprecipitation method
Fe_3O_4	CdSe	TOPO, OA, OLA	PEG-PAG-PEG	nanoemulsification/ solvent-evaporation
Fe_3O_4	GQD	SA, TCA (antidepressant)	Gelatin	UA-DMSPE
Fe_3O_4	GQD	NHS/EDC, sodium citrate	Chitosan	solvothermal synthesis,

colloidal stability and minimum cytotoxicity to CCRF CEM cells. CEM cells bonded with MQAPs possessed high QD PL intensity (32.9) of and high binding affinity.

Li et al. (2011) prepared porous carboxylic polystyrene beads by seeded swelling polymerization and MNPs-polystyrene beads and subsequently embedded QDs to MNPs-beads. The surface of MNPs-QDs-beads were activated EDC/NHS and then coated with human IgG and blocked with bovine serum albumin (BSA) for immunoassay test. The MNPs-QDs-beads exhibited excellent binding capacity for biomolecules, uniform size distribution, great stability against environment, high separation efficiency with superparamagnetic and fluorescence properties.

Tan et al. (2011) embedded iron oxide NPs (IOs) and organic QDs inside poly(lactide)tocopherol polyethylene glycol succinate (PLA-TPGS) with biodegradable properties by a modified nanoprecipitation process for multimodal imaging and detection. Figure 2.15 showed the efficiency of cellular uptake (cancer cells) of the nanocomposites containing various Cd concentrations under different incubation times, cell viability of QDs, IO and the nanocomposites under different incubation times. The QDs and IOs-loaded PLA-TPGS NPs showed great biocompatibility, stability (-37.3 mV of zeta potential), high encapsulation efficiency, uniform size distribution, high cellular uptake efficiency (40-50% within the first 4 h), low cytotoxicity (2.42 times lower in comparison with free QDs and MNPs), significantly improved tumor imaging properties.

Ortgies et al. synthesized hybrid nanostructures (HNS) formed PbS QDs-iron oxide NPs embedded in biocompatible poly(lactic-co-glycolic-acid) (PLGA) polymer via a double emulsion route for in-vivo deep-tissue tracking and hyperthermia treatments. HNSs exhibited superparamagnetic properties, good magnetic separation, low toxicity, long-term luminescence stability, emitting in the II-BW (1000 to 1350 nm.) enabling deep tissue high-resolution anatomical imaging and subtissue thermal sensing.

Purpose	Emission/Size	QY	Magnetism	Ref.
Chemotherapeutic, photothermal cancer therapy	655 nm/193 nm			Cheng 2010
Bioimaging, anticancer therapy.	535 nm/200 ± 38 nm	2 ± 0.2%		Antoniak 2021
Novel magnetic sorbent			~1.5 emu/g	Aladaghlo 2021
fluorescence imaging, MRI	434 nm		11.4 emu/g	Wang 2021

Lou et al. carried out the self-assembling of trioctylphosphine oxide (TOPO) capped CdSe/ZnS QDs on the surface of polyethyleneimine (PEI) capped Fe3O4 NPs for cancer cell imaging. Water soluble MNP-PEI-QDs presented colloidal stability, long-term fluorescence stability, high luminescence, great magnetic and fluorescence performance. MNP-PEI-QDs functionalized with TAT peptide exhibited cancer cell separation and emission of orange fluorescence from cell surface.

Koktysh et al. (2011) activated iron oxide MNPs surface with the oxidized dextran and then conjugated to GSH-capped CdTe/CdS. AgInS2/ZnS QDs, and CdHgTe/CdS QDs to produce fluorescent magnetic hybrid imaging nanoprobe (HINP) for bioimaging and magnetic separation. The conjugation with CdHgTe/CdS QDs displayed emitting in NIR (~800 nm) region. AgInS2/ZnS QDs as cadmium-free multimodal imaging agents were also synthesized and provided simultaneous VIS or NIR optical and MR imaging.

Cho et al. (2010) embedded Fe_3O_4 NPs in polystyrene matrix (MNSs). Carboxyl-functionalized MNS with EDC and amine-functionalized QDs with NHS were linked with electrostatic interaction to fabricate QD-conjugated MNSs. Subsequently QD-conjugated MNS were coated biodegradable poly(lactic-co-glycolic acid) (PLGA) with and without Paclitaxel (PTX) chemotherapeutic drug. The carboxyl-functionalized PTX-PLGA-QD-MNSs were coupled with ethylenediamine and then activated with NHS. NHS-activated PTX-PLGA-QD-MNSs were conjugated with anti-prostate specific membrane antigen (anti-PSMA) for targeting. The nanocarriers with 150 nm of particle size exhibited fluorescent and magnetic properties, monitoring capabilities emissions (790 nm) in the near-infrared range, successful drug loading capacity, hyperthermia under an alternating magnetic field, enabling their utilization in in-vivo imaging, cell targeting, and drug storage (Figure 2.16). PLGA coating caused rising of cell viability and significantly decreasing toxicity of QD-MNSs, greatly improved noninvasive tracking capabilities while PTX loading lead to decrease of cell viability

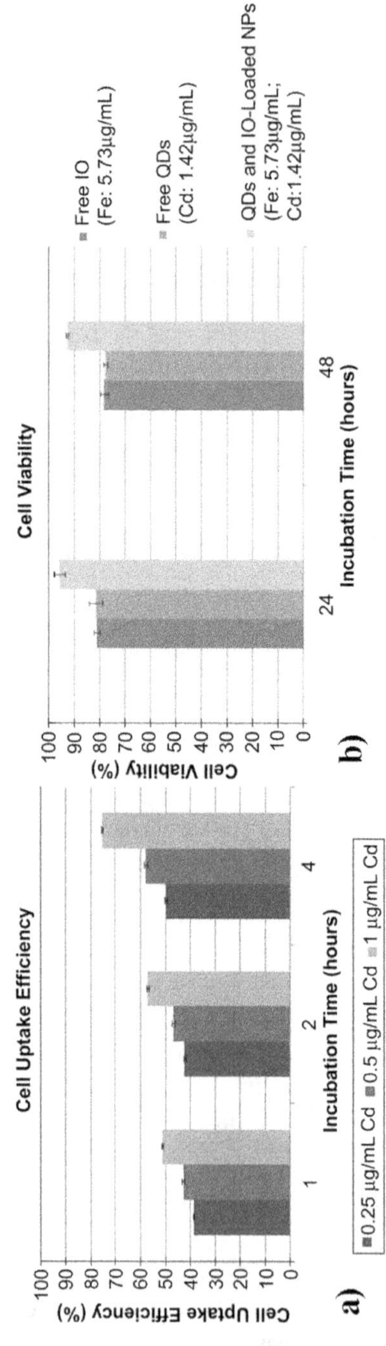

FIGURE 2.15 a) The efficiency of cellular uptake (cancer cells) of the nanocomposites containing various Cd concentrations under different incubation times, b) cell viability of QDs, IO and the nanocomposites under different incubation times (Tan 2011).

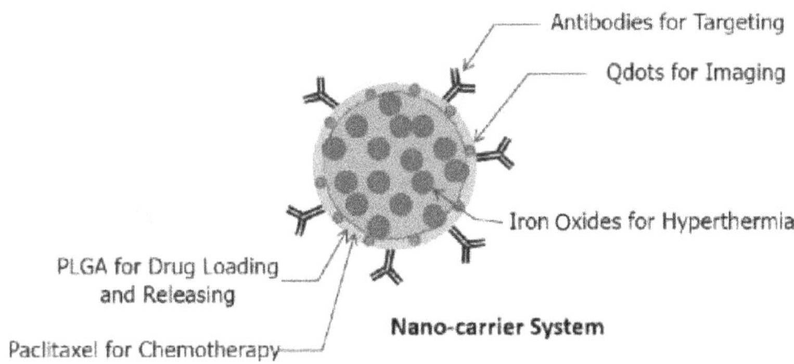

FIGURE 2.16 Schematic presentation of new multifunctional nanocarrier system. Reproduced with permission from (Cho 2010). Copyright (2010) American Chemical Society.

depending on the concentration of MNSs and thus drug loading capacity and drug release rate (Shi 2009). Anti-PSMA-conjugated nanocarrier with binding activity in cancer cells exhibited specific targeting of cancer cells.

Chen et al. (2010) functionalized the surface of multi-walled carbon nanotubes (MWCNTs) with poly(allylamine hydrochloride) (PAH) and then consecutively activated with EDC/NHS mixture solution containing iron oxides, PAH solution and EDC/NHS mixture solution containing CdTe by layer-by-layer (LBL) assembly route. CNT-SPIO-CdTe nanohybrids exhibited a strong emission band at 734 nm, superparamagnetic properties with higher magnetic saturation value from SPIOs, strong MRI contrast and great cellular fluorescence marker performance. It was reported that the toxicity of nanohybrids depend on Cd concentration and incubation time.

Liu et al. (2012) coated Fe3O4 with chitosan (CS) to produce CS– Fe3O4 MNPs and then synthesized ZnS:Mn on the surface of CS–Fe3O4 MNPs nanoparticles (Figure 2.17). The aqueous dispersion of CS–Fe3O4@ZnS:Mn NPs displayed orange fluorescence, superparamagnetic properties with 10.45 emu/g of magnetic saturation and magnetic separation properties.

Walia et al. (2017) added various concentrations of CdS-iron oxide NPs together with or without anticancer drug (podophyllotoxin-PD) to acidic solution of CS to fabricate CS/CdS-IONP/PD and CS/CdS-IONP. CS/CdS-IONP/PD exhibited minimal toxicity and successful internalization for specific cells with positive fluorescence response. It was also reported that the drug release ability of CS microspheres was low and CS/CdS-IONP/PD existed in blood and urine.

Ding et al. (2017) synthesized carboxymethyl chitosan (CMCS)-modified Fe_3O_4 NPs (Fe_3O_4@CMCS) and GSH-stabilized CdTe@ZnS QDs and subsequently attached them with different concentration of MNPs and QDs and presence of glutaraldehyde and anticancer drug doxorubicin (DOX) in chitosan matrix to fabricate drug carrier magnetic fluorescence nanocomposites (DCMFNPs). It was reported that the fluorescence intensity of CMFNPs increased while QDs:Fe3O4 molar ratios were increasing, and CS concentration was decreasing. DCMFNPs displayed excellent magnetic (36.8 emu/g of magnetic saturation) and fluorescence properties (71.2%

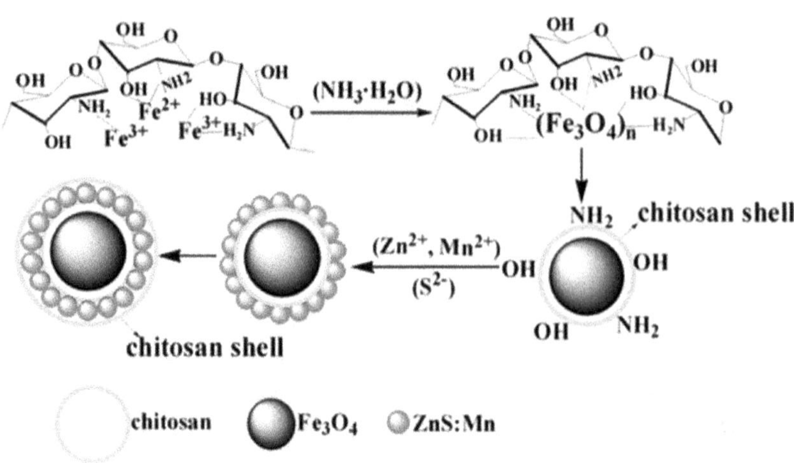

FIGURE 2.17 Schematic illustration of the synthesize of CS–Fe3O4@ZnS:Mn NPs (Liu 2012).

of PLQY) and good stability (-30mV of zeta potential), drug release with solubility of chitosan depending on pH and electrostatic interaction between different charged components, low toxicity, cell labelling features and cellular uptake depending on time. Maximum loading efficiency and encapsulation efficiency was achieved CS:MNPs:DOX = 4: 1: 5 of mass ratio.

Cheng et al. (2010) attached NH-Fe_3O_4 and QDs with Taxol (paclitaxel)-encapsulated PLGA NPs and then poly(styrenesulfonate) (PSS)-coated Au nanorods (NRs) for chemotherapeutic and in-vitro and in-vivo NIR photothermal cancer therapy. Au NR/QD/Fe_3O_4/PLGA NPs modified with 1-ethyl-3-(3-dimethylaminopropyl)-carbodiimide (EDC) were conjugated with anti-Her2 antibodies for targeted binding of HeLa cancer cells. Fluorescence intensity of QD/Fe_3O_4/PLGA NPs decreased ~30% by coating with Au NRs. Drug release from Au NR/QD/Fe_3O_4/PLGA NPs was carried out after NIR-laser irradiation by NIR absorbing and heat conversion of Au NRs and destroying of PLGA, thus releasing Taxol. Au NR/QD/Fe_3O_4/Taxol loaded PLGA NPs possessed low toxicity but their cell viability after NIR irradiation at high PLGA concentration decreased to ~10%. Au NR/QD/Fe_3O_4/Taxol-loaded PLGA NPs were suggested as contrast agent for MR imaging with 14.05 s-1 mM-1 of r^2 relaxivities.

Pei et al. (2018) electrostatically linked water soluble CdHgTe QD and dextran-magnetic layered double hydroxide ([FeII xFeIII2(OH)2(x+2)](Ay-)2/y·(1~2x)H2O)-fluorouracil(drug) (DMF) (Huang 2013) to produce fluorescent magnetic QD@DMF for magnetic targeted therapy and bioimaging.

Yin et al. (2011) synthesized superparamagnetic, magnetoresponsive and fluorescent Janus supraballs formed from one hemisphere containing poly(methyl methacrylate- co -2-hydroxyethyl methacrylate) [poly(MMA- co -HEMA)]/CdS QD–polymer hybrids and the other hemisphere containing modified Fe_3O_4 NPs with poly(MMA- co -HEMA)/cadmium acrylate Cd(AA) ionomers by microfluidic synthesis. The balls possessed good monodispersity, high performance and fluorescent stability.

Dincer (2022) firstly precipitated Fe_3O_4 nanoparticles the surface of GO nanosheet by coprecipitation process and then coated polypyrrole polymer (PPy) on GO-Fe_3O_4 by in situ polymerization (Getiren 2019) and subsequently loaded DOX to Fe_3O_4 @PPy-NGQDs. The nanocomposite exhibited superparamagnetic and high photothermal efficiency properties and drug releasing with NIR irradiation.

Antoniak (2021) encapsulated trioctylphosphine oxide-capped CdSeCdSe QDs and Fe_3O_4 in (poly(ethylene glycol)-poly(propylene glycol)-poly(ethylene glycol) (PEG-PAG-PEG) to produce NPsCdSe/Fe_3O_4 nanocapsules (NCs). The NCs demonstrated strong one- or two-photon emission at 535 nm, improved biocompatibility and potential application for bioimaging and anticancer therapy.

Aladaghlo (2021) encapsulated Fe_3O_4@GQDs nanohybrid with gelatin using glutaraldehyde as crosslinking agent via the w/o emulsion method to produce Fe_3O_4@GQDs@gelatin microspheres (GM) as magnetic sorbent to detect tricyclic antidepressants (TCA) in plasma and urine samples.

Chen et al. (2022) encapsulated OA-modified CdSe QDs and hydrophilic Fe3O4 particles in liposome to develop magnetic fluorescence liposomes (MFL) with near 82.8 nm average particle size for the treatment of hepatocellular carcinoma and drug release. The MFL exhibited good biocompatibility, low toxicity, labelling with HepG2 cells with good fluorescence images, magnetic separation properties.

Wang et al. (2021) investigated the synthesis of Fe_3O_4@CS-GQDs nanocomposites for dual-modal imaging properties. First, Fe_3O_4 NPs were coated with chitosan and subsequently GQD, and then NHS/EDC were added to Fe_3O_4@chitosan aqueous solutions. to fabricate Fe_3O_4@CS-GQDs NCs with great magnetic and optical performance (Figure 2.18).

2.2.3 Doping of Paramagnetic Transition Metal Ions or Iron Ions into QDs

Doping of transition metal ions to semiconductor quantum dots (QDs) presents excellent way for adjusting electronic, optical and magnetic properties of MQDS by interaction of the dopant with the host is another approach to synthesize MQDs (Tufani 2021; Makkar 2017; Pradeep 2020). Figure 2.19 illustrated MQDs synthesis approach with doping of paramagnetic transition metal ions into QDs. The composition, synthesis route, analysis techniques, purposes and optical and magnetic performance of MQDs synthesized by doping of paramagnetic transition metal ions or iron ions into QDs were given in Table 2.4.

Saha et al. (2016) synthesized Fe_3O_4 core- CdS shell MQD by diffusion the CdS matrix inside the Fe_3O_4 core by SILAR techniques.

Part et al. (2018) fabricated core doped and shell doped CdTe/ZnS quantum dots with ferrous ions and capped with N-acetyl-L-cysteine (NAC) to improve water dispersibility and colloidal stability (Figure 2.20).

Saha et al. (2013) successfully synthesized Fe doped CdTeS MQDs with water dispersibility, high quantum yield and NIR emitting by aqueous hydrothermal technique for bioimaging applications. The capping of MQDs with NAC provides

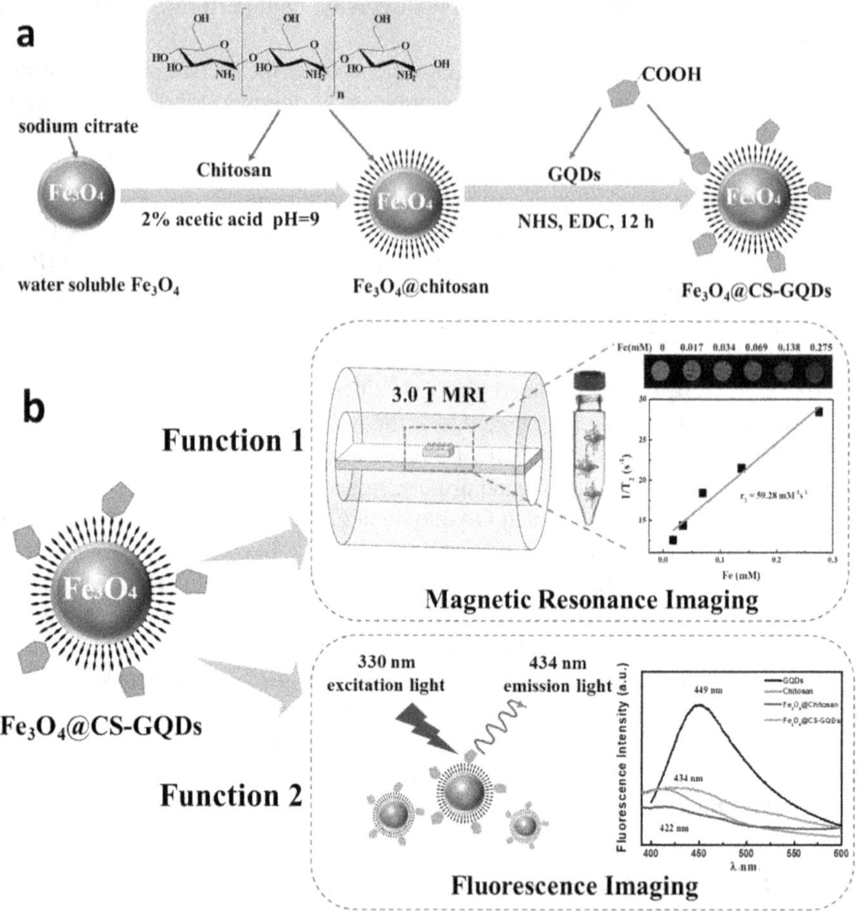

FIGURE 2.18 The synthesis and imaging properties of Fe_3O_4@CS-GQDs (Wang 2021).

stabilization of aqueous dispersion, enables bioconjugation, protects cells from cytotoxicity induced by MQD, and hinders oxidation due to antioxidant property of NAC.

Vyshnava et al. (2022) fabricated Ni-doped CdSe/ZnS by a hydrothermal method and chemical precipitation for bioimaging and sorting applications.

Doping with Mn2+ ions to QDs such as CdS (Zhao 2016), PbS (Turyanska 2016); ZnS (Gaceur 2012; Jahanbin 2015) and ZnSe (Sharma 2014; Wang 2016; Zhou 2018); In–Zn–S (Lai 2017), Cu–In–Zn–S (CIZS) and Cu–In–Zn–Se (CIZSe) QDs (Ding 2014; Sitbon 2014; Chetty 2019); Ag–In–Ga–Zn–S (AIGZS) (Galiyeva 2021) have successfully been carried out for bioimaging applications.

Galiyeva et al. (2021) incorporated Mn^{2+} dopant into Ag–In–Ga–Zn–S QDs as host material (Mn: AIGZS) by one-pot synthesis, that caused a PL QY of up to 41.3% and a high PL lifetime of up to 887.9 μs.

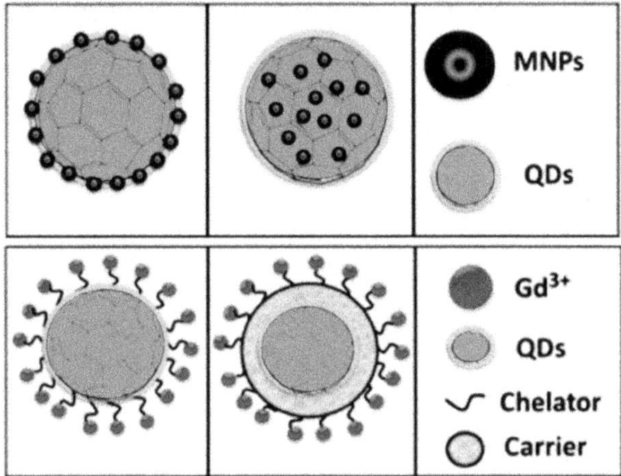

FIGURE 2.19 MQDs synthesis approach with doping of paramagnetic transition metal ions into QDs. Reproduced with permission from Das (2021); published by The Royal Society of Chemistry, 2022. Licensed by Creative Common CC BY 3.0.

Nadtochenko (2022) reported the synthesis of Mn^{2+}-doped ZnCdS/ZnS core/shell quantum dots by one-pot and two-step synthesis. They showed quantum yield (QY) of ~ 60% in anhydrous and ~ 30% in aqueous media for Mn2+ photoluminescence (PL).

Nguyen (2022) synthesized Mn ZnSe quantum dots by using a green method of precipitation in aqueous solutions containing MPA, PEG, or starch as capping agents and investigated the effect of capping agent on photoluminescence quantum yield (PLQY)of MQDs.

Sukkabot (2022) reported that optical band gaps of Mn-doped ZnSe/CdSe@Thin core/shell nanocrystals are greater than those of Mn-doped ZnSe/CdSe@Thick core/shell nanocrystals. The optical properties in Mn-doped ZnSe/CdSe@Thick core/shell nanocrystals improved under high magnetic fields.

Mulder et al. (2006) fabricated paramagnetic quantum dots (pQD)s by covering CdSe/ZnS core-shell QDs with a PEG lipids (PEG-DSPE (1,2-distearoyl-sn-glycero-3-phosphoethanolamine-N-[methoxy-(poly(ethylene glycol))-2000])) and paramagnetic lipid, Gd-DTPA-BSA (Gd-DTPA-bis(stearylamide) with high relaxivity for biomolecular imaging (Figure 2.21). The pQDs were also functionalized by Mal-PEG-DSPE (1,2-distearoyl-sn-glycero-3-phosphoethanolamine-N-[maleimide(poly(ethylene glycol))-2000]) to coupling cyclic RGD peptides for the detection of tumor cells.

2.2.4 Miscellaneous Methods for MQD Synthesis

Table 2.5 presented composition, synthesis route, analysis techniques, purposes and optical and magnetic performance of MQDs synthesized by miscellaneous methods.

TABLE 2.4
The Composition, Synthesis Route, Analysis Techniques, Purposes, and Optical and Magnetic Performance of MQDs Synthesized by Doping of Paramagnetic Transition Metal Ions or Iron Ions

Modifying Agent	Composition	Method	Purpose	Emission/Size-Zeta Potential	QY	Magnetism	r2 Relaxivity	Ref.
OLA, GSH, TMAH	Mn^{2+}-Ag–In–Ga–Zn–S	one-pot synthesis	bioimaging	611 nm/~2nm- −58.5 ± 1.2 mV	41.3%		0.57 mM−1 s−1 at 298 K	Galiyeva 2021
TOPO, TBP, NHS, MPA, TMAH, STA, OLA	Ni-CdSe/ZnS core/shell	hydrothermal method and chemical precipitation	cellular imaging and sorting	620 nm/ 9.0 ± 2.0 nm-(-24.7 mV)	10.6%			Vyshnava 2022
MPA, PEG 1500, starch	Mn^{2+}-ZnSe	a green method of precipitation in aqueous solutions	Sensors, bacteria detectors	580 nm/ PEG capped: 10-20 nm MPA capped: 25-30 nm Starch capped: 10-20 nm	PLQYstarch 26%, PLQYmla: 32%, PLQYpeg: 45%			Nguyen 2022
OLA, ODE, TMAH, STA	Mn^{2+}-ZnCdS/ ZnS core/shell	One-pot and two separate steps	bioimaging	One-pot: 654 nm/ `~7.6nm Tow-separete step: 537 nm/ ~6.7 nm	One-pot: 60% Tow-separete step: 39% in cyclohexane			Nadtochenko 2022

NAC	Fe-CdTe/ZnS, CdTe/Fe-ZnS	facile aqueous synthesis route	Bioimaging	Core doped: 562 nm Shell doped: 573 nm/2.9 nm	Core doped: 3% Shell doped: 5%	Core doped: 23 emu/g Shell doped: 4.4 emu/g at room temp.		Part 2018
NAC	Fe-CdTeS	aqueous hydrothermal technique	Bioimaging	711 nm/2.8-12.2 nm	27.2%	71 emu/g at 300K		Saha 2013
OLA, ODE, OA	Fe-CdS	SILAR technique	spintronics device	5.4nm- 15.4 nm		80 memu/g at 300 K		Saha 2016
TOPO, TOP, SA, HDA, OLA, PEG lipid	Paramagnetic lipid (Gd)-PEG lipid-CdSe/ZnS		targeted to human endothelial cells in-vitro imaging	550 nm			~ 2000 $mM^{-1}s^{-1}$	Mulder 2006

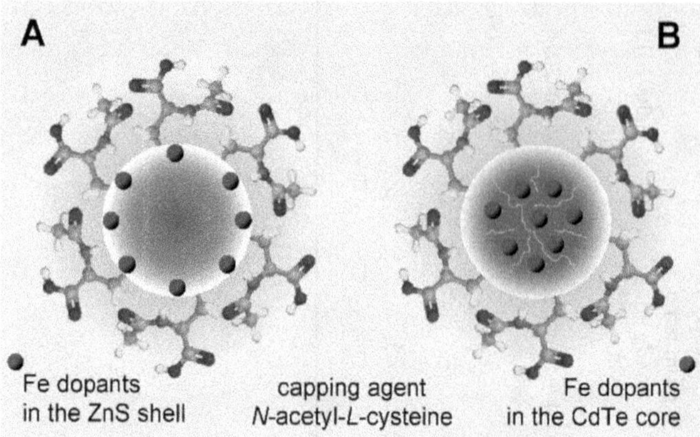

FIGURE 2.20 Schematic illustration of core iron-doped and shell iron-doped CdTe core/ZnS shell quantum dots capped with N-acetyl-L-cysteine (NAC). Reproduced with permission from (Part 2018). Copyright (2018) American Chemical Society.

Cao et al. (2013) reported the synthesis of one-dimensional ZnS:Mn^{2+} nanowires (NWs)/ Fe_3O_4 QDs/SiO_2 core-shell heterostructures by the Stober process to produce the water-soluble, FMNCs for bioimaging applications. The highest fluorescence intensity and 2.5 emu/g of magnetic saturation was determined at 5 h of hydrolysis time with TEOS.

Li et al. (2015) modified silicon oxide microspheres with (3-mercapto propyl) trimethoxy silane (MPMS) and assembled Fe3O4 NPs on MPMS-modified silicon oxide microspheres (SiMS) and subsequently modified SiMS@FeNP with TEOS and 1,6-dimercapto hexane (DMH) to produce SiMS@FeNP@SiO2. MPMS-modified SiMS@FeNPs@@SiO2 CPs were linked to thioglycolic acid (TGA)-stabilized CdTe QDs. SiMS@FeNP@SiO2@CdTe displayed great superparamagnetic and luminescence properties, magnetic separation performance.

Liu et al. (2016) fabricated nitrogen-doped carbon-Fe_3O_4 QDs (C-Fe_3O_4 QDs) by one-pot hydrothermal method. It was achieved covalently bonding of carbon quantum dots (CQD) on Fe_3O_4, doping of N atoms to CQDs. C-Fe_3O_4 QDs exhibited superior water dispersibility, good chemical- and photostability, tunable and strong FL emission peak at 464 nm, short FL lifetime, high quantum yield, excellent superparamagnetic properties, effective magnetic separation, great biocompatibility and blood compatibility, high accumulation of C-Fe_3O_4 QDs in tumor, MRI, FL and CT contrast performance.

Molaei et al. (2022) attached alginate-coated Fe3O4@SiO2 and N-doped carbon quantum dots (CQDs) produced by the hydrothermal route to obtain multifunctional magnetofluorescent Fe_3O_4@SiO_2@alginate/CQDs nanocomposites for simultaneous bioimaging and drug release (Figure 2.22).

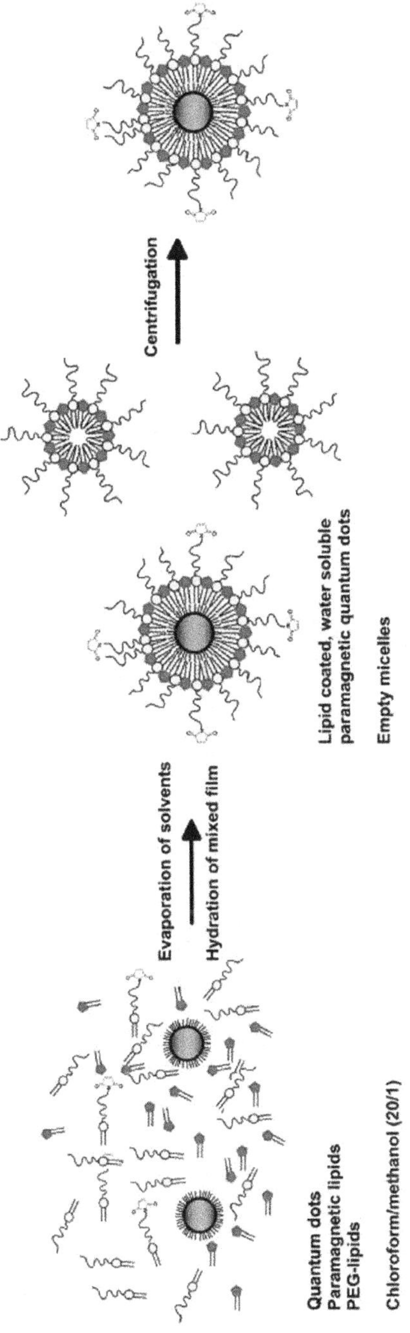

FIGURE 2.21 Schematic illustration of the fabrication of MQDs. Reproduced with permission from Mulder (2006). Copyright (2006) American Chemical Society.

TABLE 2.5
The Composition, Synthesis Route, Analysis Techniques, Purposes, and Optical and Magnetic Performance of MQDs Synthesized by Miscellaneous Methods

Composition	Modifying Agent	Method	Purpose	Emission/Size-Zeta Potential	QY	Magnetism	Cell / Drug	Ref.
ZnS:Mn2+ NW/ Fe3O4 QD/ SiO2	CA for QDs, TGA for ZnS:Mn2NWs, TEOS	stöber method for MQDs, chemical coprecipitation method for QD	biological imaging, magnetic guiding and separation	574 nm/18 nm	20%	~2.5emu/g		Cao 2013
SiMS@FeNP@ SiO2@CdTe	DMH, MPMS, TGA	directly assembled LBL	magnetodisplay application	651 nm		30.5 emu/g at 298 K		Li 2015
C-Fe3O4	γ-PGA	green and facile one-pot hydrothermal approach	triple-modal bioimaging, magnetic separation	464 nm/3 nm- (-43.2 mV)	21.6%	62 emu/g at room temperature	HeLa cells	Liu 2016
Fe3O4@SiO2@ alginate/ CQDs	SA, TEOS, alginate, PEI	coprecipitation method,	Bioimaging, cancer therapy	~530 nm/42 nm		42.46 emu/g	DOX	Molaei 2022

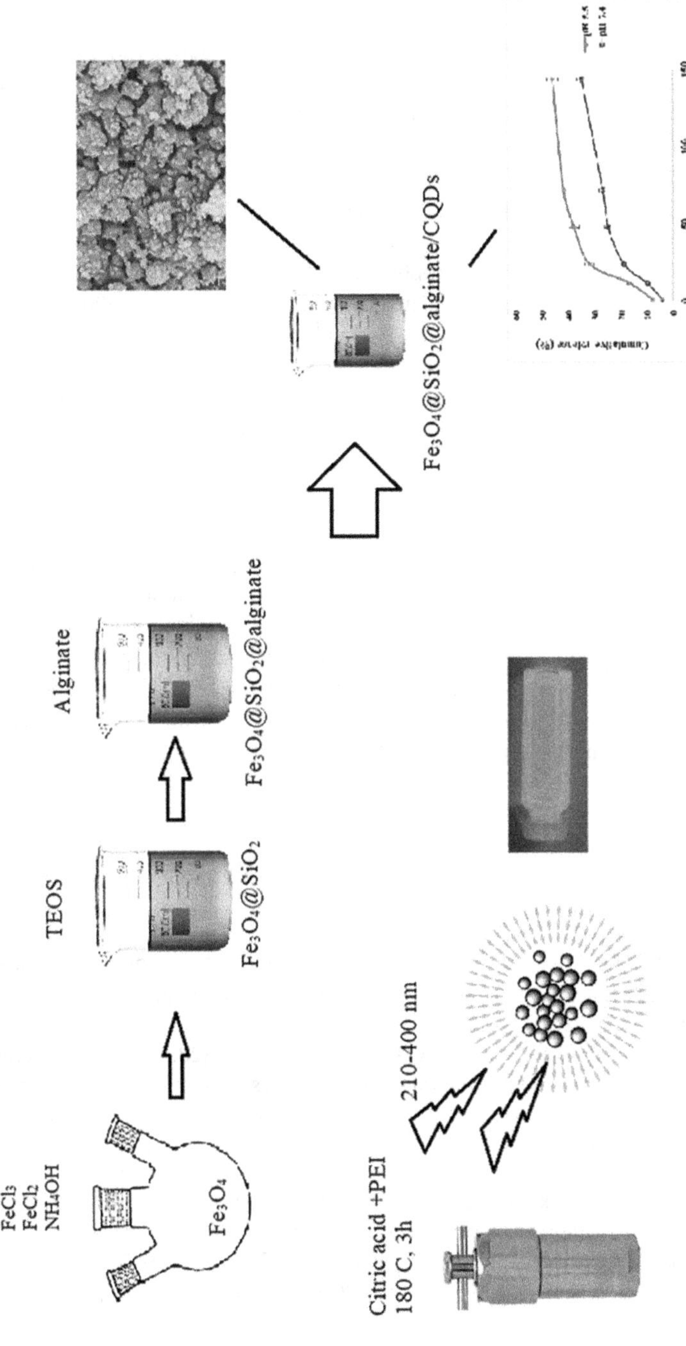

FIGURE 2.22 The synthesis route of Fe3O4@SiO2@alginate/CQDs (Molaei 2022).

2.3 CONCLUSIONS AND FUTURE OUTLOOK

MQDs formed of QDs and MNPs exhibit unique multimodality and multifunctionality because their great combination of magnetic, electrical and optical properties would be an advantage for in-vivo and in-vitro bioimaging applications.

In this chapter, we presented comprehensive strategies for synthesis approaches for integration of MNPs and QDs. Recent approaches to synthesizing MQDs with multifunctionality and multimodality could play a significant role in their biomedical applications with less or no toxicity. The synthesis of less-toxic $CuInSe_2$, $CuInS_2$, InP, Ag_2Se or AgS, and $AgInSe_2$ alternatives to cadmium and lead-containing QDs due to the toxicity of cadmium and lead heavy metals have sparsely been reported. Binding hydrophilic ligands to hydrophobic QDs to passivation of the hydrophobic surface of QDs or QDs synthesis in water or other hydrophilic solvents to solubility or dispersibility of MQDs in biological fluids are other inspiring approaches.

For bioimaging applications, the synthesis of water soluble MQDs and MQDCs with lower size than 20 nm or to enhance water dispersibility of MQD by their surface functionalization with some polymers have great importance, and limited research has been reported up until now.

The effect of some factors such as their size, shape, surface charge, composition, morphology, ligands and surface coating materials on biocompatibility, multifunctional, multimodality cytotoxicity, water dispersibility properties of MQDs are still investigated. Their long-term stability, biodegradability, elimination from the body and dosage limit to body are some challenges in biomedical applications. It was supposed to rise the imaging, cancer diagnosis, therapeutic and treatment, biomolecular separation, targeted drug delivery applications in the biomedical fields of MQDs due to their multimodal with powerful magnetic susceptibility and strong fluorescence intensity and multifunctional properties of MQDs.

REFERENCES

Acharya, A. (2013). Luminescent magnetic quantum dots for in vitro/in vivo imaging and applications in therapeutics. *Journal of Nanoscience and Nanotechnology*, *13*(6), 3753–3768.

Ahmed, S. R., Dong, J., Yui, M., Kato, T., Lee, J., & Park, E. Y. (2013). Quantum dots incorporated magnetic nanoparticles for imaging colon carcinoma cells. *Journal of Nanobiotechnology*, *11*(1), 1–9.

Aladaghlo, Z., Javanbakht, S., Fakhari, A. R., & Shaabani, A. (2021). Gelatin microsphere coated Fe3O4@ graphene quantum dots nanoparticles as a novel magnetic sorbent for ultrasound-assisted dispersive magnetic solid-phase extraction of tricyclic antidepressants in biological samples. Microchimica Acta, 188(3), 1–9.

Antoniak, M. A., Pązik, R., Bazylińska, U., Wiwatowski, K., Tomaszewska, A., Kulpa-Greszta, M., ... & Nyk, M. (2021). Multimodal polymer encapsulated CdSe/Fe3O4 nanoplatform with improved biocompatibility for two-photon and temperature stimulated bioapplications. Materials Science and Engineering: C, 127, 112224.

Armăşelu, A. (2017). Quantum dots and fluorescent and magnetic nanocomposites: Recent investigations and applications in biology and medicine. *Nonmagnetic and Magnetic Quantum Dots* (2017): 221–229.

Bhandari, S., Khandelia, R., Pan, U. N., & Chattopadhyay, A. (2015). Surface complexation-based biocompatible magnetofluorescent nanoprobe for targeted cellular imaging. *ACS Applied Materials & Interfaces*, *7*(32), 17552–17557.

Brett, M. W., Gordon, C. K., Hardy, J., & Davis, N. J. (2022). The rise and future of discrete organic–inorganic hybrid nanomaterials. *ACS Physical Chemistry Au* (2017): 221–229.

Cao, J., Niu, H., Du, J., Yang, L., Wei, M., Liu, X., ... & Yang, J. (2018). Fabrication of P (NIPAAm-co-AAm) coated optical-magnetic quantum dots/silica core-shell nanocomposites for temperature triggered drug release, bioimaging and in vivo tumor inhibition. *Journal of Materials Science: Materials in Medicine*, *29*(11), 1–13.

Cao, J., Wang, B., Han, D., Yang, S., Yang, J., Wei, M., ... & Wang, T. (2013). Effects of surface modification and SiO 2 thickness on the optical and superparamagnetic properties of the water-soluble ZnS: Mn 2+ nanowires/Fe 3 O 4 quantum dots/SiO 2 heterostructures. CrystEngComm, *15*(35), 6971–6978.

Cao, J., Wang, B., Han, D., Yang, S., Yang, J., Wei, M., ... & Wang, T. (2013). Effects of surface modification and SiO 2 thickness on the optical and superparamagnetic properties of the water-soluble ZnS: Mn 2+ nanowires/Fe 3 O 4 quantum dots/SiO 2 heterostructures. CrystEngComm, *15*(35), 6971–6978.

Chen, B., Zhang, H., Zhai, C., Du, N., Sun, C., Xue, J., ... & Wu, Y. (2010). Carbon nanotube-based magnetic-fluorescent nanohybrids as highly efficient contrast agents for multimodal cellular imaging. *Journal of Materials Chemistry*, *20*(44), 9895–9902.

Chen, G., Chen, P., Xu, D., & Min, W. (2022). Preparation and microwave absorbtion performance of composite hollow Carbon/Fe3O4 magnetic quantum dots. *Chinese Journal of Materials Research*, *36*(1), 29–39.

Chen, M., Huang, H., Pan, Y., Li, Z., Ouyang, S., Ren, C., & Zhao, Q. (2022). Preparation of layering-structured magnetic fluorescent liposomes and labeling of HepG2 cells. Bio-Medical Materials and Engineering, (Preprint), 1–12.

Cheng, F. Y., Su, C. H., Wu, P. C., & Yeh, C. S. (2010). Multifunctional polymeric nanoparticles for combined chemotherapeutic and near-infrared photothermal cancer therapy in vitro and in vivo. *Chemical Communications*, *46*(18), 3167–3169.

Chetty, S. S., Praneetha, S., Vadivel Murugan, A., Govarthanan, K., & Verma, R. S. (2019). Human umbilical cord wharton's jelly-derived mesenchymal stem cells labeled with Mn2+ and Gd3+ Co-doped CuInS2–ZnS nanocrystals for multimodality imaging in a tumor mice model. *ACS Applied Materials & Interfaces*, *12*(3), 3415–3429.

Cho, H. S., Dong, Z., Pauletti, G. M., Zhang, J., Xu, H., Gu, H., ... & Shi, D. (2010). Fluorescent, superparamagnetic nanospheres for drug storage, targeting, and imaging: a multifunctional nanocarrier system for cancer diagnosis and treatment. *ACS Nano*, *4*(9), 5398–5404.

Corr, S. A., Rakovich, Y. P., & Gun'ko, Y. K. (2008). Multifunctional magnetic-fluorescent nanocomposites for biomedical applications. *Nanoscale Research Letters*, *3*(3), 87–104.

Das, P., Ganguly, S., Margel, S., & Gedanken, A. (2021). Tailor made magnetic nanolights: Fabrication to cancer theranostics applications. *Nanoscale Advances* (2017): 221–229.

Demillo, V. G., Liao, M., Zhu, X., Redelman, D., Publicover, N. G., & Hunter Jr, K. W. (2015). Fabrication of MnFe2O4–CuInS2/ZnS magnetofluorescent nanocomposites and their characterization. *Colloids and Surfaces A: Physicochemical and Engineering Aspects*, *464*, 134–142.

Deng, S., Ruan, G., Han, N., & Winter, J. O. (2010). Interactions in fluorescent-magnetic heterodimer nanocomposites. *Nanotechnology*, *21*(14), 145605.

Dincer, C. A., Getiren, B., Gokalp, C., Ciplak, Z., Karakecili, A., & Yildiz, N. (2022). An anticancer drug loading and release study to ternary GO-Fe3O4-PPy and Fe3O4@

PPy-NGQDs nanocomposites for photothermal chemotherapy. *Colloids and Surfaces A: Physicochemical and Engineering Aspects*, *633*, 127791.

Ding, K., Jing, L., Liu, C., Hou, Y., & Gao, M. (2014). Magnetically engineered Cd-free quantum dots as dual-modality probes for fluorescence/magnetic resonance imaging of tumors. *Biomaterials*, *35*(5), 1608–1617.

Ding, Y., Yin, H., Shen, S., Sun, K., & Liu, F. (2017). Chitosan-based magnetic/fluorescent nanocomposites for cell labelling and controlled drug release. *New Journal of Chemistry*, *41*(4), 1736–1743.

Evans, C. W., Raston, C. L., & Iyer, K. S. (2010). Nanosized luminescent superparamagnetic hybrids. *Green Chemistry*, *12*(7), 1175–1179.

Fan, H. M., Olivo, M., Shuter, B., Yi, J. B., Bhuvaneswari, R., Tan, H. R., ... & Ding, J. (2010). Quantum dot capped magnetite nanorings as high performance nanoprobe for multiphoton fluorescence and magnetic resonance imaging. *Journal of the American Chemical Society*, *132*(42), 14803–14811.

Frey, N. A., Peng, S., Cheng, K., & Sun, S. (2009). Magnetic nanoparticles: synthesis, functionalization, and applications in bioimaging and magnetic energy storage. *Chemical Society Reviews*, *38*(9), 2532–2542.

Gaceur, M., Giraud, M., Hemadi, M., Nowak, S., Menguy, N., Quisefit, J. P., ... & Ammar, S. (2012). Polyol-synthesized Zn0. 9Mn0. 1S nanoparticles as potential luminescent and magnetic bimodal imaging probes: synthesis, characterization, and toxicity study. *Journal of Nanoparticle Research*, *14*(7), 1–15.

Galiyeva, P., Rinnert, H., Bouguet-Bonnet, S., Leclerc, S., Balan, L., Alem, H., ... & Schneider, R. (2021). Mn-doped quinary Ag–In–Ga–Zn–S quantum dots for dual-modal imaging. *ACS Omega*, *6*(48), 33100–33110.

Getiren, B., Çıplak, Z., Gökalp, C., & Yıldız, N. (2020). Novel approach in synthesizing ternary GO-Fe3O4-PPy nanocomposites for high Photothermal performance. *Journal of Applied Polymer Science*, *137*(26), 48837.

Goryacheva, O. A., Wegner, K. D., Sobolev, A. M., Häusler, I., Gaponik, N., Goryacheva, I. Y., & Resch-Genger, U. (2022). Influence of particle architecture on the photoluminescence properties of silica-coated CdSe core/shell quantum dots. *Analytical and Bioanalytical Chemistry*, *414*(15), 1–13.

Gui, R., Wang, Y., & Sun, J. (2014). Encapsulating magnetic and fluorescent mesoporous silica into thermosensitive chitosan microspheres for cell imaging and controlled drug release in vitro. *Colloids and Surfaces B: Biointerfaces*, *113*, 1–9.

Huang, J., Gou, G., Xue, B., Yan, Q., Sun, Y., & Dong, L. E. (2013). Preparation and characterization of "dextran-magnetic layered double hydroxide-fluorouracil" targeted liposomes. *International Journal of Pharmaceutics*, *450*(1–2), 323–330.

Hsu, J. C., Huang, C. C., Ou, K. L., Lu, N., Mai, F. D., Chen, J. K., & Chang, J. Y. (2011). Silica nanohybrids integrated with CuInS2/ZnS quantum dots and magnetite nanocrystals: multifunctional agents for dual-modality imaging and drug delivery. *Journal of Materials Chemistry*, *21*(48), 19257–19266.

Ishikawa, M., & Biju, V. (2011). Luminescent quantum dots, making invisibles visible in bioimaging. *Progress in Molecular Biology and Translational Science*, *104*, 53–99.

Jahanbin, T., Gaceur, M., Gros-Dagnac, H., Benderbous, S., & Merah, S. A. (2015). High potential of Mn-doped ZnS nanoparticles with different dopant concentrations as novel MRI contrast agents: synthesis and in vitro relaxivity studies. *Journal of Nanoparticle Research*, *17*(6), 1–12.

Karimi, S., & Namazi, H. (2021). A photoluminescent folic acid-derived carbon dot functionalized magnetic dendrimer as a pH-responsive carrier for targeted doxorubicin delivery. *New Journal of Chemistry*, *45*(14), 6397–6405.

Koc, K., Karakus, B., Rajar, K., & Alveroglu, E. (2017). Synthesis and characterization of ZnS@ Fe3O4 fluorescent-magnetic bifunctional nanospheres. *Superlattices and Microstructures*, *110*, 198–204.

Koktysh, D., Bright, V., & Pham, W. (2011). Fluorescent magnetic hybrid nanoprobe for multimodal bioimaging. *Nanotechnology*, *22*(27), 275606.

Koole R, Mulder WMJ, Van Schooneveld MM, Strijkers GJ, Meijerink A, Nicolay K. (2009) Magnetic quantum dots for multimodal imaging. *Wiley Interdisciplinary Reviews Nanomedicine and Nanobiotechnology*, *1*(5):475–491. DOI: 10.1002/wnan.14

Koole, R., Mulder, W. J., Van Schooneveld, M. M., Strijkers, G. J., Meijerink, A., & Nicolay, K. (2009). Magnetic quantum dots for multimodal imaging. *Wiley Interdisciplinary Reviews: Nanomedicine and Nanobiotechnology*, *1*(5), 475–491

Kumar, D. S., Kumar, B. J., & Mahesh, H. M. (2018). Quantum nanostructures (QDs): an overview. *Synthesis of Inorganic Nanomaterials*, 59–88.

Lai, P. Y., Huang, C. C., Chou, T. H., Ou, K. L., & Chang, J. Y. (2017). Aqueous synthesis of Ag and Mn co-doped In2S3/ZnS quantum dots with tunable emission for dual-modal targeted imaging. *Acta Biomaterialia*, *50*, 522–533.

Lee, J. S., Bodnarchuk, M. I., Shevchenko, E. V., & Talapin, D. V. (2010). "Magnet-in-the-semiconductor" FePt– PbS and FePt– PbSe nanostructures: magnetic properties, charge transport, and magnetoresistance. *Journal of the American Chemical Society*, 132(18), 6382–6391.

Li, C. L., Huang, B. R., Chang, J. Y., & Chen, J. K. (2015). Bifunctional superparamagnetic–luminescent core–shell–satellite structured microspheres: Preparation, characterization, and magnetodisplay application. *Journal of Materials Chemistry C*, *3*(18), 4603–4615.

Li, C. L., Huang, B. R., Chang, J. Y., & Chen, J. K. (2015). Bifunctional superparamagnetic–luminescent core–shell–satellite structured microspheres: preparation, characterization, and magnetodisplay application. *Journal of Materials Chemistry C*, *3*(18), 4603–4615.

Li, Y. H., Song, T., Liu, J. Q., Zhu, S. J., & Chang, J. (2011). An efficient method for preparing high-performance multifunctional polymer beads simultaneously incorporated with magnetic nanoparticles and quantum dots. *Journal of Materials Chemistry*, *21*(33), 12520–12528.

Li, Z., Wang, G., Shen, Y., Guo, N., & Ma, N. (2018). DNA-templated magnetic nanoparticle-quantum dot polymers for ultrasensitive capture and detection of circulating tumor cells. *Advanced Functional Materials*, *28*(14), 1707152.

Lin, A. W., Ang, C. Y., Patra, P. K., Han, Y., Gu, H., Le Breton, J. M., ... & Ying, J. Y. (2011). Seed-mediated synthesis, properties and application of γ-Fe2O3–CdSe magnetic quantum dots. *Journal of Solid State Chemistry*, *184*(8), 2150–2158.

Liu, L., Xiao, L., & Zhu, H. Y. (2012). Preparation and characterization of CS–Fe3O4@ ZnS: Mn magnetic-fluorescent nanoparticles in aqueous media. *Chemical Physics Letters*, *539*, 112–117.

Liu, X., Jiang, H., Ye, J., Zhao, C., Gao, S., Wu, C., ... & Wang, X. (2016). Nitrogen-doped carbon quantum dot stabilized magnetic iron oxide nanoprobe for fluorescence, magnetic resonance, and computed tomography triple-modal in vivo bioimaging. *Advanced Functional Materials*, *26*(47), 8694–8706.

Liu, X., Jiang, H., Ye, J., Zhao, C., Gao, S., Wu, C., ... & Wang, X. (2016). Nitrogen-doped carbon quantum dot stabilized magnetic iron oxide nanoprobe for fluorescence, magnetic resonance, and computed tomography triple-modal in vivo bioimaging. *Advanced Functional Materials*, *26*(47), 8694–8706.

Lou, L., Yu, K., Zhang, Z., Li, B., Zhu, J., Wang, Y., ... & Zhu, Z. (2011). Functionalized magnetic-fluorescent hybrid nanoparticles for cell labelling. *Nanoscale*, *3*(5), 2315–2323.

Ma, Q., Nakane, Y., Mori, Y., Hasegawa, M., Yoshioka, Y., Watanabe, T. M., ... & Jin, T. (2012). Multilayered, core/shell nanoprobes based on magnetic ferric oxide particles and quantum dots for multimodality imaging of breast cancer tumors. *Biomaterials*, *33*(33), 8486–8494.

Ma, X., Yang, W., Ge, X., Wang, C., Wei, M., Yang, L., ... & Liu, W. (2021). Design a novel multifunctional (CsPbBr3/Fe3O4)@ MPSs@ SiO2 magneto-optical microspheres for capturing circulating tumor cells. *Applied Surface Science*, *551*, 149427.

Mahajan, K. D., Fan, Q., Dorcéna, J., Ruan, G., & Winter, J. O. (2013). Magnetic quantum dots in biotechnology–synthesis and applications. *Biotechnology Journal*, *8*(12), 1424–1434.

Mahato, R. (2017). Multifunctional Micro-and Nanoparticles. In *Emerging Nanotechnologies for Diagnostics, Drug Delivery and Medical Devices* (pp. 21–43). Amsterdam: Elsevier.

Makkar, M., & Viswanatha, R. (2017). Recent advances in magnetic ion-doped semiconductor quantum dots. *Current Science*, *112*(7), 1421–1429.

Martynenko, I. V., Kusic, D., Weigert, F., Stafford, S., Donnelly, F. C., Evstigneev, R., ... & Resch-Genger, U. (2019). Magneto-fluorescent microbeads for bacteria detection constructed from superparamagnetic Fe3O4 nanoparticles and AIS/ZnS quantum dots. *Analytical Chemistry*, *91*(20), 12661–12669.

Molaei, M. J., & Salimi, E. (2022). Magneto-fluorescent superparamagnetic Fe3O4@ SiO2@ alginate/carbon quantum dots nanohybrid for drug delivery. *Materials Chemistry and Physics*, *288*, 126361.

Mulder, W. J., Koole, R., Brandwijk, R. J., Storm, G., Chin, P. T., Strijkers, G. J., ... & Griffioen, A. W. (2006). Quantum dots with a paramagnetic coating as a bimodal molecular imaging probe. *Nano Letters*, *6*(1), 1–6.

Nadtochenko, V., Cherepanov, D., Kochev, S., Motyakin, M., Kostrov, A., Golub, A., ... & Rtimi, S. (2022). Structural and optical properties of Mn2+-doped ZnCdS/ZnS core/shell quantum dots: New insights in Mn2+ localization for higher luminescence sensing. *Journal of Photochemistry and Photobiology A: Chemistry*, *429*, 113946.

Nguyen, V. K., Pham, D. K., Tran, N. Q., Dang, L. H., Nguyen, N. H., Nguyen, T. M., Viet, N. T., Oh, J. W., ... & Luong, B. T. (2022). Effect of stabilizers on Mn ZnSe quantum dots synthesized by using green method. *Green Processing and Synthesis*, *11*(1), 327–337.

Ortgies, D. H., de la Cueva, L., Del Rosal, B., Sanz-Rodríguez, F., Fernandez, N., Iglesias-de la Cruz, M. C., ... & Martin Rodriguez, E. (2016). In vivo deep tissue fluorescence and magnetic imaging employing hybrid nanostructures. *ACS Applied Materials & Interfaces*, *8*(2), 1406–1414.

Park, J. H., von Maltzahn, G., Ruoslahti, E., Bhatia, S. N., & Sailor, M. J. (2008). Micellar hybrid nanoparticles for simultaneous magnetofluorescent imaging and drug delivery. *Angewandte Chemie*, *120*(38), 7394–7398.

Part, F., Zaba, C., Bixner, O., Grünewald, T. A., Michor, H., Küpcü, S., ... & Ehmoser, E. K. (2018). Doping Method Determines Para-or Superparamagnetic Properties of Photostable and Surface-Modifiable Quantum Dots for Multimodal Bioimaging. *Chemistry of Materials*, *30*(13), 4233–4241.

Pei, Q. Y., Wang, R., Jin, X. Q., & Gou, G. J. (2018). Recombination of CdHgTe quantum dot and "dextran-magnetic layered double hydroxide-fluorouracil" system for cell imaging. In *Materials Science Forum* (Vol. 914, pp. 11–18). (2017): 221–229 Trans Tech Publications Ltd.

Pradeep, KR, & Viswanatha, R. (2020). Mechanism of Mn emission: Energy transfer vs charge transfer dynamics in Mn-doped quantum dots. *APL Materials*, *8*(2), 020901.

Qiu, Y., Palankar, R., Echeverría, M., Medvedev, N., Moya, S. E., & Delcea, M. (2013). Design of hybrid multimodal poly (lactic-co-glycolic acid) polymer nanoparticles for neutrophil labeling, imaging and tracking. *Nanoscale*, *5*(24), 12624–12632.

Qureshi, A., Tufani, A., Corapcioglu, G., & Niazi, J. H. (2020). CdSe/CdS/ZnS nanocrystals decorated with Fe3O4 nanoparticles for point-of-care optomagnetic detection of cancer biomarker in serum. *Sensors and Actuators B: Chemical*, *321*, 128431.

Ruan, G., Vieira, G., Henighan, T., Chen, A., Thakur, D., Sooryakumar, R., & Winter, J. O. (2010). Simultaneous magnetic manipulation and fluorescent tracking of multiple individual hybrid nanostructures. *Nano Letters*, *10*(6), 2220–2224.

Saha, A. K., Sharma, P., Sohn, H. B., Ghosh, S., Das, R. K., Hebard, A. F., ... & Moudgil, B. M. (2013). Fe doped CdTeS magnetic quantum dots for bioimaging. Journal of Materials Chemistry B, 1(45), 6312–6320.

Saha, A., Shetty, A., Pavan, A. R., Chattopadhyay, S., Shibata, T., & Viswanatha, R. (2016). Uniform doping in quantum-dots-based dilute magnetic semiconductor. *The Journal of Physical Chemistry Letters*, *7*(13), 2420–2428.

Scaria, J., Karim, A. V., Divyapriya, G., Nidheesh, P. V., & Kumar, M. S. (2020). Carbon-supported semiconductor nanoparticles as effective photocatalysts for water and wastewater treatment. In *Nano-Materials as Photocatalysts for Degradation of Environmental Pollutants* (pp. 245–278). Amsterdam: Elsevier.

Selvan, S. T. (2010). Silica-coated quantum dots and magnetic nanoparticles for bioimaging applications (Mini-Review). *Biointerphases*, *5*(3), FA110–FA115

Selvan, S. T., Patra, P. K., Ang, C. Y., & Ying, J. Y. (2007). Synthesis of silica-coated semiconductor and magnetic quantum dots and their use in the imaging of live cells. *Angewandte Chemie*, *119*(14), 2500–2504.

Sharma, V. K., Gokyar, S., Kelestemur, Y., Erdem, T., Unal, E., & Demir, H. V. (2014). Manganese doped fluorescent paramagnetic nanocrystals for dual-modal imaging. *Small*, *10*(23), 4961–4966.

Shi, D., Cho, H. S., Chen, Y., Xu, H., Gu, H., Lian, J., ... & Dong, Z. (2009). Fluorescent polystyrene–Fe3O4 composite nanospheres for in vivo imaging and hyperthermia. *Advanced Materials*, *21*(21), 2170–2173.

Shibu, E. S., Ono, K., Sugino, S., Nishioka, A., Yasuda, A., Shigeri, Y., ... & Biju, V. (2013). Photouncaging nanoparticles for MRI and fluorescence imaging in vitro and in vivo. *Acs Nano*, *7*(11), 9851–9859.

Sitbon, G., Bouccara, S., Tasso, M., Francois, A., Bezdetnaya, L., Marchal, F., ... & Pons, T. (2014). Multimodal Mn-doped I–III–VI quantum dots for near infrared fluorescence and magnetic resonance imaging: from synthesis to in vivo application. *Nanoscale*, *6*(15), 9264–9272.

Song, E., Han, W., Li, J., Jiang, Y., Cheng, D., Song, Y., ... & Tan, W. (2014). Magnetic-encoded fluorescent multifunctional nanospheres for simultaneous multicomponent analysis. *Analytical Chemistry*, *86*(19), 9434–9442.

Su, H., Wang, Z., & Liu, G. (2014). Near-infrared fluorescence imaging probes for cancer diagnosis and treatment. In *Cancer Theranostics* (pp. 55–67). Cambridge, Massachusetts: Academic Press.

Sun, L., Zang, Y., Sun, M., Wang, H., Zhu, X., Xu, S., ... & Shan, Y. (2010). Synthesis of magnetic and fluorescent multifunctional hollow silica nanocomposites for live cell imaging. *Journal of Colloid and Interface Science*, *350*(1), 90–98.

Sun, X., Ding, K., Hou, Y., Gao, Z., Yang, W., Jing, L., & Gao, M. (2013). Bifunctional superparticles achieved by assembling fluorescent CuInS2@ ZnS quantum dots and amphibious Fe3O4 nanocrystals. *The Journal of Physical Chemistry C*, *117*(40), 21014–21020.

Swierczewska, M., Lee, S., & Chen, X. (2011). Inorganic nanoparticles for multimodal molecular imaging. *Molecular Imaging*, *10*(1), 7290–2011.

Tan, Y. F., Chandrasekharan, P., Maity, D., Yong, C. X., Chuang, K. H., Zhao, Y., ... & Feng, S. S. (2011). Multimodal tumor imaging by iron oxides and quantum dots formulated in poly (lactic acid)-d-alpha-tocopheryl polyethylene glycol 1000 succinate nanoparticles. *Biomaterials*, *32*(11), 2969–2978.

Tufani, A., Qureshi, A., & Niazi, J. H. (2021). Iron oxide nanoparticles based magnetic luminescent quantum dots (MQDs) synthesis and biomedical/biological applications: A review. *Materials Science and Engineering: C*, *118*, 111545.

Turyanska, L., Moro, F., Patanè, A., Barr, J., Köckenberger, W., Taylor, A., ... & Thomas, N. R. (2016). Developing Mn-doped lead sulfide quantum dots for MRI labels. *Journal of Materials Chemistry B*, *4*(42), 6797–6802.

Vyshnava, S. S., Pandluru, G., Kumar, K. D., Panjala, S. P., Paramasivam, K., Banapuram, S., ... & Dowlatabad, M. R. (2022). Biocompatible Ni-doped CdSe/ZnS semiconductor nanocrystals for cellular imaging and sorting. *Luminescence* 37, no. 3 (2022): 490–499.

Walia, S., Sharma, S., Kulurkar, P. M., Patial, V., & Acharya, A. (2016). A bimodal molecular imaging probe based on chitosan encapsulated magneto-fluorescent nanocomposite offers biocompatibility, visualization of specific cancer cells in vitro and lung tissues in vivo. *International Journal of Pharmaceutics*, *498*(1-2), 110–118.

Wang, K., Ruan, J., Qian, Q., Song, H., Bao, C., Zhang, X., ... & Cui, D. (2011). BRCAA1 monoclonal antibody conjugated fluorescent magnetic nanoparticles for in vivo targeted magnetofluorescent imaging of gastric cancer. *Journal of Nanobiotechnology*, *9*(1), 1–12.

Wang, K., Xu, X., Li, Y., Rong, M., Wang, L., Lu, L., ... & Jiang, Y. (2021). Preparation Fe3O4@ chitosan-graphene quantum dots nanocomposites for fluorescence and magnetic resonance imaging. *Chemical Physics Letters*, 783, 139060.

Wang, Y., Wu, B., Yang, C., Liu, M., Sum, T. C., & Yong, K. T. (2016). Synthesis and characterization of Mn: ZnSe/ZnS/ZnMnS sandwiched QDs for multimodal imaging and theranostic applications. *Small*, *12*(4), 534–546.

Wang, Z., Jiang, X., Liu, W., Lu, G., & Huang, X. (2019). A rapid and operator-safe powder approach for latent fingerprint detection using hydrophilic Fe3O4@ SiO2-CdTe nanoparticles. *Science China Chemistry*, *62*(7), 889–896.

Wegner, K. D., & Hildebrandt, N. (2015). Quantum dots: bright and versatile in vitro and in vivo fluorescence imaging biosensors. *Chemical Society Reviews*, *44*(14), 4792–4834.

Wu, W., Wu, Z., Yu, T., Jiang, C., & Kim, W. S. (2015). Recent progress on magnetic iron oxide nanoparticles: synthesis, surface functional strategies and biomedical applications. *Science and technology of advanced materials* 16, no. 2 (2015): 023501.

Xu, Y., Karmakar, A., Wang, D., Mahmood, M. W., Watanabe, F., Zhang, Y., ... & Biris, A. S. (2010). Multifunctional Fe3O4 cored magnetic-quantum dot fluorescent nanocomposites for RF nanohyperthermia of cancer cells. *The Journal of Physical Chemistry C*, *114*(11), 5020–5026.

Yi, D. K., Selvan, S. T., Lee, S. S., Papaefthymiou, G. C., Kundaliya, D., & Ying, J. Y. (2005). Silica-coated nanocomposites of magnetic nanoparticles and quantum dots. *Journal of the American Chemical Society*, *127*(14), 4990–4991.

Yin, N., Wang, X., Yang, T., Ding, Y., Li, L., Zhao, S., ... & Zhu, L. (2021). Multifunctional Fe3O4 cluster@ quantum dot-embedded mesoporous SiO2 nanoplatform probe for cancer cell fluorescence-labelling detection and photothermal therapy. *Ceramics International*, *47*(6), 8271–8278.

Yin, N., Wu, P., Liang, G., & Cheng, W. (2016). A multifunctional mesoporous Fe3O4/SiO2/CdTe magnetic-fluorescent composite nanoprobe. *Applied Physics A*, *122*(3), 1–7.

Yin, S. N., Wang, C. F., Yu, Z. Y., Wang, J., Liu, S. S., & Chen, S. (2011). Versatile bifunctional magnetic-fluorescent responsive janus supraballs towards the flexible bead display. *Advanced Materials*, *23*(26), 2915–2919.

Yu, K. O., Grabinski, C. M., Schrand, A. M., Murdock, R. C., Wang, W., Gu, B., ... & Hussain, S. M. (2009). Toxicity of amorphous silica nanoparticles in mouse keratinocytes. *Journal of Nanoparticle Research*, *11*(1), 15–24.

Zhang, L., Yang, H., Tang, Y., Xiang, W., Wang, C., Xu, T., ... & Zhang, J. (2022). High-performance CdSe/CdS@ ZnO quantum dots enabled by ZnO sol as surface ligands: A novel strategy for improved optical properties and stability. *Chemical Engineering Journal*, *428*, 131159.

Zhao, B., Huang, P., Rong, P., Wang, Y., Gao, M., Huang, H., ... & Li, W. (2016). Facile synthesis of ternary CdMnS QD-based hollow nanospheres as fluorescent/magnetic probes for bioimaging. *Journal of Materials Chemistry B*, *4*(7), 1208–1212.

Zhou, R., Sun, S., Li, C., Wu, L., Hou, X., & Wu, P. (2018). Enriching Mn-doped ZnSe quantum dots onto mesoporous silica nanoparticles for enhanced fluorescence/magnetic resonance imaging dual-modal bio-imaging. *ACS Applied Materials & Interfaces*, *10*(40), 34060–34067.

Zou, W. S., Yang, J., Yang, T. T., Hu, X., & Lian, H. Z. (2012). Magnetic-room temperature phosphorescent multifunctional nanocomposites as chemosensor for detection and photo-driven enzyme mimetics for degradation of 2, 4, 6-trinitrotoluene. *Journal of Materials Chemistry*, *22*(11), 4720–4727.

3 Optical Properties of Magnetic Quantum Dots

Varsha Lisa John and Vinod T. P.
Department of Chemistry, CHRIST (Deemed to be University), Bangalore, India

CONTENTS

3.1 INTRODUCTION

Quantum dots (QDs) are zero-dimensional nanomaterials with unique optical properties administered by the effect of quantum confinement (QCE) [1]. Sharp emission, broad absorption and high photostability are some of the attractive characteristics of QDs [2]. Many of the QD systems featuring useful photophysical responses are prepared as a core-shell structure where a lower bandgap material is capped with a higher bandgap material [3–6]. Semiconductor nanocrystals are considered artificial atoms due to their discrete electronic structure similar to that of an atom, arising due to the QCE. For QDs, the electron motion is confined along the three dimensions (QCE) and thus can be regarded as artificial atoms [7]. Significant differences exist between real and artificial atoms in terms of (i) the effective electron mass and free electron mass, (ii) the atomic potential having a singular character that becomes periodic in QDs, and (iii) the screening due to electron-electron interactions. However,

DOI: 10.1201/9781003319870-3

there exists an agreement with the Zeeman effect, the electron-hole correlation, and the required magnetic field relevant for exhibiting optical and electronic properties [7].

Research conducted by Alivisatos and Nie demonstrated the efficiency of QDs for bio labelling and sensing applications for the first time [8]. Resistance to photobleaching, unblinking photoluminescence, small size, and water solubility are the advantages of QDs over organic dyes [9]. Harnessing iron-dextran magnetic nanoparticles for immunolabelling applications is the first evidence of applying magnetic nanoparticles for biological applications [10]. Magnetic nanoparticles such as Fe_3O_4 or γ-Fe_2O_3 with very small sizes exhibit superparamagnetism [9,11]. After removing the external magnetic field, these nanoparticles do not display residual magnetization. Large magnetization induced in these particles facilitates their easy manipulation with the application of an external magnetic field [12]. The possibilities of aggregation or coagulation of these particles are little without applying an external magnetic field, which helps to develop an "action at a distance" strategy for many applications. The magnetic nanoparticles [13] along with QDs are thus considered nanomaterials with pivotal roles in biomedical applications [14].

Merging fluorescent and magnetic entities within a single nanomaterial opens up opportunities for synthesizing new nanocomposites beneficial for multi-functional, multitargeting, and multi-theranostic tools [15]. Combining magnetic and fluorescent components can bring out novel and versatile "two-in-one" nanohybrid materials with varied applications [11, 12]. The advancements in magnetic resonance imaging (MRI) [17], bioimaging [18], biolabelling [19] etc., due to the incorporation of these nanomaterials over the years are commendable. Combining both magnetic and optical properties within a single nanohybrid would concurrently facilitate enhanced bio labelling/bioimaging and cell sorting/separation applications [20]. The possibility led to synthesizing a new variety of compounds, the magnetic quantum dots (MQD), consisting of semiconductors and magnetic particles. The CdS-FePt heterodimer synthesized by R. Zheng and colleagues is the first of its kind [21]. Thus, MQDs fall under the category of Janus nanoparticles [22, 23].

This chapter discusses the optical properties of QDs consisting of magnetic impurities. The optical properties discussed in this chapter are refractive index, absorption coefficient and oscillator strength with their theoretical explanation. An account of the applicability of MQDs in a range of biomedical applications such as MRI, theranostics, cell detection, and cell separation.

3.2 THEORETICAL BACKGROUND

Core-shell structured QDs typically have cores consisting of a lower bandgap II-IV (CdTe, CdSe), III-V (GaN, GaAs, InAs), and IV-VI (PbSe) semiconductor enclosed within a shell of a wider bandgap semiconductor such as ZnS, for minimizing the surface deficiency and enhancing the quantum yield [24]. When incorporated with magnetic impurities (Mn, rare earth elements, etc.) QDs undergo consequent changes in optical and electronic properties by lifting their degeneracies, known as the Zeeman effect [25]. The spatial extent of wave functions in QDs is different from atoms due to two major reasons: (i) the orbital wave function of a QD is usually 100 times more comprehensive than that of an atom, and (ii) the additional symmetry in its Bloch function within the wave function of QDs result in a denser energy spectrum than that of an atom. For these reasons, the cyclotron energy ($\hbar\omega_c$), the electron-hole exchange energy (Δ_{ex}), and the confinement energy (E_0) govern the magnetic field effect of QDs.

The Δ_{ex} is related to the energetic ordering of optically allowed and forbidden exciton states, the $\hbar\omega_c$ is connected with the number of allowed transitions in QDs, and the E_0 is related to the spatial confinement of electron-hole pairs are important parameters pertinent to a QD [26]. The E_0 can either be the exciton's binding energy or the energy required for the confinement of potential. The angular momentum of the hole (F) and electron (f) are contained in the total angular momentum (J). The contribution of Coulomb interaction is avoided to assume spherical approximation.

Zeeman splitting of QDs due to the influence of magnetic field on electronic energy levels is verified under two circumstances ($\hbar\omega_c \leq E_0$ and $\hbar\omega_c \geq E_0$) and are explained in the following sections. When $\hbar\omega_c \leq E_0$, the spin states are split due to the impact of magnetic field. There exist two possibilities namely the cases where $\hbar\omega_c \gg \Delta_{ex} < E_0$ and $\hbar\omega_c \ll \Delta_{ex} < E_0$. Consider the system where the exciton state consists of a conduction electron (Γ_7) and a valence hole (Γ_9) whose spins are $\frac{1}{2}$. It has the J values of 2 and 1, respectively, for the lowest and the second-lowest exciton states at 0T. When $\hbar\omega_c \ll \Delta_{ex} < E_0$, the exciton states perturbed by the magnetic field are split into the following levels as $(J, J_z) = (1, \pm1), (1,0), (2, \pm1), (2,0)$, and $(2, \pm2)$. Here, the optically active states are $(1, \pm1)$ and $(2, \pm1)$. The quantum-confined Zeeman effect is established by this splitting. The exciton states with the labelling of J define the electron-hole states better. Under the condition $\hbar\omega_c \gg \Delta_{ex} < E_0$, the cyclotron energy is larger and thus the energy concerned with the influence of electron-hole exchange is ignored. The energy states are labelled by f, f_z, F, and F_z that are split into $(f, f_z, F, F_z) = (\frac{1}{2},\frac{1}{2},\frac{3}{2},-\frac{1}{2})$, $(\frac{1}{2},-\frac{1}{2},\frac{3}{2},\frac{1}{2})$, $(\frac{1}{2},\frac{1}{2},\frac{3}{2},\frac{1}{2})$, $(\frac{1}{2},-\frac{1}{2},\frac{3}{2},-\frac{3}{2})$, $(\frac{1}{2},\frac{1}{2},\frac{3}{2},-\frac{3}{2})$, $(\frac{1}{2},-\frac{1}{2},\frac{3}{2},-\frac{1}{2})$, $(\frac{1}{2},\frac{1}{2},\frac{3}{2},\frac{3}{2})$, and $(\frac{1}{2},-\frac{1}{2},\frac{3}{2},-\frac{3}{2})$, of which the optically active states are $(\frac{1}{2},\frac{1}{2},\frac{3}{2},\frac{1}{2})$, $(\frac{1}{2},-\frac{1}{2},\frac{3}{2},\frac{3}{2})$, $(\frac{1}{2},\frac{1}{2},\frac{3}{2},-\frac{3}{2})$, and $(\frac{1}{2},-\frac{1}{2},\frac{3}{2},-\frac{1}{2})$. Here, the electron-hole state is considered as the sum of a hole and an electron with the labelling of their angular momenta.

The opposite limit of $\hbar\omega_c \geq E_0$ applies to QDs of large Bohr radius (GaAs, InSb). A few studies exist on the Zeeman splitting in these systems, but they are not well studied as the former systems [27, 28]. However, the diamagnetic shifts, as well as the electron-hole transitions existing within the states having a non-zero orbital angular momenta observed by Bayer et al. during the measurement of the photoluminescence spectra of InGaAs/GaAs QDs, has a radius of 100-400 Å at the magnetic field strength of 12T [29].

Merging QDs with magnetic materials has impending applications in bioimaging, separation of cells, separation of molecules, and theranostics [13]. Rare earth elements (e.g., Gd), iron phosphide, superparamagnetic iron oxide nanoparticles (SPIONs), and transition metals with half-filled d orbitals (Mn) are the commonly investigated magnetic materials [15]. The paramagnetic behavior exhibited by these nanoparticles in the solution reduces the aggregation tendency considerably. There exist methods such as doping, high-temperature decomposition, encapsulation, crosslinking, etc., for introducing magnetism to the QDs. However, the doping strategy is the most popular due to the comparative ease of synthesis.

There has been a surge in the research (experimental and theoretical) related to Mn-doped QDs [30, 31]. The case where magnetic impurities like Mn atoms that consist of a different number of mobile carriers, when doped on a QD, represent an artificial atom. The spins of doped impurities act through mobile carriers (electron/hole) and are capable of generating a ferromagnetic state in MQDs. Magnetic polaron is the stable state formed due to the interaction of carriers with magnetic impurities. The single Mn atom on MQD has a spin that lasts longer and may be thus called a qubit [32]. The characteristics of the electronic state produced after doping the Mn atom depend on the system's material. The Mn atom is an isoelectronic impurity in the case of II-VI semiconductors such as CdTe and CdSe. With III-V materials (GaAs/InAs), an acceptor state is created due to Mn doping.

3.2.1 Mn Doping in II-VI QDs

Instead of creating an electric potential for the carriers (electron/hole), an exchange interaction dependent on the spins between the carriers is created by the Mn impurity [33]. A disc-like, self-assembled QD within the vertical (z) and in-plane (x,y) directions (Figure 3.1a) has to be assumed to study the effect of doping in the case of II-VI QDs.

When the mixing of the heavy and the light holes are neglected and when the states related to the heavy holes alone are considered, the effective Hamiltonian becomes:

$$\hat{H}_{eff}^{spin} = A_e\left(\hat{\vec{S}}_e.\vec{I}_{Mn}\right)\left|\varphi\left(R_{Mn}\right)\right|^2 + A_h\left(\hat{J}_{h,z}.\hat{I}_{Mn,z}\right) + \hat{H}_{exchange}^{e-h}, \quad (1)$$

where $A_e = -\alpha\left|\varphi_e\left(R_{Mn}\right)\right|^2$ and $A_h = \frac{\beta}{3}\left|\varphi_h\left(R_{Mn}\right)\right|^2$. When an Mn ion is confined within CdTe, the parameters $\alpha N_0 = 0.29eV$ and $\beta N_0 = -1.4eV$ are significant, where α and β denote the strengths of the carrier-Mn exchange interaction, and $N_0 = 15\ nm^{-3}$gives an account of the spatial density of cations within the CdTe lattice. The signs of the exchange constant indicate the formation of the ferromagnetic state between Mn and electron, whereas the complexes of Mn atom and hole prefer an antiferromagnetic alignment. Equation (1) depicts the anisotropic behavior of Mn-hole interaction. While exploring the emission from a QD that consists of a single Mn ion (Figure 3.1b), a six-line exciton structure is observed.

Consider "switching on" the interaction between the electron and Mn atom and "switching off" the coupling between the hole and Mn atom as in Figure 3.2. In the case where $A_h \gg A_e$, each of the line structures of the exciton is split. The term "d" represents the degeneracy of a single level. When $A_h \sim A_e$, the observed spectrum consists of 12 lines which are equally spaced between the neighboring lines. While applying the condition of $A_h \ll A_e$, two groups of lines are detected with degeneracies

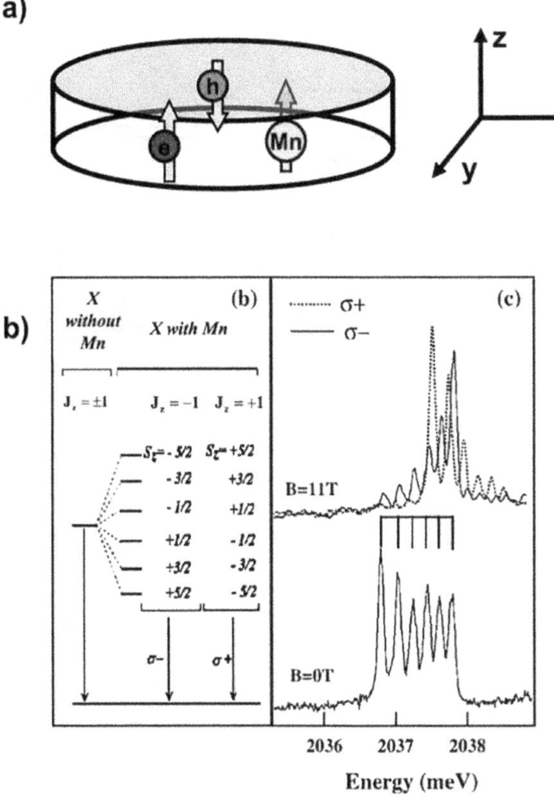

FIGURE 3.1 (a) Scheme depicting the geometry of a QD consisting of Mn [31], Reproduced with permission [31], Copyright 2008, Elsevier. (b, c) Interband transitions of the exciton contained in a QD, Spectrum of emission obtained experimentally [33], Reproduced with permission [33], Copyright 2004, American Physical Society.

of 5 and 7 acquired by the spectrum. Only dark states remain under the conditions when $A_e \neq 0$, $A_h \neq 0$. Incorporation of electron-hole exchange interaction, $\hat{H}^{e-h}_{exchange}$, distorts the spectrum and leads to the photon emission from all states. During the Mn-carrier interactions, this distortion affects the optical activity of all exciton states. The advancement of the exciton-Mn atom spectrum as a function of the position of Mn-impurity with respect to the centre of the QD is depicted in Figure 3.3. When $R_{Mn} \rightarrow \infty$, there exist two levels, upper and lower, depicting the brighter and darker states [34].

3.2.2 Mn Doping in III-V QDs

Here, the Mn impurity induces an acceptor state that is strongly bound, perturbing the orbital motion of the electrons or holes. The spin and orbital motions of the exciton in a III-V QD with a single Mn atom (impurity) are represented by the Hamiltonian,

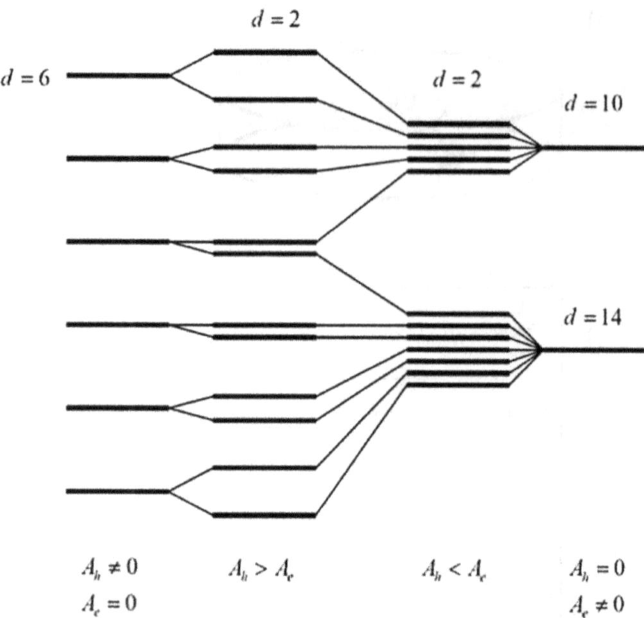

FIGURE 3.2 The advancement of exciton spectra as the interactions of Mn atoms and electrons as well as the Mn atom and hole are switched OFF and ON [31], Reproduced with permission [31], Copyright 2008, Elsevier.

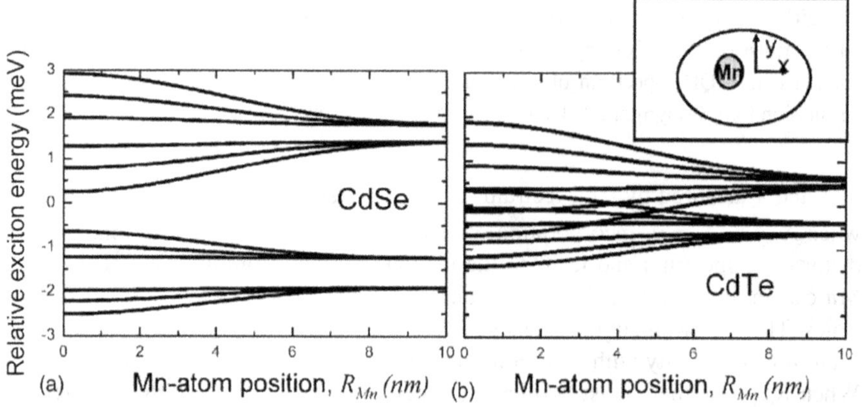

FIGURE 3.3 Excitation energy spectrum conforming to the position of Mn atom for CdSe and CdTe QDs [32], Reproduced with permission [32], Copyright 2004, American Physical Society.

$$\widehat{H}_{exciton} = \widehat{H}_h^{QD}(r_h) + \widehat{H}_e^{QD}(r_e) + U_h^{imp}(r_h) + U_e^{imp}(r_e)$$
$$+ A_{imp}\left(\overrightarrow{\hat{j}_h}.\overrightarrow{\hat{I}_{Mn}}\right)\delta\left(r_h - R_{Mn}\right) + H_{exchange}^{e-h} + U_{Coul}\left(r_e, r_h\right) \tag{2}$$

where the operators $\widehat{H}_h^{QD}(r_h)$, and $\widehat{H}_e^{QD}(r_e)$ refer to the motion of the electron and hole in the absence of an Mn atom, $U_h^{imp}(r_h)$ and $U_e^{imp}(r_e)$ are the localizing potentials of holes and electrons, $A_{imp}\left(\overrightarrow{\hat{j}_h}.\overrightarrow{\hat{I}_{Mn}}\right)\delta\left(r_h - R_{Mn}\right)$ depicts the Coulomb interaction due to the existing exchange interaction amid the impurity (Mn) and hole, $U_{Coul}\left(r_e, r_h\right)$.

When the size of the Mn atom is lesser than the dimensions of QD, and when an Mn atom is located in the centre of the QDs, the corresponding spacing existing within the potential maximum in the valence band and impurity level accounts for the energy. Figure 3.4 portrays the schematic of the process of photon absorption to the exciton state ($X_{QD+Mn}^{-} = A^0 + e_{CB}^{-}$) from the initial state ($A^{-} = A^0 + e_{imp}^{-}$).

Here, A^0 depicts the neutral state in an Mn acceptor when the hole is bound to the attractive potential, e_{imp}^{-} is the electron residing in the Mn-acceptor orbital, and e_{CB}^{-} is the electron confided in the states of the conduction band. The initial state consists of the electron originating from the Fermi sea. During the absorption of a photon, a bound electron (e_{imp}^{-}) is stimulated to the conduction band to become e_{CB}^{-}. Thus the final state will have an Mn atom bound to the hole. The potentials of QD and the interactions amid electron and hole result in the splitting of the triplet state of the hole (J_{imp}=1), while the exciton (X_{QD+Mn}^{-}) acquires a fine structure (Figure 3.5).

The optical emissions occurring within an InGaAs QD that consists of a single Mn atom as an impurity are described in Figures 3.6 and 3.7. The $A^0 + X_{QD}^0$ and $A^0 + X_{QD}^{-}$ are the exciton states of the spectra. The exciton states corresponding to the s-s transitions between electronic energy levels of a QD are portrayed, which occur via exchange interaction. The transitions within the s-Mn atom have lesser intensities

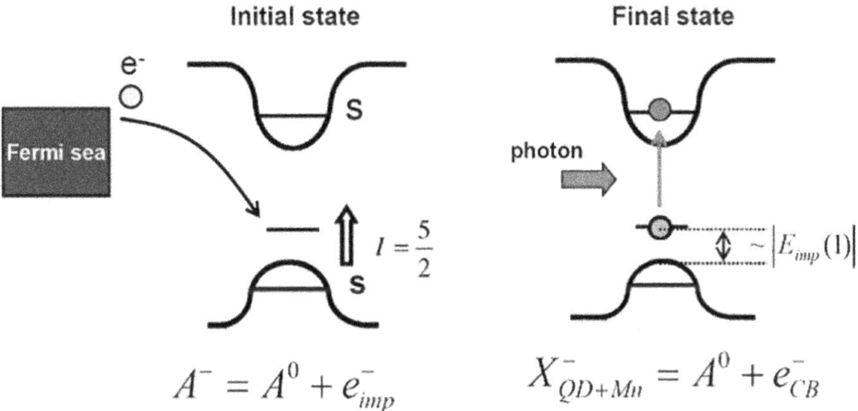

FIGURE 3.4 Scheme representing the optical absorption process within a QD having an Mn-acceptor level [31], Reproduced with permission [31], Copyright 2008, Elsevier.

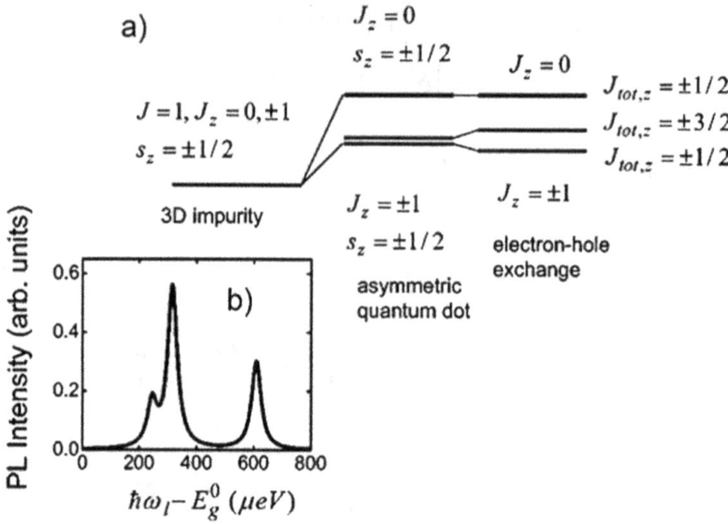

FIGURE 3.5 (a) The lowest states of an exciton in an InGaAs dot consisting of a single Mn atom as an impurity (b) The emission spectrum (calculated)[35], Reproduced with permission [35], Copyright 2004, American Physical Society.

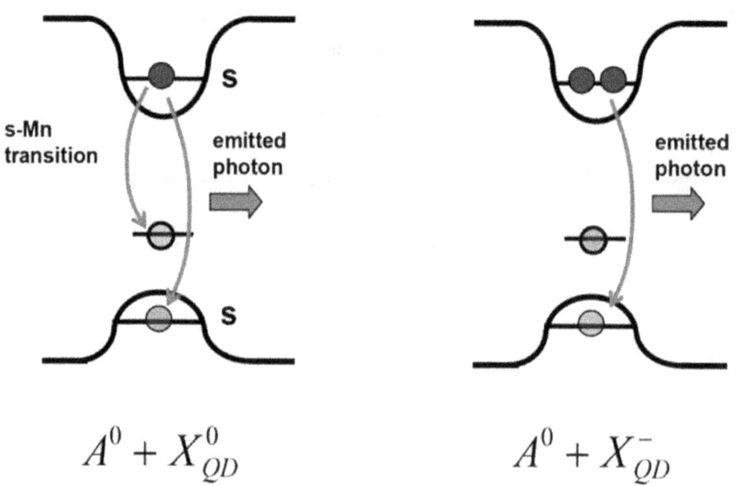

FIGURE 3.6 Scheme depicting the process of optical emissions in an InGaAs QD containing a single Mn impurity [31], Reproduced with permission [31], Copyright 2008, Elsevier.

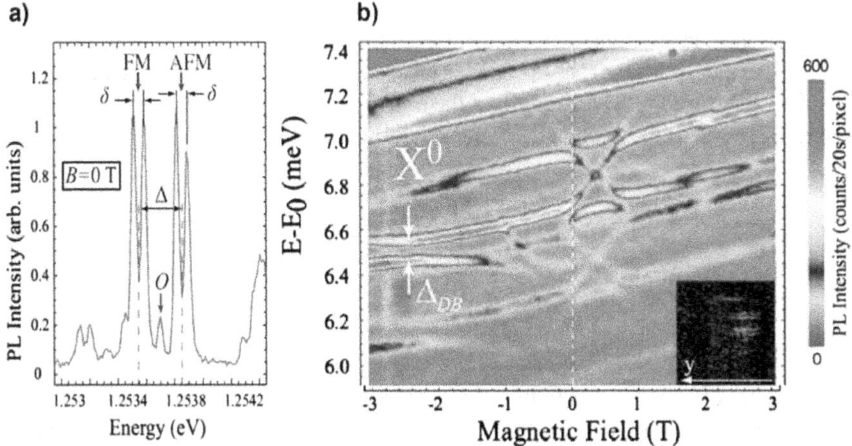

FIGURE 3.7 (a, b) The magneto-photoluminescence spectra of the exciton and Mn spins were determined experimentally [36]. Reproduced with permission [36], Copyright 2005, American Physical Society.

due to a relatively smaller overlap of the wave functions of the Mn-bound state and conduction-band electron wave function.

3.3 OPTICAL PROPERTIES OF MAGNETIC QUANTUM DOTS

The sections below unravel the optical properties of MQDs, such as refractive index, absorption coefficient, and oscillator strength, beneficial for biomedical applications.

3.3.1 Oscillator Strength

Oscillator strength is one of those key factors for investigating the optical properties of nanostructures. The dimensionless quantity that measures the probability of emission or absorption of electromagnetic radiations during the transitions between the energy levels is called oscillator strength [3]. The bright transitions are associated with large oscillator strength; with small oscillator strengths, non-radiative decay outstrips radiative decay. From the normalized wave functions obtained from diagonalized Hamiltonian and the energy spectrum, the oscillator strength of transitions from the initial state $|i = | n, l, m\rangle$ to the final state $|f\rangle = |n', l', m'\rangle$ in the dipole approximation is expressed as,

$$P_{i \to f} = E_{if} \left| M_{if} \right|^2 \tag{3}$$

where P is the polarization of the applied radiation,

$$E_{if} = E_f - E_i \tag{4}$$

$$M_{if} = \langle n', l', m' \mid \hat{p}.r \mid n, l, m \rangle \tag{5}$$

Here, $\hat{p}$ is the unit vector of polarization of the applied radiation. The type of polarization of the chosen light source depends on the final state.

The Zeeman splitting from the energy levels in QDs is observed by the far-infrared (Far-IR) spectroscopy. The strength of the magnetic field applied perpendicular to the surface gives the relative change of measured transmittance.

$$t = -\frac{\left[T(V_g) - T(V_t)\right]}{T(V_t)} \tag{6}$$

3.3.2 Refractive Index

The measure of bending of a light ray during its passage from one medium to another is known as the refractive index (RI). It is a fundamental parameter for a material used for optical applications [37]. The optical properties of QDs can be elucidated through the density matrix formalism. When a monochromatic field of electromagnetic radiations of a frequency ω excites the QDs corresponding to the following equation,

$$E(t) = \tilde{E}e^{i\omega t} + \tilde{E}e^{-i\omega t} \tag{7}$$

The linear and third-order non-linear RI is

$$\frac{\Delta n_r^{(1)}(\omega)}{n_r} = \frac{\sigma_v |M_{21}|^2}{2n_r^2 \varepsilon_0} \frac{(E_{21} - \hbar\omega)}{(E_{21} - \hbar\omega)^2 + (\hbar\Gamma_{21})^2} \tag{8}$$

$$\frac{\Delta n_r^{(3)}(\omega)}{n_r} = \frac{\mu I c \sigma_v |M_{21}|^4}{n_r^3 \varepsilon_0} \frac{(E_{21} - \hbar\omega)}{[(E_{21} - \hbar\omega)^2 + (\hbar\Gamma_{21})^2]^2} \tag{9}$$

The total RI is,

$$\frac{\Delta n_r(\omega)}{n_r} = \frac{\Delta n_r^{(1)}(\omega)}{n_r} + \frac{\Delta n_r^{(3)}(\omega)}{n_r} \tag{10}$$

Here, μ, σ_v and n_r are the system's permeability, carrier intensity, and refractive index, respectively. When ħω is the energy of the incident photon,

$$\Gamma_{21} = \frac{1}{T_{21}} \tag{11}$$

where T_{21} is the time of relaxation between the energy states 1 and 2, the optical intensity of the incident wave of light is denoted by I, and c is the speed of light. The terms $E_{21} = E_2 - E_1$ represent the difference between energies of energy states 1 and 2,

$$M_{21} = \left| \langle \psi_2 | ez | \psi_1 \rangle \right| \tag{12}$$

is the transition matrix element of the dipole energy states 1 and 2.

3.3.3 Absorption Coefficient

The attenuation of the intensity of light while passing through a material is defined as the absorption coefficient [38]. When QDs are excited by an incident light according to equation (7), the Hamiltonian defining the system is given by,

$$H' = H + ezE(t) \tag{12}$$

where H denotes the Hamiltonian of the system without the magnetic field, E(t). The total absorption coefficient can be expressed as

$$\alpha(\omega, I) = \alpha^1(\omega) + \alpha^3(\omega, I) \tag{13}$$

where,

$$\alpha^1(\omega) = \sqrt{\frac{\mu}{\varepsilon_R}} \frac{\sigma_v \hbar\omega \Gamma_{fi} |M_{fi}|^2}{(E_{fi} - \hbar\omega)^2 + (\hbar\Gamma_{fi})^2} \tag{14}$$

$$\alpha^3(\omega, I) = -\sqrt{\frac{\mu}{\varepsilon_R}} \left(\frac{I}{2\varepsilon_0 n_r c}\right) \frac{4\sigma_v \hbar\omega \Gamma_{fi} |M_{fi}|^4}{[(E_{fi} - \hbar\omega)^2 + (\hbar\Gamma_{fi})^2]^2} \tag{15}$$

are linear and third-order non-linear absorption coefficients.

3.4 TAILORING OF OPTICAL PROPERTIES FOR BIOLOGICAL APPLICATIONS

MQDs find applicability in multimodal imaging, bioimaging, theranostics, molecular detection, and separation. Being a hybrid nanoparticle, both aspects of optical properties and magnetism are applied for their usability. Such utility of MQDs is summarized in the following sections and Table 3.1.

3.4.1 Multimodal Imaging

When MQDs are integrated as magnetic resonance imaging (MRI) contrast agents, the preoperative diagnosis is enabled with the magnetic domain and an upright determination during in-vivo surgery is facilitated with the fluorescent signal [16]. The use of MQDs thus materializes the multimodal fluorescent-magnetic system. Merging QDs with other functional nanoparticles is expected to render multimodality to the imaging probes. Relaxation is the occurrence of nuclear spins' release due to the

TABLE 3.1
Representative Examples of MQDs Useful for Biomedical Applications

Quantust	Magnetic Material (Dopant)	Applications	ref.
CdTe	Fe_3O_4	Bioimaging (Colon carcinoma cells)	[49]
CdSe (coated with Si)	Fe_2O_3	Bioimaging (Live cells)	[50]
CdTe	Fe_3O_4	Bioimaging (HeLa cells)	[20]
CdTe (embedded with Si)	Fe_3O_4	Drug delivery (Adriamycin), Bioimaging (HepG2 cells)	[48]
CdSe/ZnS	Fe_2O_3	Detection and separation of the cells of leukaemia and prostate cancer	[51]
CdSe/ZnS	Gd	MRI contrast agent	[14]
CdSe/ZnS (with paramagnetic lipidic coating)	Gd	MRI contrast agent	[52]

energy assimilated from the radiofrequency pulse. The relaxations can be categorized into two, such as the T_1 (longitudinal relaxation/ spin-lattice relaxation) and T_2 (transverse relaxation/spin-spin relaxation). There exist T_2^* relaxations as well that reflect both T_2 relaxation and inhomogeneities of the magnetic field affect the MRI signal. The MR signals recuperate with the time constants T_1 or T_2 according to the direction of decay, longitudinal or in the transverse plane. The differences in the relaxation taking place within different tissues are regarded as the primary reason for image contrast, observed as the signal-to-noise ratio in MRI [39]. The longitudinal relaxivity (r_1) determines the efficacy of an MRI contrast agent [40]. The r_1 accounts for the change of relaxation rates (r_i, where i = 1 and 2) of a solution and is expressed as a function of concentration (Figure 3.8).

Metals accelerate the T1 of water protons by exerting bright distinctions or contrasts in the regions confined by the nanoprobe. The half-filled orbitals containing ground states with s symmetry take up decisive roles in MR imaging, and that is how the metal atoms like Mn (II) and Gd (III) fit into the scenario [41]. The greatest advantage of the MRI technique is its high spatial resolution. The maximization of dipole-dipole interaction between hydrogens of the QD and Gd(III) results in high r_1, indicating the effectiveness of doping [40–44]. The high r_1 is also a sign of the effectiveness of doping, thereby improving their detection sensitivity by which Mn(II) and Gd(III) are considered as potential contrast agents [43,45]. Thus, the half-filled metal ions of Mn(II) and Gd(III) render MRI modality to QDs through their magnetic properties that result in nuclear spin relaxation.

The synthetic approaches of nanoprobes with multimodality have to be less toxic and robust for their applicability in nanomedicine. A multi-layered core-shell

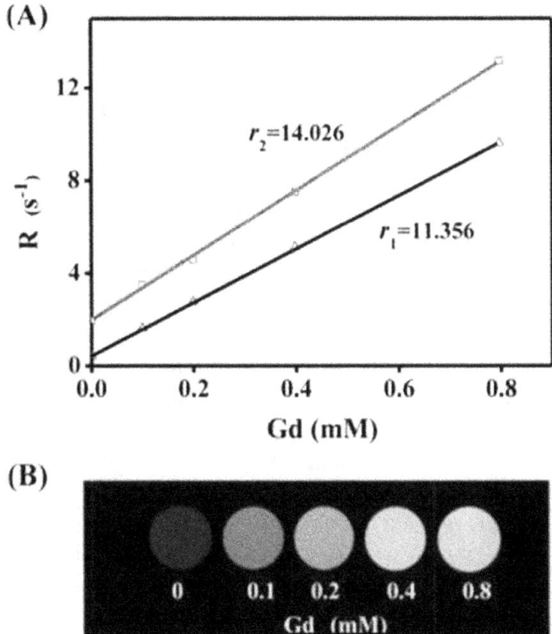

FIGURE 3.8 (A) The linear plot of rate of relaxation against the concentration of Gadolinium. (B) T_1 weighted MR images of water and serial dilutions of QDs doped with Gadolinium [40]. Reproduced with permission [40], Copyright 2014, American Chemical Society.

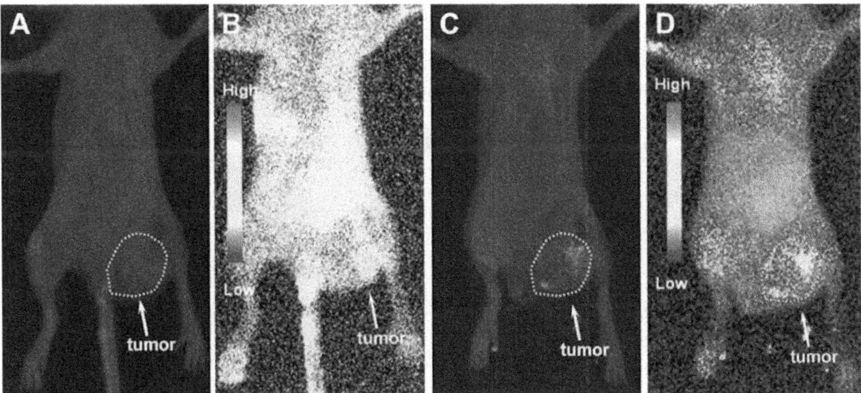

FIGURE 3.9 The in-vivo near-IR fluorescence imaging of the probe of HER2-MQQ was injected into mice that bear the human breast tumors at various intervals. Here (A, B) is observed after 0.1 h, and (C, D) is observed 48 h after injection of the HER2-MQQ-probe [46]. Reproduced with permission [46], Copyright 2012, Elsevier.

(MQQ) probe that involves MNPs having high T_2 relaxivity and QDs (CdSe/ZnS) that integrates all the benefits of NIR-fluorescence imaging techniques and MRI were developed by Ma et al. [46]. These MQQ probes utilized a mouse model that bore human breast cancer tumors for multimodal imaging (Figure 3.9).

3.4.2 Molecular Detection and Separation

The advantage of nanoparticle-based detection is their size similarity with biomarkers. MQDs are efficient for this purpose because their fluorescence attribute can quantify the concentration of the target, and the magnetism contributes to the isolation of the target. Rare cells helpful for disease diagnosis and drug screening have been isolated with the help of MQDs. The fluorescence functionality of MQDs favors the visualization of the separated cells helping the elucidation of the molecular mechanism of the disease at the level of genes and proteins. The activity of biomolecules such as DNA, RNA, and proteins is accompanied by minute changes in their structure, which bulk measurements like absorbance cannot detect [15]. These minuscule structural changes disturb the overall activity of biomolecules; hence, a precise diagnosis is imperative. The "MagDots" developed by Jessica and coworkers [47] were used on a lab-on-chip platform employing MQDs for the sensing as well as separation of biomarkers and cells (Figure 3.10). A versatile platform for the sensing, segregation and quantifying of cell surface markers, nucleic acid sequences, and proteins has been demonstrated with MagDots. The analytes such as cells, proteins, or DNA have been quantified by the additive nature of fluorescence.

3.4.3 Theranostics

MQDs function as a delivery agent of therapeutic drugs and nucleic acids to cells. They are used as gene delivery vectors facilitated by their surface functionalities. The

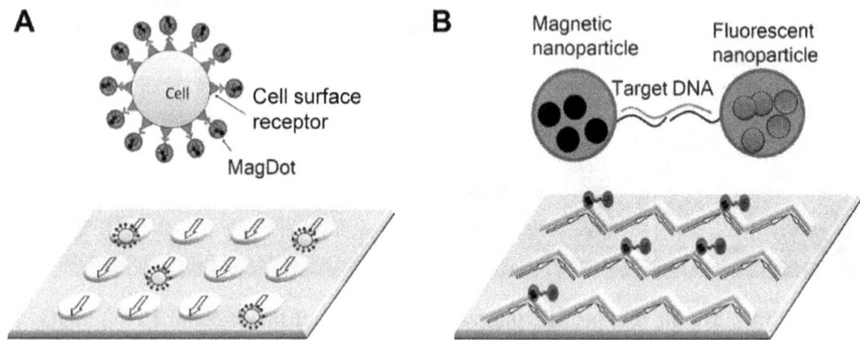

FIGURE 3.10 MQDs for the captured cells and molecules. (A) MQDs conjugate with antibodies for the recognition of specific cells. (B) Target DNA forms a sandwich amidst magnetic and fluorescent nanoparticles and is isolated on micropatterned nanowires [15]. Reproduced with permission [15], Copyright 2013 Wiley-VCH Verlag GmbH & Co. KGaA, Weinheim.

presence of surface functionalities and the magnetic fields determines the transfection ability of MQDs. The cellular uptake is enhanced by the presence of negatively charged carboxyl groups or positively charged poly-L-lysine.

Developing magnetic carriers of drugs that consist of fluorescent markers (organic dyes or QDs) has been considered ideal for site-specific, targeted drug delivery with reduced side effects [48]. Hence, these nanoparticles with superparamagnetic nature (SPMNPs), such as Fe_3O_4 with extraordinary magnetization and better biocompatibility, are utilized as high-quality drug delivery systems (DDS) (Figure 3.11). Functional groups should reform the SPMNPs to avoid being washed away by the macrophages or reticuloendothelial system before being received at preferred sites. Biodegradable polymers functionalize an ideal DDS as shells and SPMNPs as cores. Thus, anticancer drugs are shielded by polymer shells during the drug delivery process and thus minimize the possible side effects of drugs.

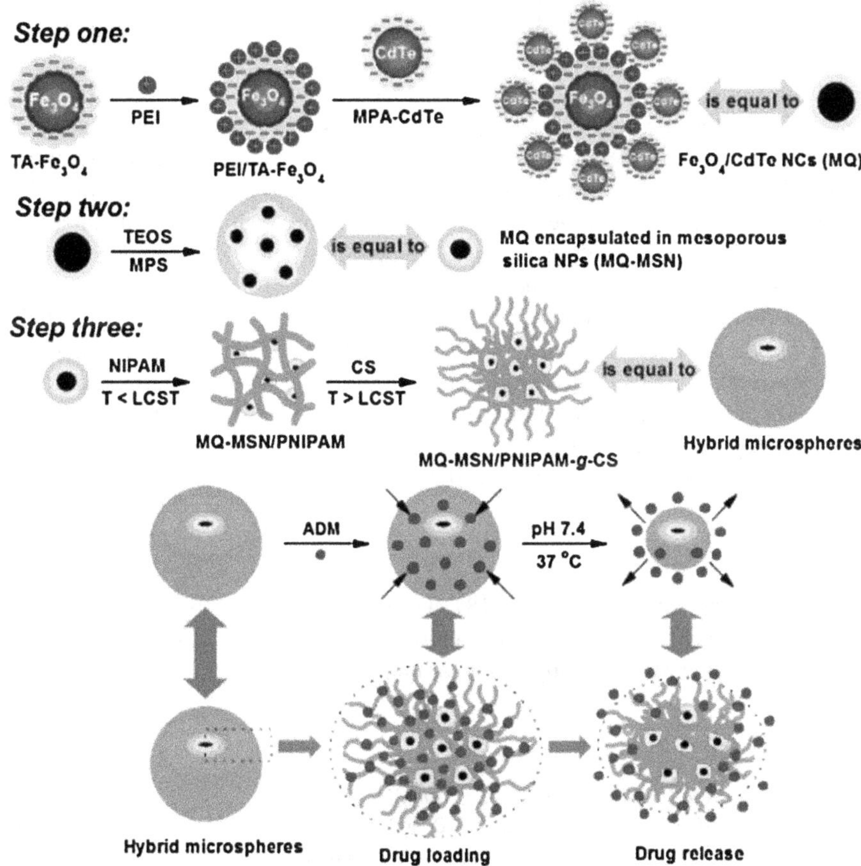

FIGURE 3.11 Schematic illustration of the drug (Adriamycin) loading and microspheres release [48]. Reproduced with permission [48], Copyright 2014, Elsevier.

3.5 TOXICITY

Incorporating magnetic ions or SPIONs into semiconductor QDs invites potential biological system toxicity. There have been reports on both acute and chronic toxicity. The toxicity mechanism involves the following steps, (i) the generation of reactive oxidation species and (ii) the heavy metal photooxidation from the QDs. However, QDs have proved to be nontoxic for up to 90 days in the model of the primate or cleared via the renal system if maintained under appropriate conditions. Developing nontoxic QDs or alternative QD materials [53–57] has to be encouraged for in-vivo applications. Novel fluorescent probes containing inorganic, nontoxic, or less toxic nanoparticles must be developed [16].

3.6 CONCLUSION

Confining the electronic motion along the three dimensions in QDs because of QCE makes them regarded as artificial atoms. The QDs undergo significant changes in optical and electronic properties by lifting their degeneracies (Zeeman effect) when incorporated with magnetic impurities such as Mn atom, rare earth elements etc. The combination of fluorescent and magnetic entities can bring out novel "two-in-one" multi-functional nanohybrid materials applicable for multi-functional, multitargeting, and multi-theranostic tools. The MQDs thus formed are suitable for multimodal imaging, bioimaging, theranostics, molecular detection, and separation. As contrast agents, MQDs enable the preoperative diagnosis with the magnetic domain and an upright resolution during surgery in-vivo with the fluorescent signal. The molecular detection and separation are materialized with MQDs when the fluorescence attribute quantifies the concentration of the target, and the magnetism contributes to the isolation of the target. The presence of surface functionalities and the magnetic fields determine the transfection ability of MQDs and facilitates drug delivery and gene therapy efficiently. Novel fluorescent probes consisting of inorganic nanoparticles that are nontoxic or less toxic have to be developed. Synthetic strategies based on self-assembly than following conventional approaches have to be developed to minimize the complexity of the existing methods.

ACKNOWLEDGEMENTS

Varsha Lisa John is thankful to CHRIST (Deemed to be University) for the research fellowship. Vinod T. P. is thankful to Centre for Research, CHRIST (Deemed to be University) for the research support through mono-graph project MNGDSC-1806.

REFERENCES

[1] Reed M A 1993 Quantum dots *Sci. Am.* 268 118–23

[2] Bayal M, Chandran N, Pilankatta R and Nair S S 2021 Semiconductor quantum dots and core shell systems for high contrast cellular/bio imaging *Nanomaterials for Luminescent Devices, Sensors, and Bio-imaging Applications* (Singapore: Springer Singapore) pp 27–38

[3] Jafari A R, Naimi Y and Davatolhagh S 2013 Optical properties of nano-multi-layered quantum dot: Oscillator strength, absorption coefficient and refractive index *Opt. Quantum Electron.* 45 517–27

[4] Hohler M E G, Muller T, Ruckenstein A, Steiner F, Trfirnper J and Steiner F 2003 *Springer Tracts in Modern Physics* 184 1–255

[5] Bailey R E, Smith A M and Nie S 2004 Quantum dots in biology and medicine *Phys. E Low-dimensional Syst. Nanostructures* 25 1–12

[6] Jacak L, Wójs A and Hawrylak P 1998 *Quantum Dots* (Berlin, Heidelberg: Springer Berlin Heidelberg)

[7] Rinaldi R 1998 Artificial atoms in magnetic field: Electronic and optical properties *Int. J. Mod. Phys. B* 12 471–502

[8] Bruchez M, Moronne M, Gin P, Weiss S and Alivisatos A P 1998 Semiconductor nanocrystals as fluorescent biological labels *Science (80-.).* 281 2013–6

[9] Wang G and Su X 2011 The synthesis and bio-applications of magnetic and fluorescent bifunctional composite nanoparticles *Analyst* 136 1783–98

[10] Molday R S and Mackenzie D 1982 Immunospecific ferromagnetic iron-dextran reagents for the labeling and magnetic separation of cells *J. Immunol. Methods* 52 353–67

[11] Sukumaran S, Neelakandan M S, Shaji N, Prasad P and Yadunath V K 2018 Magnetic nanoparticles: Synthesis and potential biological applications *JSM Nanotechnol Nanomed* 6 1068

[12] Akbarzadeh A, Samiei M and Davaran S 2012 Magnetic nanoparticles: preparation, physical properties, and applications in biomedicine *Nanoscale Res. Lett.* 7 144

[13] Cho H, Dong Z, Pauletti G M, Zhang J, Xu Ḱ H, Gu H, Wang L, Ewing Ḱ R C, Huth Ḱ C, Wang F and Shi D 2010 Multifunctional nanocarrier system for cancer diagnosis and treatment *ACS Nano* 4 5398–404

[14] Mulder W J M, Koole R, Brandwijk R J, Storm G, Chin P T K, Strijkers G J, De Mello Donegá C, Nicolay K and Griffioen A W 2006 Quantum dots with a paramagnetic coating as a bimodal molecular imaging probe *Nano Lett.* 6 1–6

[15] Mahajan K D, Fan Q, Dorcéna J, Ruan G and Winter J O 2013 Magnetic quantum dots in biotechnology--synthesis and applications. *Biotechnol. J.* 8 1424–34

[16] Quarta A, Di Corato R, Manna L, Ragusa A and Pellegrino T 2007 Fluorescent-magnetic hybrid nanostructures: Preparation, properties, and applications in biology *IEEE Trans. Nanobioscience* 6 298–308

[17] Tufani A, Qureshi A and Niazi J H 2021 Iron oxide nanoparticles based magnetic luminescent quantum dots (MQDs) synthesis and biomedical/biological applications: A review *Mater. Sci. Eng. C* 118 111545

[18] Chandran N, Janardhanan P, Bayal M, Unniyampurath U, Pilankatta R and Nair S S 2020 Label free, nontoxic Cu-GSH NCs as a nanoplatform for cancer cell imaging and subcellular pH monitoring modulated by a specific inhibitor: Bafilomycin A1 *ACS Appl. Bio Mater.* 3 1245–57

[19] Yao J, Li P, Li L and Yang M 2018 Biochemistry and biomedicine of quantum dots: from biodetection to bioimaging, drug discovery, diagnostics, and therapy *Acta Biomater.* 74 36–55

[20] Sun P, Zhang H, Liu C, Fang J, Wang M, Chen J, Zhang J, Mao C and Xu S 2010 Preparation and characterization of Fe3O4/CdTe magnetic/fluorescent nanocomposites and their applications in immuno-labeling and fluorescent imaging of cancer cells *Langmuir* 26 1278–84

[21] Gu H, Zheng R, Zhang X X and Xu B 2004 Facile one-pot synthesis of bifunctional heterodimers of nanoparticles: A conjugate of quantum dot and magnetic nanoparticles *J. Am. Chem. Soc.* 126 5664–5

[22] Xing Z, Dong K, Pavlopoulos N, Chen Y and Amirav L 2021 Photoinduced self-assembly of carbon nitride quantum dots *Angew. Chemie – Int. Ed.* 60 19413–8

[23] Ren B 2014 Magnetic janus particles and their applications *CUNY Thesis* 161

[24] Liu L, Jin S, Hu Y, Gu Z and Wu H C 2011 Application of quantum dots in biological imaging *J. Nanomater.* 2011

[25] Chang K and Xia J B 2002 Magneto-optical properties of diluted magnetic semiconductor quantum dots *J. Phys. Condens. Matter* 14 13661–5

[26] Nomura S, Segawa Y, Misawa K, Kobayashi T, Zhao X, Aoyagi Y and Sugano T 1996 Optical properties of semiconductor quantum dots in magnetic fields *J. Lumin.* 70 144–57

[27] Leonard D, Krishnamurthy M, Reaves C M, Denbaars S P and Petroff P M 1993 Direct formation of quantum-sized dots from uniform coherent islands of InGaAs on GaAs surfaces *Appl. Phys. Lett.* 63 3203–5

[28] Eaglesham D J and Cerullo M 1990 Dislocation-free Stranski-Krastanow growth of Ge on Si(100) *Phys. Rev. Lett.* 64 1943–6

[29] Bayer M, Schmidt A, Forchel A, Faller F, Reinecke T L, Knipp P A, Dremin A A and Kulakovskii V D 1995 Electron-hole transitions between states with nonzero angular momenta in the magnetoluminescence of quantum dots *Phys. Rev. Lett.* 74 3439–42

[30] Mendes U C, Korkusinski M, Trojnar A H and Hawrylak P 2013 Optical properties of charged quantum dots doped with a single magnetic impurity *Phys. Rev. B* 88 115306

[31] Govorov A O 2008 Optical and electronic properties of quantum dots with magnetic impurities *Comptes Rendus Phys.* 9 857–73

[32] Govorov A O and Kalameitsev A V 2005 Optical properties of a semiconductor quantum dot with a single magnetic impurity: photoinduced spin orientation *Phys. Rev. B* 71 35338

[33] Besombes L, Léger Y, Maingault L, Ferrand D, Mariette H and Cibert J 2004 Probing the spin state of a single magnetic ion in an individual quantum dot *Phys. Rev. Lett.* 93 207403

[34] Bayer M, Kuther A, Forchel A, Gorbunov A, Timofeev V B, Schäfer F, Reithmaier J P, Reinecke T L and Walck S N 1999 Electron and Hole *g* Factors and Exchange Interaction from Studies of the Exciton Fine Structure in $In_{0.60}Ga_{0.40}$ As Quantum Dots *Phys. Rev. Lett.* 82 1748–51

[35] Govorov A O 2004 Optical probing of the spin state of a single magnetic impurity in a self-assembled quantum dot *Phys. Rev. B – Condens. Matter Mater. Phys.* 70 1–5

[36] Kudelski A, Lemaître A, Miard A, Voisin P, Graham T C M, Warburton R J and Krebs O 2007 Optically probing the fine structure of a single Mn atom in an InAs quantum dot *Phys. Rev. Lett.* 99 1–4

[37] Chen J, Chen X, Xu R, Zhu Y, Shi Y and Zhu X 2008 Refractive index of aqueous solution of CdTe quantum dots *Opt. Commun.* 281 3578–80

[38] Rahimi F, Ghaffary T, Naimi Y and Khajehazad H 2021 Effect of magnetic field on energy states and optical properties of quantum dots and quantum antidots *Opt. Quantum Electron.* 53 1–16

[39] Shah B, Anderson S W, Scalera J, Jara H and Soto J A 2011 Quantitative MR imaging: physical principles and sequence design in abdominal imaging. *Radiographics* 31 867–80

[40] Gong N, Wang H, Li S, Deng Y, Chen X, Ye L and Gu W 2014 Microwave-assisted polyol synthesis of gadolinium-doped green luminescent carbon dots as a bimodal nanoprobe *Langmuir* 30 10933–9

[41] Pan Y, Yang J, Fang Y, Zheng J, Song R and Yi C 2017 One-pot synthesis of gadolinium-doped carbon quantum dots for high-performance multimodal bioimaging *J. Mater. Chem. B* 5 92–101

[42] Bourlinos A B, Bakandritsos A, Kouloumpis A, Gournis D, Krysmann M, Giannelis E P, Polakova K, Safarova K, Hola K and Zboril R 2012 Gd(III)-doped carbon dots as a dual fluorescent-MRI probe *J. Mater. Chem.* 22 23327–30

[43] Yu C, Xuan T, Chen Y, Zhao Z, Liu X, Lian G and Li H 2016 Gadolinium-doped carbon dots with high quantum yield as an effective fluorescence and magnetic resonance bimodal imaging probe *J. Alloys Compd.* 688 611–9

[44] Ding H, Wang D, Sadat A, Li Z, Hu X, Xu M, de Morais P C, Ge B, Sun S, Ge J, Chen Y, Qian Y, Shen C, Shi X, Huang X, Zhang R-Q and Bi H 2021 Single-atom gadolinium anchored on graphene quantum dots as a magnetic resonance signal amplifier *ACS Appl. Bio Mater.* 4 2798–809

[45] Chen H, Wang L, Fu H, Wang Z, Xie Y, Zhang Z and Tang Y 2016 Gadolinium functionalized carbon dots for fluorescence/magnetic resonance dual-modality imaging of mesenchymal stem cells *J. Mater. Chem. B* 4 7472–80

[46] Ma Q, Nakane Y, Mori Y, Hasegawa M, Yoshioka Y, Watanabe T M, Gonda K, Ohuchi N and Jin T 2012 Multilayered, core/shell nanoprobes based on magnetic ferric oxide particles and quantum dots for multimodality imaging of breast cancer tumors *Biomaterials* 33 8486–94

[47] Ruan G, Vieira G, Henighan T, Chen A, Thakur D, Sooryakumar R and Winter J O 2010 Simultaneous magnetic manipulation and fluorescent tracking of multiple individual hybrid nanostructures *Nano Lett.* 10 2220–4

[48] Gui R, Wang Y and Sun J 2014 Encapsulating magnetic and fluorescent mesoporous silica into thermosensitive chitosan microspheres for cell imaging and controlled drug release in vitro *Colloids Surfaces B Biointerfaces* 113 1–9

[49] Ahmed S R, Dong J, Yui M, Kato T, Lee J and Park E Y 2013 Quantum dots incorporated magnetic nanoparticles for imaging colon carcinoma cells *J. Nanobiotechnology* 11 1–9

[50] Selvan S T, Patra P K, Ang C Y and Ying J Y 2007 Synthesis of silica-coated semiconductor and magnetic quantum dots and their use in the imaging of live cells *Angew. Chemie – Int. Ed.* 46 2448–52

[51] Song E Q, Hu J, Wen C Y, Tian Z Q, Yu X, Zhang Z L, Shi Y B and Pang D W 2011 Fluorescent-magnetic-biotargeting multi-functional nanobioprobes for detecting and isolating multiple types of tumor cells *ACS Nano* 5 761–70

[52] Van Tilborg G A F, Mulder W J M, Chin P T K, Storm G, Reutelingsperger C P, Nicolay K and Strijkers G J 2006 Annexin A5-conjugated quantum dots with a paramagnetic lipidic coating for the multimodal detection of apoptotic cells *Bioconjug. Chem.* 17 865–8

[53] Mangalath S, Saneesh Babu P S, Nair R R, Manu P M, Krishna S, Nair S A and Joseph J 2021 Graphene quantum dots decorated with boron dipyrromethene dye derivatives for photodynamic therapy *ACS Appl. Nano Materials* 4, no. 4 (2021): 4162–4171.

[54] Lisa John V, Joy F, Jose Kollannoor A, Joseph K, Nair Y and Vinod T P 2022 Amine functionalized carbon quantum dots from paper precursors for selective binding and fluorescent labelling applications *J. Colloid Interface Sci.* 617 730–44

[55] John V L, Nair Y and Vinod T P 2021 Doping and surface modification of carbon quantum dots for enhanced functionalities and related applications *Part. & Part. Syst. Charact.* 2100170 1–28

[56] A. R G, John V L, P. S A, Krishnan K. A A and T. P V Carbon dots from natural sources for biomedical applications *Part. & Part. Syst. Charact.* **n/a** 2200017

[57] Nandi S, Kolusheva S, Malishev R, Trachtenberg A, Vinod T P and Jelinek R 2015 Unilamellar vesicles from amphiphilic graphene quantum dots *Chem. – A Eur. J.* 21 7755–9

4 Characterization Techniques of Magnetic Quantum Dots

Dipti Rawat and Ragini Raj Singh
Nanotechnology Laboratory, Department of Physics and Materials Science, Jaypee University of Information Technology, Waknaghat, Solan, India

CONTENTS

4.1 INTRODUCTION

Nanostructures with numerous components, such as metallic, metal oxides, semiconducting, magnetic or core/shell topologies provide structures, material characteristics, and functionalities distinct from those of the individual bulk components [1-5]. The characterization of magnetic and magnetic-luminescent core-shell nanostructures with dual functions is covered in this chapter. Due to their unusual spatial organization,

DOI: 10.1201/9781003319870-4

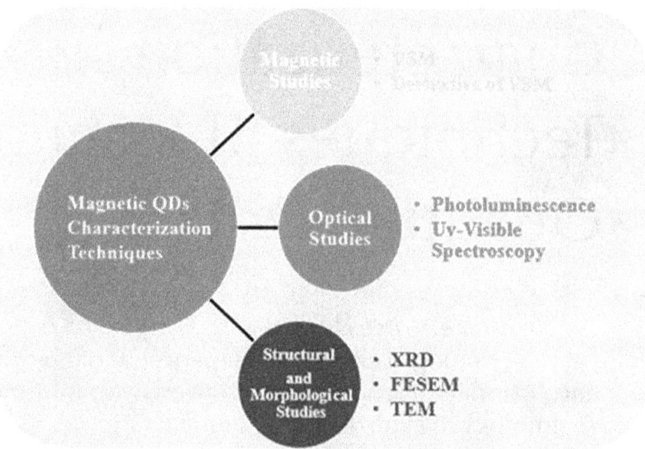

FIGURE 4.1 Different characterization tools for magnetic quantum dots.

numerous capabilities, and improved characteristics compared to their single-component counterparts, smart nanoparticles with cores and shells present intriguing potential in fundamental studies and cutting-edge technological applications [6–9]. Materials used in the core and shell can be semiconducting, magnetic, organic, inorganic, metallic, or dielectric. Magnetic and semiconducting materials conjugated systems are the focus of researchers these days. The magnetic component of core-shell nanoparticles is anticipated to offer a variety of potential functionality in areas including catalysts, ferrofluids, and biological applications. The surface properties of such shell layers include catalysis, improved photoluminescence, and tailored band structures in the case where semiconductors constitute the shell. Despite extensive study on sequential reduction strategies for assembling magnetic and core-shell nanoparticles, real multi-functional examples are few, and substantial research is ongoing. Figure 4.1 represents the different tools and characterization techniques for magnetic quantum dots.

Due to its distinct physicochemical characteristics, magnetite is preferred because it gives an edge over other magnetic materials. In addition, it demonstrates several intriguing features, such as charge ordering, mixed valence, and the metal-insulator transition known as the Verwey transition, which helps analyze the synergistic effects of surface modification and the spatial restrictions imposed by a core-shell arrangement [10-13]. CdS is a particularly effective luminous semiconducting quantum-dot substantial in the optical arena. It is expected that by combining ferrites and QDs into a core-shell nanoarchitecture, a new nanostructured material will be created that holds the optical and magnetic properties of the original constituents while offering synergistically improved performance and functionalities that may go beyond the original components. This chapter describes how to characterize magnetic and multifunctional magnetic-luminescent nanostructures, which are made up of a semiconductor shell and a magnetic core. Using a variety of tools and procedures, we can characterize these nanostructures. The magnetic and core-shell nanostructures were first subjected to basic characterization techniques, after which the computer modelling was successfully implemented.

4.2 TOOLS AND TECHNIQUES FOR CHARACTERIZATION

4.2.1 Scattering Analysis

One of the most used approaches for the characterization of nanoparticles in colloidal or powder form is an electron, neutron, or light-scattering from the sample. Another common approach for measuring direct particle size in nanoparticle solutions is dynamic light scattering (DLS) [14-17]. The shell thickness can be determined by measuring the particle size before and after coating. In addition, gauging the potential of the magnetic core particles in the solution can provide unintended mark of the extent of magnetic core surface alteration. This method determines the particle's hydrodynamic diameter.

4.2.2 X-Ray Characterization

X-ray powder diffraction is a technique for structural study. The three fundamental parts of XRD are an X-ray tube, a detector, and sample holder. A Shimadzu (XRD 6000) X-ray diffractometer was utilized for the XRD testing. A popular non-destructive technique for characterizing crystalline materials is XRD. Data on crystal structure, atomic spacing, phases, preferred crystal orientations (texture), and several other aspects like crystallite size, strain, and crystal defects are provided by this method. Constructive interference of dispersed X-ray photons from lattice planes at specific angles results in the formation of XRD peaks.

XRD is the most frequently used method for characterizing NPs. Typically, XRD is used to determine the crystalline structure, phase nature, lattice parameters, and crystalline grain size. The latter parameter is computed using the broadening of the peak with the highest intensity in the XRD measurement for a specific sample and Scherrer's equation. The advantage of the XRD methods is the generation of statistically representative, volume-averaged data. The particle's composition can be resolved by linking the position and intensity of the peaks with the reference patterns in the Joint Committee on Powder Diffraction Standards (JCPDS). But, for amorphous samples, it is not appropriate because of the extreme broadening of the XRD peaks for particles smaller than 3 nm [18].

We investigated the X-ray diffraction spectrum of $NiZnFe_2O_4$ core, CdS QDs, and$NiZnFe_2O_4$/CdS core-shell with $NiZnFe_2O_4$annealed at 900°C temperature in the present Figure 4.2 [19]. As seen in Figure 4.2 (a–c), the (022), (113), (222), (004), (333), and (044) hkl planes for $NiZnFe_2O_4$ (Figure 4.2(a)) indicate the creation of single-phase spinel cubic structure for$NiZnFe_2O_4$ (JCPDS No. 52-0277). There is no impurity phase in the nanoparticles. Peaks indexed as (111), (220), and (311) hkl planes show the cubic structure of CdS QDs (JCPDS No. 80-0019) with a crystallite size of approximately 1.83nm in CdS QDs spectra (Figure 4.2(b)). The spectra for core-shell nanostructures with a $NiZnFe_2O_4$ loading of 0.05 g are shown in Figure 4.2(c). Peaks from both the CdS QDs and the $NiZnFe_2O_4$/CdS $NiZnFe_2O_4$/CdS $NiZnFe_2O_4$/CdS $NiZnFe_2O_4$/CdS $NiZnFe_2O_4$/CdS $NiZnFe_2O_4$/ CdS $NiZnFe_2O_4$/C As a result, we can state that the core-shell system is more intact at higher $NiZnFe_2O_4$ annealing temperatures. There was a shift in the peaks

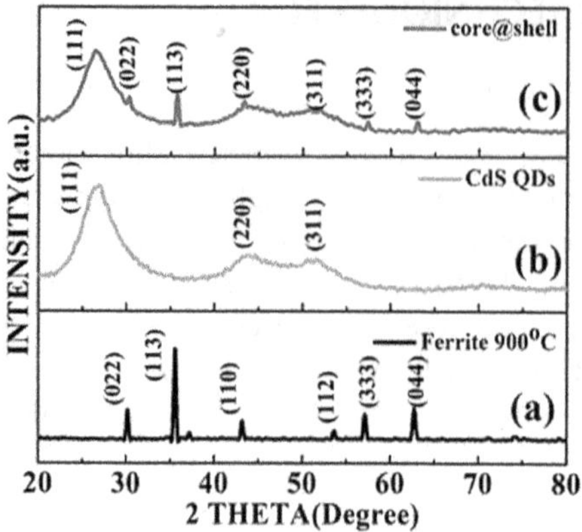

FIGURE 4.2 XRD spectra of core, shell and the core/shell of $NiZnFe_2O_4$ with CdSquantum dots.

when core-shell nanostructures were compared to bare core and separate shell structures, showing that core-shell nanostructures were successfully created. There was no alloying, and all of the nanostructures were confirmed to have pure crystalline phases in the samples. To determine the particle sizes for core-shell nanostructures, XRD could not be employed since NiZnFe2O4, and CdS QDs have distinctive peaks in their spectra. According to the literature, various magnetic core and semiconductor shell core-shell nanostructures exhibit a similar pattern in their XRD spectra [20–22].

4.2.2.1 Rietveld Refinement

X-ray diffractograms were refined using the Rietveld method to determine the nanoparticle's lattice parameter, bond lengths, atomic coordinates, and site occupancy. For methodology, a succinct summary of Rietveld refinement was provided. Using the FULLPROF tool from the WINPLOTR family of applications, global parameters like zero and background were tuned first, and then the XRD patterns were Rietveld refined [23–24]. Figure 4.3 shows the Rietveld analysis of the NiZnFe2O4 X-ray diffraction spectra after it has been annealed at various temperatures using magnetofluorescent magnetic quantum dots.

4.2.3 Spectroscopic Analysis

The magnetic nanoparticle surface can be inferred indirectly from optical properties since they are sensitive to changes to the nanocrystal surface. A popular spectroscopic method for analyzing different kinds of nanoparticles is UV-vis spectroscopy. Those in particular with an absorbance spectrum and energy absorption capacity in the UV-vis range. Another helpful method for characterizing materials with fluorescence properties, such as semiconductors, is fluorescence or photoluminescence spectroscopy. Following coating, both the peak wavelengths in the UV and PL spectra and the intensity (light absorbed or emitted) change. The intensity and peak wavelengths

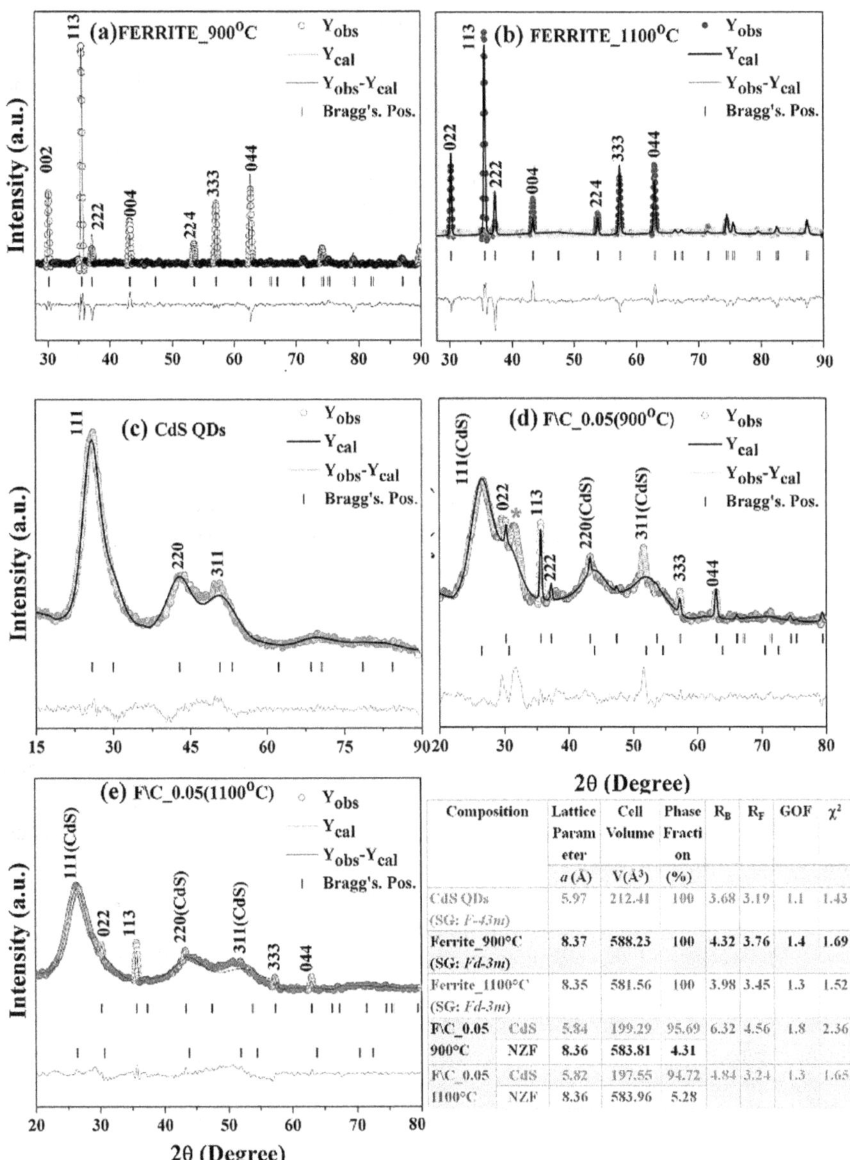

Composition		Lattice Parameter	Cell Volume	Phase Fraction	R_B	R_F	GOF	χ^2
		a (Å)	V(Å³)	(%)				
CdS QDs (SG: *F-43m*)		5.97	212.41	100	3.68	3.19	1.1	1.43
Ferrite_900°C (SG: *Fd-3m*)		**8.37**	**588.23**	**100**	**4.32**	**3.76**	**1.4**	**1.69**
Ferrite_1100°C (SG: *Fd-3m*)		8.35	581.56	100	3.98	3.45	1.3	1.52
F\C_0.05 900°C	CdS	5.84	199.29	95.69	6.32	4.56	1.8	2.36
	NZF	**8.36**	**583.81**	**4.31**				
F\C_0.05 1100°C	CdS	5.82	197.55	94.72	4.84	3.24	1.3	1.65
	NZF	8.36	583.96	5.28				

FIGURE 4.3 The Rietveld refinement pattern of (**a-b**) $NiZnFe_2O_4$; annealed at different temperatures) CdS QDs; and (**d-e**) $NiZnFe_2O_4$/CdS core@shell nanostructures from reference [28].

also move closer to those of pure shell materials as the thickness of the shell material increases.

On the other hand, indirect evidence for the coating of shell material on the core surface is primarily provided by UV or PL spectroscopy. X-ray photoelectron spectroscopy (XPS) [25] is another powerful spectroscopic technique for revealing surface information. This method can calculate the number of electrons leaving a 110 nm thick surface layer and their kinetic energy (KE). The two main drawbacks of XPS are that characterization can only go down 10 nm from the particle surface and that it needs an ultrahigh vacuum (UHV) chamber.

4.2.3.1 UV-Visible Spectroscopy

UV-vis spectroscopy is a well-liked depiction method for nanoscale samples that is equally simple and affordable (UV-vis). In comparison to the light intensity reflected from reference material, it compares the light intensity reflected from a sample. UV-vis spectroscopy is a helpful tool for identifying, characterizing, and researching these materials as well as evaluating the stability of NP colloidal solutions because the optical properties of NPs depend on their size, shape, concentration, agglomeration state, and refractive index close to the NP surface. To detect and categorize the four various types of gold nanostructures they created, the researchers used the UV-vis technique. This NP system and type classification method is quick, non-destructive, and reasonably priced [26].

4.2.3.2 Photoluminescence Spectroscopy

Photoluminescence spectroscopy (PL), which measures the light emitted by atoms or molecules that have absorbed photons, is another tool for researching nanoscale materials [27]. PL is frequently used as a characterization tool for metal nanoclusters and fluorescent nanoparticles like quantum dots. Recently, research on the intrinsic PL of metallic NPs has gained popularity. In spite of the low quantum efficiency of the emission mechanism, this inefficiency is made up for by the enormous excitation cross sections at the plasmon resonances. Additionally, the PL of metal NPs is devoid of photo blinking and photobleaching. As a result, PL can be viewed as a superior substitute for fluorescent molecules in optical labelling applications. Plasmonic nanostructures of various types have been used to produce PL for single-photon and multi-photon stimulation. The PL behavior of a single Au nanoflower, with several branches, was studied by Gong and colleagues. ZnO NPs and other metal oxide nanoparticles are photoluminescent [27]. Saliba et al. created zinc oxide nanoparticles. For the constituents, three emissions were seen depending on the excitation wavelength. Surface defects were one of the factors cited as the cause of the emissions (e.g., oxygen vacancies). Figure 4.4 displays the PL spectra of the following $NiZnFe_2O_4/CdS$ nanostructures: (a) 0.2 g $NiZnFe_2O_4/CdS$ nanostructures; (b) 0.1 g $NiZnFe_2O_4/CdS$ nanostructures; and (c) 0.05 g $NiZnFe_2O_4/CdS$ nanostructures with regard to naked CdS QDs. The semiconductor shell's emission profile trend is also in the magnetic/semiconducting core/shell nanostructure. The substrate impact of the magnetic core causes a slight shift in the emission peak profile in magnetofluorescent core/shell nanostructures.

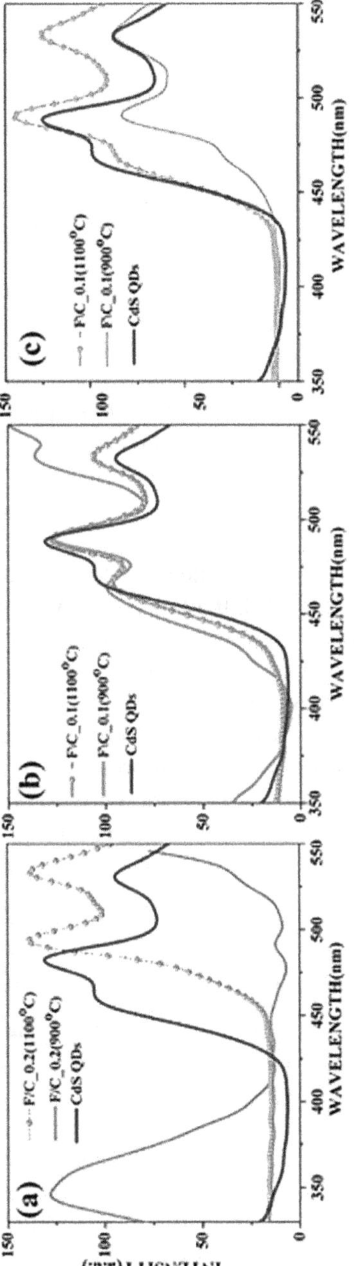

FIGURE 4.4 Photoluminescence spectra of (a) $NiZnFe_2O_4$/CdS nanostructures with 0.2 g loading of magnetic nanoparticles; (b) $NiZnFe_2O_4$/ CdS nanostructures with 0.1 g loading; (c) $NiZnFe_2O_4$/CdS nanostructures with 0.05 g loading w.r.t., bare CdS QDs from reference [28].

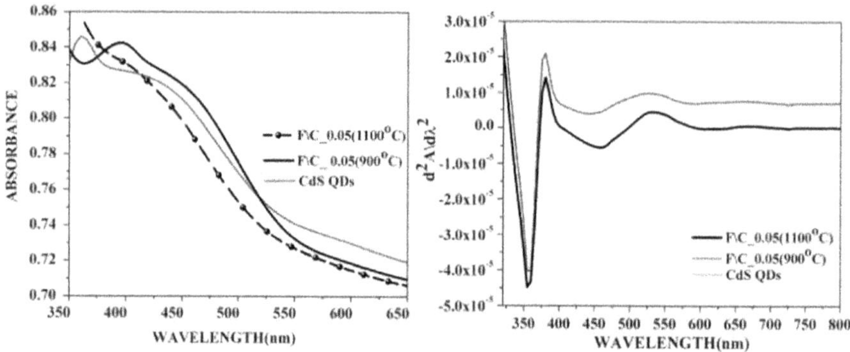

FIGURE 4.5 Absorbance and derivative curve of $NiZnFe_2O_4/CdS$ nanostructures from reference [28].

4.2.3.3 Optical Analysis Using Derivative Spectroscopy

Derivative spectroscopy uses the original or complex derivative of absorbance w.r.t., wavelength for a qualitative and quantitative study: "The first recommendations for derivatizing spectra data were made in the 1950s, when it was discovered to offer many advantages [29–31]."

The first-order derivative is used to calculate the rate at which absorbance changes in relation to wavelength. At 0, the first-order derivative starts and stops. The strong negative or positive band in the even-order derivatives has a minimum or extreme at the same wavelength as the extreme of absorbance band. It is essential to keep in mind that the number of bands observed in spectra is always equal to the derivative order plus one. We calculated the detailed absorbance sites for each sample using the second derivative of absorbance for samples. The bands seen in the graphs of double derivatives are equal to the derivative order plus one. Figure 4.5 illustrates how precisely the exact absorbance peak position may be produced with specific features by deriving the absorbance spectra of $NiZnFe_2O_4$/CdS core/shell nanostructures.

4.2.4 Microscopic Techniques for Nanostructure Characterization

Microscopic examination is the most common and effective method for observing diverse categories of nanoparticles up close. Scanning electron microscopy (SEM) is the supreme microscopic procedure for analyzing the size and structure of nanoparticles. However, it is challenging to distinguish between the core and shell materials in core/shell nanoparticles because they can only produce a surface image. When combined with energy-dispersive X-ray spectroscopy, SEM can be used for fundamental research on the magnetic nanoparticle surface (EDX). Recent research has shown the value of FESEM (field-emission SEM), which can achieve significantly higher magnification than traditional SEM. High-magnification FESEM and HRTEM images and SAED patterns can be used to study the evidence of magnetic core and luminous shell nanostructures [32].

4.2.4.1 FESEM Analysis

High-resolution surface imaging is frequently used for scanning electron microscopy (SEM), which may also be used to examine materials at the nanoscale. The SEM uses electrons in the same way a light microscope uses visible light for imaging. Mazzaglia et al. coupled field emission SEM (FESEM) and XPS experiments to investigate supramolecular colloidal systems of Au NPs/amphiphilic cyclodextrin [32]. Due to their small size within the range of nanometers, this approach can also be used to analyze magnetic nanoparticles.

This technique guarantees the observation of NPs using high-resolution SEM (HRSEM), and sample preparation are quick and straightforward. However, in the case of magnetic nanoparticles, a metal coating may be required in order to reduce magnetic effects. The benefit of HRSEM over other imaging methods is its ability to scale down and examine the configurations of nanometric elements in a broader environment. Furthermore, it enables the investigation of potential interactions between NPs by allowing the study of the precise spatial arrangement of NPs. The findings of that investigation suggested that HRSEM has the potential to be a very straightforward method for qualitatively identifying the elements. It can be seen as a strong and versatile tool for researching how biological systems interact with metallic nanostructures.

Through the T-SEM technology, SEM can be operated in transmission mode. By assembling in-depth data and analyzing ensembles of NPs, advanced NP analysis can be done in transmission mode. Figure 4.6 below includes a few SEM pictures of zinc ferrites, zinc cobalt magnesium ferrites, and cobalt magnesium ferrites nanoparticles.

4.2.4.2 HRTEM Analysis

Because X-ray diffraction and other bulk characterization methods have limits, more complex strategies, including spatially resolved micro-analytical methods, must be used to determine the actual structure of individual particles. Such a spatially resolved research platform, as offered by transmission electron microscopy (TEM), is ideally suited to characterize multi-layered nanostructures [34]. Based on lattice fringes and contrast fluctuations, bright field TEM and high-resolution TEM (HRTEM) can be utilized to distinguish between various compositions. High angle annular dark field (HAADF), a type of dark field imaging, is particularly informative in this latter area since, unlike brilliant field imaging, picture production now primarily relies on Rutherford scattered electrons rather than Bragg scattered electrons [35].

Additionally, single crystal and polycrystalline anisotropic NPs can be differentiated using HRTEM [36] despite sharing the same optical properties. One example of a structural change that can be identified by HRTEM is the temperature transition in iron-platinum nanoparticles from disordered face-centered cubic to ordered L10 [37]. The NPs produced by this thermally induced event have improved magneto crystalline anisotropy and greater coercivity, which are essential for making permanent magnets (Figure 4.7).

HRTEM observations can also reveal information about NP growth and qualities that are related to structure. For instance, Zhang et al. investigated in-situ HRTEM's role in CuO NP production [38]. They discovered that coalescence, which occurs

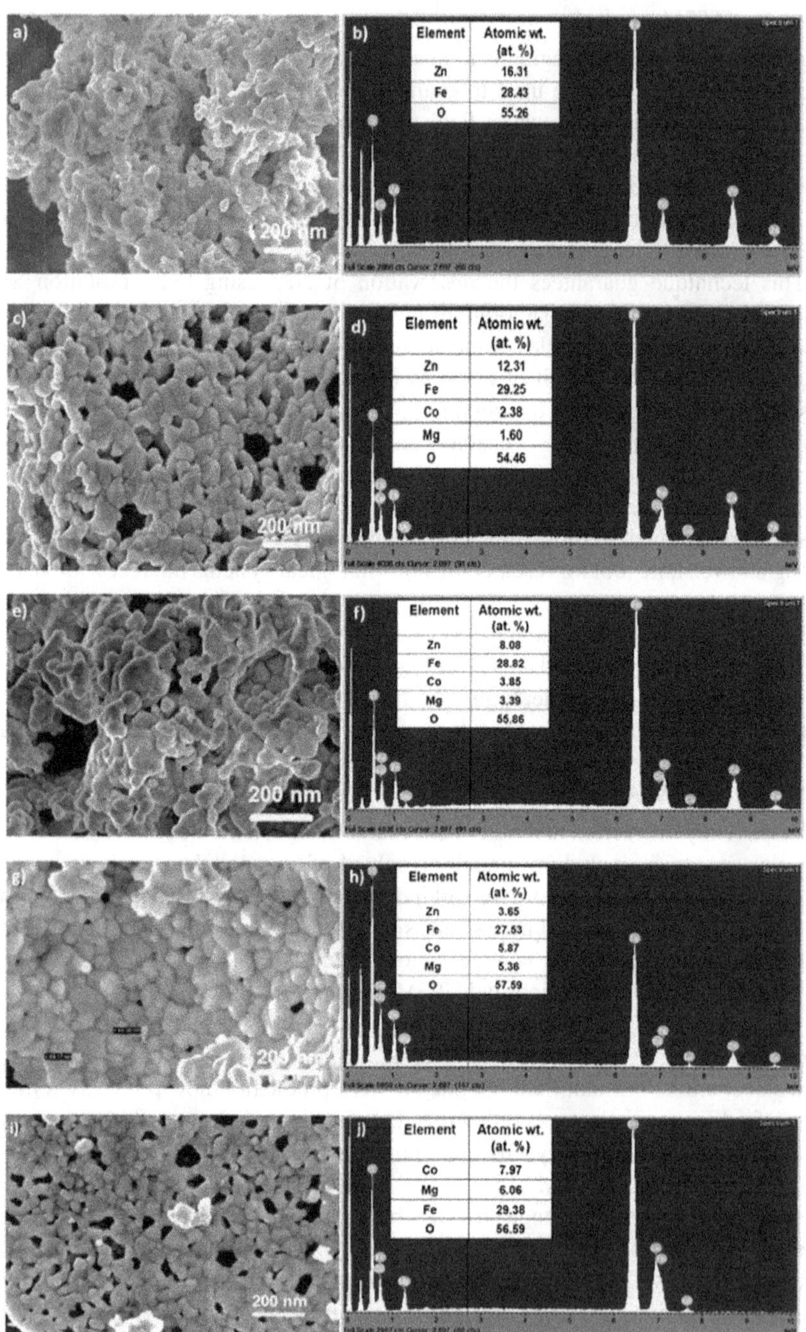

FIGURE 4.6 FESEM and EDS images of (a and b) ZF, (c and d) ZCMF1, (e and f) ZCMF2, (g and h) ZCMF3, ald (i and j) CMF, individually from reference [33].

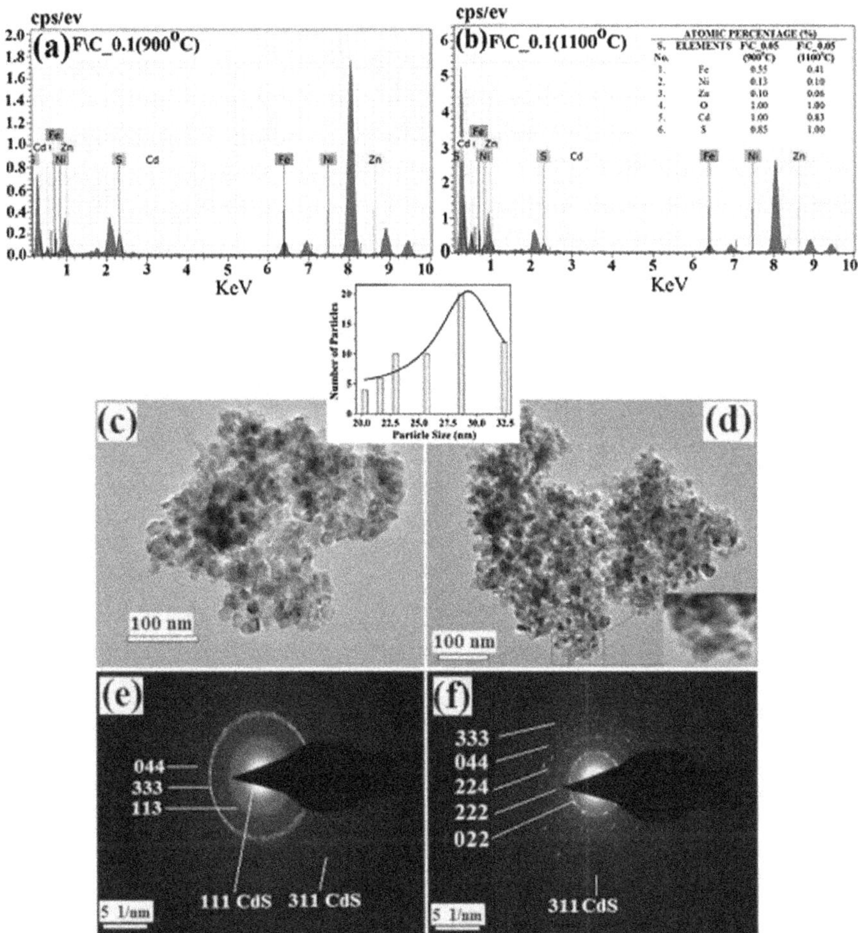

FIGURE 4.7 EDX spectra of core/shell nanostructures (a) and (b); (c) and (d) show the TEM micrograph of nanostructures with their respective SAD patterns (e-f) from reference [28].

far more quickly than other processes like nanocrystal reshaping, was the dominant mechanism. In addition, they saw that single crystal NPs were produced when the colloids were oriented before merging. Additionally, HRTEM has been used to explain how substrates affect the characteristics of metal NPs.

4.2.5 Magnetic Characterization

4.2.5.1 VSM Analysis

Another technique for capturing the M-H loops and obtaining constraints like the Ms and the Mr is vibrating sample magnetometry (VSM). The superparamagnetic FeCo@SnO_2 NPs on graphene-polyaniline were studied in terms of their magnetic

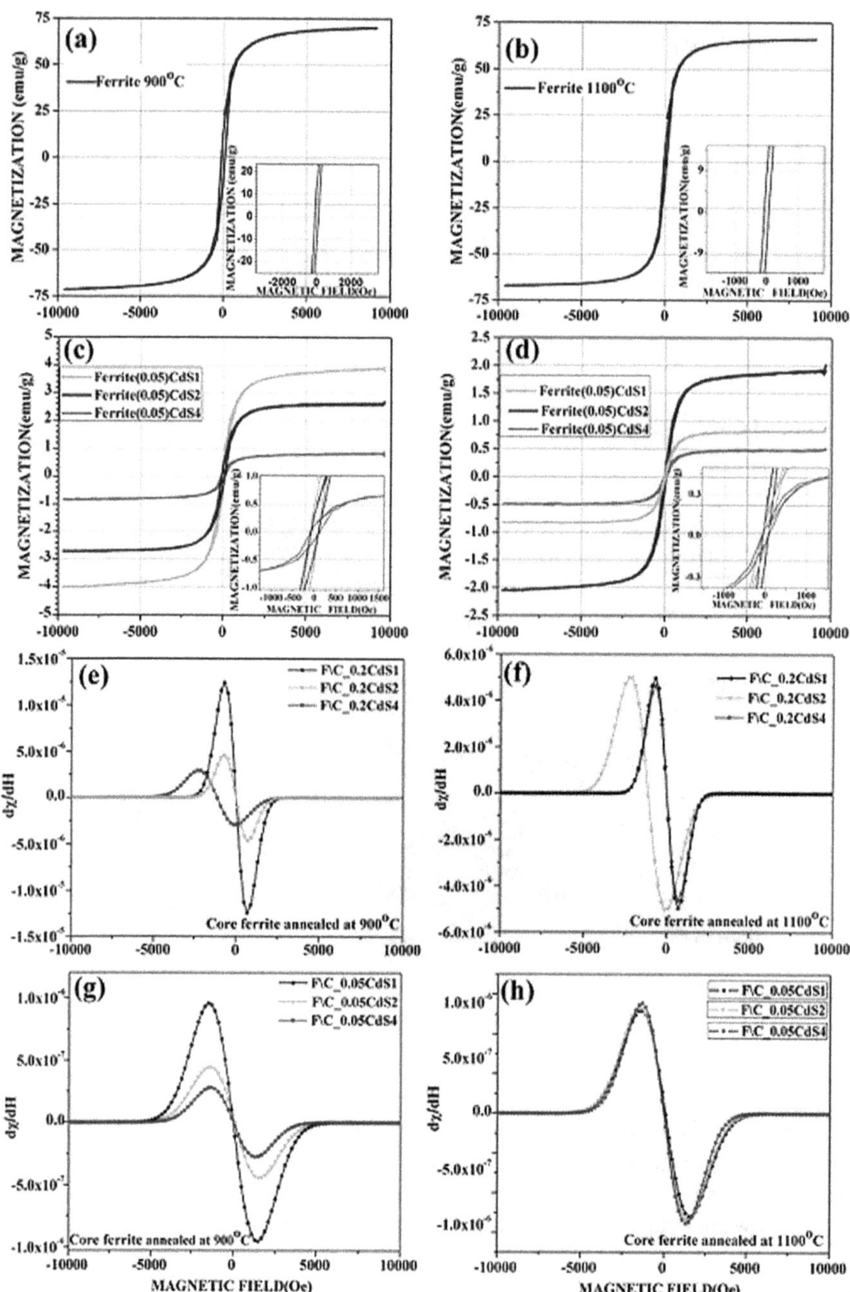

FIGURE 4.8 Hysteresis loops for $NiZnFe_2O_4$ (a) 900 °C, (b) 1100 °C; and $NiZnFe_2O_4$/CdS with $NiZnFe_2O_4$ (c) 900 °C, (d) 1100 °C;. dχ/dH curves of $NiZnFe_2O_4$/CdS nanostructures with changing CdS QDs layers with 0.2 g $NiZnFe_2O_4$loIng (e) 900 °C, (f) 1100 °C; and with 0.05 g $NiZnFe_2O_4$ loading (g) 900 °C, (h) 1100 °C. From reference [28].

properties by the FeCo NPs. We looked into their improved electromagnetic wave absorption capabilities.

VSM was used to describe several FeCo, FeCo@SnO2, and FeCo@SnO2@graphene@PANI composites. When an external magnetic field is applied, the magnetic moments of the FeCo NPs exhibit strong magnetic dipolar interactions and would point in the same direction as the field [39].

Magnetization measurements have been performed on $NiZnFe_2O_4$ that has been annealed at two temperatures, 900°C and 1100°C, as well as $NiZnFe_2O_4$/CdS core/shell nanoparticles. The M-H loop for core and core/shell nanostructures is depicted in Figure 4.8 (a–d): "The saturation magnetization (Ms) of $NiZnFe_2O_4$ NPs was found to be 70.16 emu/g and 66.35 emu/g, respectively, when they were annealed at two different temperatures, 900 °C (Figure 4.8(a)) and 1100 °C (Figure 4.8(b))" [28]. The magnetization loss resulting from raising the annealing temperature is caused by the high temperature that accelerates ion diffusion and causes pores to form. The decreased magnetization may have been caused by how challenging it was to shift the domain wall. Samples lose saturation magnetization at higher annealing temperatures due to pore development. Additionally, "samples undergo a spin-reorientation transition that is reduced by an external magnetic field and caused by a change in temperature" [40].

4.2.5.2 Deep Insight into Magnetic Results Using Derivative Spectroscopy

The as-obtained hysteresis curves have been subjected to spectroscopy for additional investigation of the magnetic character derivative. The M-H loop of bare ferrite and the produced core/shell nanostructures are depicted by the second derivative curve in Figures 4.8 (e–h). To accurately describe the magnetic characteristics of magnetic and magneto-optical core/shell systems, SFDs are crucial. One can learn about spin orientation at the intersection of high and low magnetic fields, thanks to SFD. This information is vital when comparing the core/shell nanostructure to the raw core and shell nanoparticles. On shifting from MPs to the core/shell, a significant variance in the SFD was observed: "By means of the equation SFD= H/H_C [41], where H is the FWHM of the first order differentiated curve of the simple hysteresis loop, the SFD has been calculated." The SFD loops for $NiZnFe_2O_4$magnetic nanoparticles are shown in Figure 4.9.

A distribution function for the quantity of units reversing at a specific field is the SFD from dM/dH. The SFD and particle size distribution are closely related for a nanoparticle medium [42]. High H_C and low SFD materials are of outstanding quality. The width and sharpness of derivative curves are significant because small SFDs are favorable [42]. The curves provide useful information on the magnetic confirmation of the material, which is related to the microstructure and chemical inhomogeneities in the system because a steady switching transition produces a modest SFD.

4.2.6 FT-IR Analysis

The term "Fourier-transformed infrared spectroscopy" (FTIR) refers to a method that measures the wavelength-dependent absorption of electromagnetic radiation. A molecule's dipole moment is altered as it absorbs IR light, making it IR active.

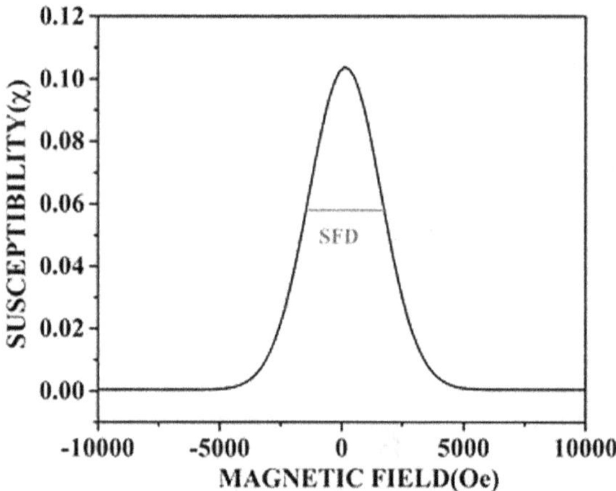

FIGURE 4.9 SFD loops for $NiZnFe_2O_4$magnetic nanoparticle.

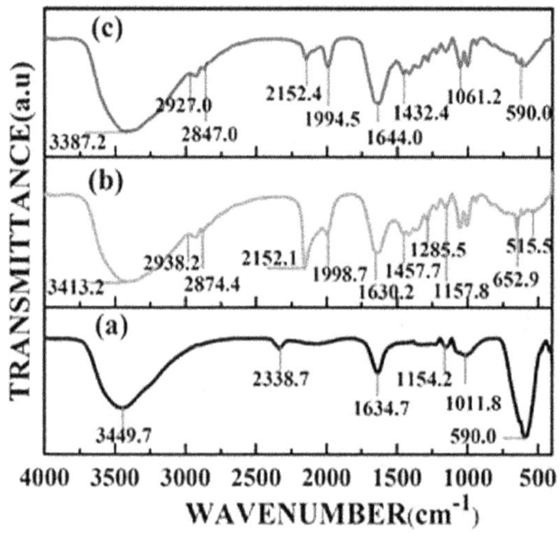

FIGURE 4.10 FTIR spectra for the $NiZnFe_2O_4$, CdS and $iZnFe_2O_4$/CdS nanostructure.

According to a recorded spectrum, specific functional groups, bands related to binding strength and type, their positions, and information about molecular structures and interactions are revealed [43].

The less intense vibrational band at 3500 cm^{-1} and the intense vibrational band near 2500 $cm^{-1}$are predicted to be caused by the free and absorbed water molecules maintained in the samples during synthesis [44, 45]. Since FTIR is a surface phenomenon in core-shell nanostructures, where the core and shell are entirely distinct from one another, only the functional groups present at the surface, i.e., of the shell, should be observable. Figure 4.10 shows the FTIR spectra for the core, shell, and core-shell magneto fluorescent nanostructures.

4.3 KEY NOTES

- **Structural characterization:** From the structural characterization of core-shell nanostructures, information regarding the shell growth over the core, shifting in crystallite peaks with the core-shell formation, if alloying is taking place at any point during the core-shell formation, generation of defects, diffusion of the shell with the core materials etc., can be revealed. By refining the XRD spectra through Rietveld refinement, we can take the idea about the phase fraction of the core-shell separately.
- **Optical analysis:** Optical analysis became an essential tool for the nanostructure analysis for the magneto fluorescent core-shell nanostructures. UV-visible and the pL spectroscopy give a brief idea about the band gap of fluorescent shell and their emission profile. Sometimes during the core-shell synthesis, leakage of shell constituents to the core might be possible. Using PL spectroscopy, defect peaks occurring due to the alloying of core and shell can be detected; hence the intactness of the core-shell structure can be checked. Particle size distribution and the intensity of emission can also be studied using the optical analysis tool. Also, with derivative spectroscopy, it is possible to precisely determine peak positions and the change in peak positions with enhanced spectral resolution.
- **Microscopic techniques for nanostructure characterization:** By doing morphological studies such as FESEM and HRTEM, the formation of core-shell nanostructures is confirmed. Clear and sharp boundaries are seen amid the interface of core and shell. Sharper the interface boundaries much intact the nanostructure system considered. By using the dark field SAED pattern, the polycrystallanity of the system is confirmed.
- **Hysteresis curve and the interface study using magnetic analysis:** Magnetic study is another essential tool for the magneto fluorescent core-shell nanostructures. The hard/soft magnetic character of the core is supported by magnetic research. Additionally, it was established that the luminous shell perfectly covers the core as the magnetic saturation degrades following shell creation. This is another solid proof that the core and shell have formed.

By taking the fundamental hysteresis loop's first-order derivative, the switching field distribution can be calculated. We can learn about spin orientation at the intersection of high and low magnetic fields due to SFD.

FT-IR analysis: The functional groups on the system were revealed by FTIR studies to be favorable for further processing of these magnetofluorescent nanostructures for biological offers. The FTIR spectra shift in the absorption band proves that the core and shell formed.

According to the findings, these nanostructures can be exploited in spintronics and for applications in the fields of drug delivery, diagnosis, and imaging in nanomedicine.

TABLE 4.1
Investigation Methods for Nanoparticle Characterizations

Entity Characterized	Characterization Techniques Suitable
Size (structural properties)	TEM, XRD, DLS, NTA, SAXS,HRTEM, SEM, AFM, EXAFS,UV-Vis MELDI, NMR, TRPS, EPLS, magnetic susceptibility.
Shape	TEM, HRTEM, AFM, EPLS, FMR, 3D-tomography
Elemental-chemical composition	XRD, ZPS, ICP-MS, ICP-OES, SEM-EDX, NMR, MFM, LEIS
Crystal stucture	XRD, EXAFS, HRTEM,electron diffraction, STEM
Size distribution	DCS, DLS, SAXS, NTA, ICP-MS, FMR,DTA,
Ligand's binding/composition/arrangement, mass, Surface composition	XPS, FTIR, NMR, SIMS, FMR, TGA, SAMS
Concentration	IC-MS, UV-Vis, RMM-MEMS, PTA, DCS, TRPS
Structural defects	HRTEM, EBSD
Detection of NPs	TEM, SEM, STEM, EBSD, magnetic susceptibility
Optical properties	UV-Vis-NIR, PL, EEL-STEM
Magnetic properties	SQUID, VSM, Mossbaues, MFM, FMR, XMCD, magnetic susceptibility

4.4 CONCLUSION AND FUTURE PERSPECTIVES

The function of numerous dissimilar practices for the nanoscale materials description is discussed in this chapter. Through our in-depth analysis of NP characterization techniques, we highlighted the benefits and drawbacks of each method while also demonstrating how to use them successfully in combination and as a complement to one another. It is common to employ more than one technique to thoroughly and accurately assess even a single attribute to provide a complete picture of the variety of features connected with a nanomaterial. Therefore, our chapter will serve as a solid reference, assisting the scientific community in comprehending the discussed topic better by comparing the role of each technique. In this approach, it will be easier for researchers to select the best practices for their characterizations and evaluate how they are used more accurately. On the other hand, the scientific community faces difficulties further enhancing several procedures' precision and clarity. We, therefore, conclude by expressing the hope that a careful reading of this chapter will assist in identifying which worthwhile strategies merit efforts for subsequent technical advancements.

REFERENCES

1. Cho, N.H., Cheong, T.C., Min, J.H., Wu, J.H., Lee, S.J., Kim, D., Yang, J.S., Kim, S., Kim, Y.K. and Seong, S.Y., 2011. A multifunctional core–shell nanoparticle

for dendritic cell-based cancer immunotherapy. *Nature nanotechnology*, *6*(10), pp.675–682.

2. Hao, R., Xing, R., Xu, Z., Hou, Y., Gao, S. and Sun, S., 2010. Synthesis, functionalization, and biomedical applications of multifunctional magnetic nanoparticles. *Advanced Materials*, *22*(25), pp.2729–2742.
3. Kim, H., Achermann, M., Balet, L.P., Hollingsworth, J.A. and Klimov, V.I., 2005. Synthesis and characterization of Co/CdSe core/shell nanocomposites: bifunctional magnetic-optical nanocrystals. *Journal of the American Chemical Society*, *127*(2), pp.544–546.
4. Wu, W., Jiang, C. and Roy, V.A., 2015. Recent progress in magnetic iron oxide–semiconductor composite nanomaterials as promising photocatalysts. *Nanoscale*, *7*(1), pp.38–58.
5. Son, J.S., Lee, J.S., Shevchenko, E.V. and Talapin, D.V., 2013. Magnet-in-the-semiconductor nanomaterials: High electron mobility in all-inorganic arrays of FePt/CdSe and FePt/CdS core–shell heterostructures. *The Journal of Physical Chemistry Letters*, *4*(11), pp.1918–1923.
6. Issa, B., Obaidat, I.M., Albiss, B.A. and Haik, Y., 2013. Magnetic nanoparticles: surface effects and properties related to biomedicine applications. *International Journal of Molecular Sciences*, *14*(11), pp.21266–21305.
7. Chen, X., Lou, Y., Samia, A.C. and Burda, C., 2003. Coherency strain effects on the optical response of core/shell heteronanostructures. *Nano Letters*, *3*(6), pp.799–803.
8. Vaidya, S., Kar, A., Patra, A. and Ganguli, A.K., 2013. Core–Shell (CS) nanostructures and their application based on magnetic and optical properties. *Reviews in Nanoscience and Nanotechnology*, *2*(2), pp.106–126.
9. Chatterjee, K., Sarkar, S., Rao, K.J. and Paria, S., 2014. Core/shell nanoparticles in biomedical applications. *Advances in Colloid and Interface Science*, *209*, pp.8–39.
10. Ghazanfari, M.R., Kashefi, M., Shams, S.F. and Jaafari, M.R., 2016. Perspective of Fe3O4 nanoparticles role in biomedical applications. *Biochemistry Research International*, *2016* (2016): 1–32.
11. McNamara, K. and Tofail, S.A., 2015. Nanosystems: the use of nanoalloys, metallic, bimetallic, and magnetic nanoparticles in biomedical applications. *Physical Chemistry Chemical Physics*, *17*(42), pp.27981–27995.
12. Liu, Y., Zhu, W., Wu, D. and Wei, Q., 2015. Electrochemical determination of dopamine in the presence of uric acid using palladium-loaded mesoporous Fe3O4 nanoparticles. *Measurement*, *60*, pp.1–5.
13. Sun, S.N., Wei, C., Zhu, Z.Z., Hou, Y.L., Venkatraman, S.S. and Xu, Z.C., 2014. Magnetic iron oxide nanoparticles: Synthesis and surface coating techniques for biomedical applications. *Chinese Physics B*, *23*(3), p.037503.
14. Chu, B., 1991. Laser Light Scattering: Basic Principles and Practice. Amsterdam: Elsevier.
15. Schmitz, K.S., 1990. *An Introduction to Dynamic Light Scattering by Macromolecules* Academic Press. *New York*.
16. Berne, B.J. and Pecora, R., 2000. *Dynamic Light Scattering: With Applications to Chemistry, Biology, and Physics*. North Chelmsford, Massachusetts: Courier Corporation.
17. Pecora, R., 2000. Dynamic light scattering measurement of nanometer particles in liquids. *Journal of Nanoparticle Research*, *2*(2), pp.123–131.
18. Mourdikoudis, S., Pallares, R.M. and Thanh, N.T., 2018. Characterization techniques for nanoparticles: comparison and complementarity upon studying nanoparticle properties. *Nanoscale*, *10*(27), pp.12871–12934.

19. Rawat, D., Barman, P.B. and Singh, R.R., 2019. Multifunctional magneto-fluorescent NiZnFe@ CdS core-shell nanostructures for multimodal applications. *Materials Chemistry and Physics*, *231*, pp.388–396.
20. Joseph, J., Mishra, N., Mehto, V.R., Banerjee, A. and Pandey, R.K., 2014. Structural, optical and magnetic characterization of bifunctional core shell nanostructure of Fe3O4/CdS synthesized using a room temperature aqueous route. *Journal of Experimental Nanoscience*, *9*(8), pp.807–817.
21. Kale, A., Kale, S., Yadav, P., Gholap, H., Pasricha, R., Jog, J.P., Lefez, B., Hannoyer, B., Shastry, P. and Ogale, S., 2011. Magnetite/CdTe magnetic–fluorescent composite nanosystem for magnetic separation and bio-imaging. *Nanotechnology*, *22*(22), p.225101.
22. Shao, H., Qi, J., Lin, T. and Zhou, Y., 2018. Preparation and characterization of Fe3O4@ SiO2@ NMDP core-shell structure composite magnetic nanoparticles. *Ceramics International*, *44*(2), pp.2255–2260.
23. Rodriguez-Carvajal, J., 2011. FULLPROF: A rietveld refinement and pattern matching analysis program. *Lab. Leon Brillouin (CEA-CNRS), Fr.*
24. Yogi, A., Bera, A.K., Maurya, A., Kulkarni, R., Yusuf, S.M., Hoser, A., Tsirlin, A.A. and Thamizhavel, A., 2017. Stripe order on the spin-1 stacked honeycomb lattice in Ba 2 Ni (PO 4) 2. *Physical Review B*, *95*(2), p.024401.
25. Mourdikoudis, S., Pallares, R.M. and Thanh, N.T., 2018. Characterization techniques for nanoparticles: comparison and complementarity upon studying nanoparticle properties. *Nanoscale*, *10*(27), pp.12871–12934.
26. *UV/Vis/IR Spectroscopy Analysis of NPs*, September 2012, NanoComposix (Nanocomposix.com).
27. Zhang, T., Lu, G., Shen, H., Shi, K., Jiang, Y., Xu, D. and Gong, Q., 2014. Photoluminescence of a single complex plasmonic nanoparticle. *Scientific Reports*, *4*(1), pp.1–7.
28. Rawat, D., Barman, P.B. and Singh, R.R., 2019. Corroboration and efficacy of magneto-fluorescent (NiZnFe/CdS) nanostructures prepared using differently processed core. *Scientific Reports*, *9*(1), pp.1–12.
29. Bridge, T.P., Fell, A.F. and Wardman, R.H., 1987. Perspectives in derivative spectroscopy Part 1-Theoretical principles. *Journal of the Society of Dyers and Colourists*, *103*(1), pp.17–27.
30. Kus, S., Marczenko, Z. and Obarski, N., 1996. Derivative UV-VIS spectrophotometry in analytical chemistry. *Chem. Anal*, *41*(6), pp.889–927.
31. Dixit, L. and Ram, S., 1985. Quantitative analysis by derivative electronic spectroscopy. *Applied Spectroscopy Reviews*, *21*(4), pp.311–418.
32. Ghosh Chaudhuri, R. and Paria, S., 2012. Core/shell nanoparticles: classes, properties, synthesis mechanisms, characterization, and applications. *Chemical Reviews*, *112*(4), pp.2373–2433.
33. Jangam, K., Patil, K., Balgude, S. Patange, S. and More, P., 2020. Magnetically separable Zn1−xCo0.5xMg0.5xFe2O4 ferrites: stable and efficient sunlight-driven photocatalyst for environmental remediation. *RSC Adv., 10*, pp.42766–42776.
34. Reimer, L., 2013. *Transmission Electron Microscopy: Physics of Image Formation and Microanalysis* (Vol. 36). New York: Springer.
35. Williams, D. B. and Carter, C. B., 1996. *Transmission Electron Microscopy*, New York: Springer.
36. Pallares, R.M., Wang, Y., Lim, S.H., K Thanh, N.N.T. and Su, X., 2016. Growth of anisotropic gold nanoparticles in photoresponsive fluid for UV sensing and erythema prediction. *Nanomedicine*, *11*(21), pp.2845–2860.

37. Dmitrieva, O., Rellinghaus, B., Kästner, J. and Dumpich, G., 2007. Quantitative structure analysis of L10-ordered FePt nanoparticles by HRTEM. *Journal of Crystal Growth*, *303*(2), pp.645–650.
38. Zhang, W.J. and Miser, D.E., 2006. Coalescence of oxide nanoparticles: In situ HRTEM observation. *Journal of Nanoparticle Research*, *8*(6), pp.1027–1032.
39. Wang, Y., Zhang, W., Luo, C., Wu, X., Wang, Q., Chen, W. and Li, J., 2016. Synthesis, characterization and enhanced electromagnetic properties of NiFe2O4@ SiO2-decorated reduced graphene oxide nanosheets. *Ceramics International*, *42*(15), pp.17374–17381.
40. Sangeetha, A., Kumar, K.V. and Kumar, G.N., 2017. Effect of annealing temperature on the structural and magnetic properties of NiFe 2 O 4 nanoferrites. *Advances in Materials Physics and Chemistry*, *7*(02), p.19.
41. Rawat, D., Sethi, J., Sahani, S., Barman, P.B. and Singh, R.R., 2021. Pioneering and proficient magneto fluorescent nanostructures: Hard ferrite based hybrid structures. *Materials Science and Engineering: B*, *265*, p.115017.
42. Kumar, S., Singh, R.R. and Barman, P.B., 2021. Reitveld Refinement and Derivative Spectroscopy of Nanoparticles of Soft Ferrites (MgNiFe). *Journal of Inorganic and Organometallic Polymers and Materials*, *31*(2), pp.528–541.
43. Mahdavi, M., Ahmad, M.B., Haron, M.J., Namvar, F., Nadi, B., Rahman, M.Z.A. and Amin, J., 2013. Synthesis, surface modification and characterization of biocompatible magnetic iron oxide nanoparticles for biomedical applications. *Molecules*, *18*(7), pp.7533–7548.
44. Waldron, R.D., 1955. Infrared spectra of ferrites. *Physical Review*, *99*(6), p.1727.
45. Nejati, K. and Zabihi, R., 2012. Preparation and magnetic properties of nano size nickel ferrite particles using hydrothermal method. *Chemistry Central Journal*, *6*(1), pp.1–6.

5 History and Techniques of Bioimaging

Astha Verma and Sanjay Kumar Gupta
Shri Rawatpura Sarkar Institute of Pharmacy, Kumhari, Durg, Chhattisgarh, India

CONTENTS

5.1 EVOLUTION OF IMAGING: BRIEF INTRODUCTION

Imaging biological phenomena, such as examining cells and microscopic organisms, has advanced from simple microscopy to computer vision technology, enabling artificial vision for any object in place of direct human vision. The discoveries have also sparked the creation of a potent new technique that looks into the intact body using photons and electromagnetic waves to diagnose diseases, with a focus on structural and anatomical imaging at the tissue or organ level that is now being extended to the cellular and molecular biology level. A brief review of microscopy's past can assist highlight specific difficulties and how more recent technology advancements have drastically altered the ways that are currently accessible for imaging procedures [1, 2].

DOI: 10.1201/9781003319870-5

Microscopy has captivated scientists for generations as a way to study cells and other microscopic organisms. Simple optical tools like lenses have been used since the time of the ancient Greeks, possibly much earlier. However, the Dutchman Antonievan Leeuwenhoek (1632–1723) and the Englishman Robert Hooke (1635–1703) are credited with conducting the first significant studies of biological systems using microscopes. Scientist Robert Hooke had a significant impact on a number of scientific fields [3]. One of many instances is Hooke's law of elasticity; Hooke is credited with coining the term "cell" to describe how the structures he observed in the bark were arranged. Antonievan Leeuwenhoek built straightforward yet effective one-lens microscopes and studied biological systems in great detail. He is known as the father of microbiology and is credited with making the first observation of bacteria. His creative design of microscope lenses accounts for a large portion of the impact of his work. Over the years, the light microscope has evolved into a vital instrument for a majority of biologists. Even though many of the basic optical principles have changed throughout time, advances in optics have significantly improved the performance of the microscope [4, 5]. The germ theory of illness, proposed by Robert Koch and Louis Pasteur in the 1870s, came from a new style of research that used the microscope as its primary observational tool. Using common laboratory equipment, German physicist Wilhelm Roentgen discovered a new form of radiation in the fall of 1895: the X-ray. The range of electromagnetic radiation was significantly increased by this finding. Additionally, this discovery sparked the creation of a potent new method that uses X-rays to examine the intact body to diagnose diseases [5,6]. The range of electromagnetic radiation was significantly increased by this finding; additionally, this discovery sparked the creation of a potent new method that uses X-rays to examine the intact body to diagnose diseases [7].

The performance of the microscope has significantly improved due to advancements in optics over time, despite the fact that many fundamental optical principles have essentially stayed constant [8]. However, breakthroughs in other areas of microscopy have radically altered what can be accomplished using the method. The processing of the sample using molecular methods, which was only developed in the 20th century, is a significant illustration of this. In addition to revealing the general shapes of particular subcellular structures, molecular biology techniques can also be used to know the locations of particular molecules within a cell. Another example is the development of microscopy modalities that, for instance, use particular illumination schemes to produce images with significantly higher information richness [9, 10]. This was made possible by the introduction of lasers as a light source. A third example is the advancement of contemporary image detectors, which can quickly and accurately record images for later extensive quantitative analysis.

One of the most fascinating developments in bioengineering is the ability to observe and understand the microscopic universe of cells in order to comprehend the biological mechanisms underlying physiological and pathological phenomena of life [10, 11]. Since researchers are concerned to find quick, accurate, and non-invasive diagnostic tools to determine the state of biological samples,

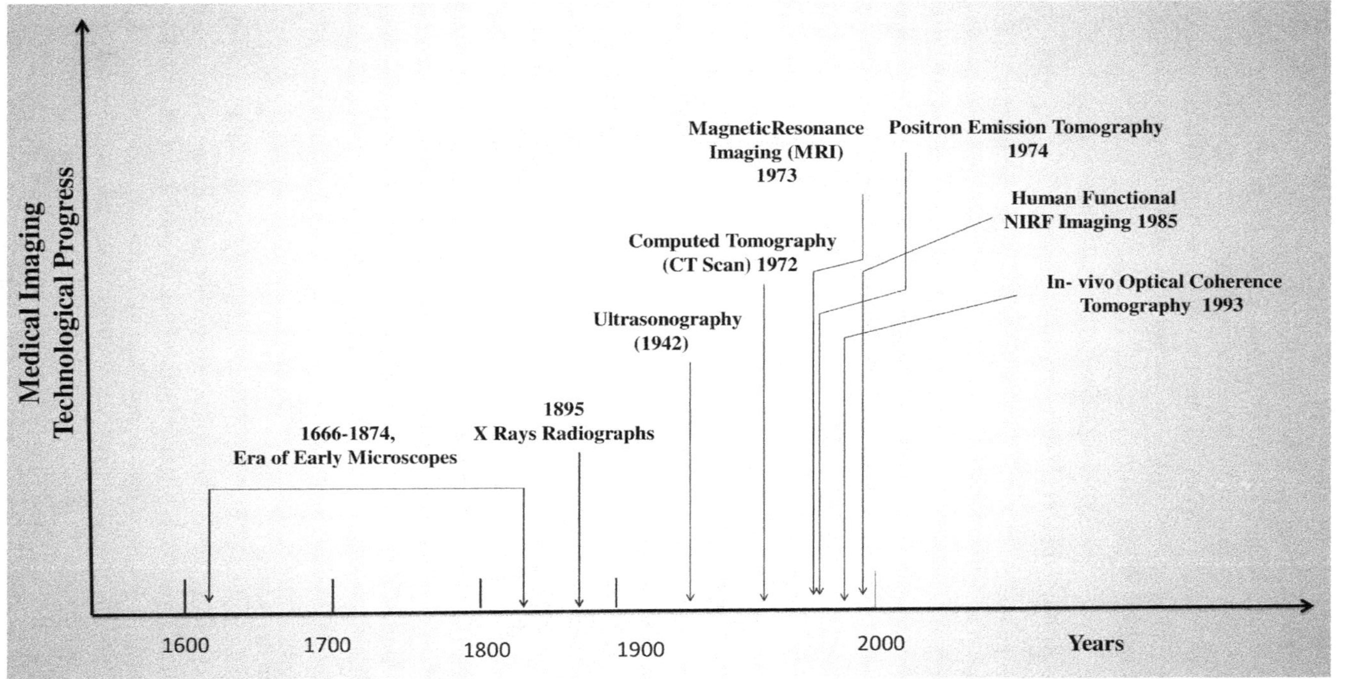

FIGURE 5.1 **Evolution of Imaging Modalities:** Imaging started with the discovery of microscope providing enhanced resolution of specimen to the use of radiography X-ray in 1895 to examine intact body. Further advancement in radiography, X-ray CT based image to development of Ultrasonography, MRI as a diagnostic tool revolutionized imaging technique. Recent uses of modalities such as PET, Optical microscopy have been applied to study diseases at subcellular level.

several micro-scale imaging techniques have recently been developed, with the goal of applying these techniques on human patients. Recent progress in optical microscopy has led to the characterization of observed living specimen across time in a non-intrusive, morphological, and functional manner [12]. The development of cutting-edge imaging technology and the introduction of new investigative techniques have expanded the boundaries of medicine and given medical professionals vital tools for analysing tissues, analysing molecules, and diagnosing diseases [13]. The evolution and progress in bioimaging modalities are summarized in Figure 5.1.

5.2 BIOIMAGING: DEFINITION

Bioimaging, also known as biological imaging, is a precise method for capturing data of biological materials utilizing a range of imaging tools and post-processing techniques. It is typically described as a visualization technique that enables the non-invasive recognition of biological processes or the recording of data from biological specimens [14]. It is frequently used to learn about the specimen's 3D structure from the outside – that is, without physical disturbance or invasion. Bioimaging also allows for the observation of subcellular structures, complete cells, tissues, and even entire multicellular animals. Additionally, bioimaging can be used to assess cellular processes, ion or metabolite levels in cell biology [15].

Three major biomedical domains where imaging is currently advancing:

(i) Single cell imaging [16].
(ii) Imaging molecular biomarkers or aiding in biomarker analysis (micro and nano imaging probes, fluorescent proteins) [17,18]
(iii) Imaging therapies [19]
    - Each domain has a very high potential for fast development.

Bioimaging is a highly interdisciplinary area that draws ideas from biology, engineering, computer science, physics, mathematics, and biomedical engineering. In translational and clinical research, the application of imaging biomarkers or imaging probes and analysis through imaging tools can identify the existence and extent of disease as well as patients' response to treatment which would help in patient selection in clinical studies [20]. Also, functional imaging technologies finds robust application in the designing, development, and testing of novel pharmaceuticals as a means of making rapid economic decisions about a drug's potential to meet therapeutic demands. Imaging helps biomedical researcher in developing a discovery-development-delivery pathway by determining the physiological responses and cellular mechanisms of drug and therapeutics through imaging techniques [22]. The engineers and physicists, designing robotic controls, laser automation, and device miniaturization, can bring technology automation and easier imaging methods and analysis [21]. Thus, the interdisciplinary conduct of biomedical imaging promotes improved imaging methods which advance the diagnosis process, therapy, and treatment through progress in imaging-driven intervention and can be summarized in Figure 5.2.

5.3 BIOIMAGING: INITIAL DEVELOPMENTS

Bioimaging has made it possible to extract biological data from deep within bodies and has completely changed how diseases are understood, identified, and treated from all angles and dimensions. With the rapid development of more precise, less intrusive, and faster equipment during the past ten years, medical imaging has undergone a revolution [23].

The basis of a medical imaging system is a sensor or energy source that can move through the human body. As the energy passes through the body, it is absorbed or reflected at intensified varying rates depending on the density and atomic number of the various tissues, producing signals. Special detectors that are compatible with the energy source pick up these signals, which are then mathematically altered to produce an image [24]. A categorization based on the energy applied to the body is made from the photos generated using the energy from human tissue, as discussed in Figure 5.3.

5.3.1 Bioimaging Modalities: An Overview

5.3.1.1 X-Ray Radiography

When Wilhelm Conrad Röntgen discovered the X-ray in 1895, radiography was born. When Rontgen viewed a photo of his wife's hand on an X-ray-created photographic plate, he realised that it could be used in medicine. A diagnostic method known as radiography uses ionizing magnetic radiation, such as X-rays, to inspect objects. A high energy electromagnetic radiation with a wavelength of between 0.01 and 10 manometers, X-rays can ionize gas and penetrate solid objects [25]. When X-rays are used for medical imaging, they move through the body and, depending on the density and atomic number of the various tissues, are absorbed or attenuated at varying rates, forming an image. These varied absorptions, or X-ray attenuation coefficients, allowed for the differentiation of the densities of bone, muscle, fat, and other soft tissues leads to the distinguished images [26]. This forms the basis for biomedical X-rays imaging technique.

X-Ray Radiography Medical Applications [27-31]

- X-ray radiography is utilized in many different sorts of examinations such as to visualize the structure of internal organs like the stomach, intestine, and colon and in dentistry.
- X-ray radiography is also used to assess the kind and severity of fractures and to envisage pathological changes in lungs.
- Fluoroscopy radiographs, which are typically used to depict the motion of internal organs like the stomach, intestine, and colon in the body.
- Mammography, which is used to examine and diagnose breast tissue.
- Bone densitometer is used to measure bone density and x ray radiography is also used in visualizing inside joint through arthrography.
- X ray radiography is used in examining uterus and fallopian tube through hysterosalpingogram.

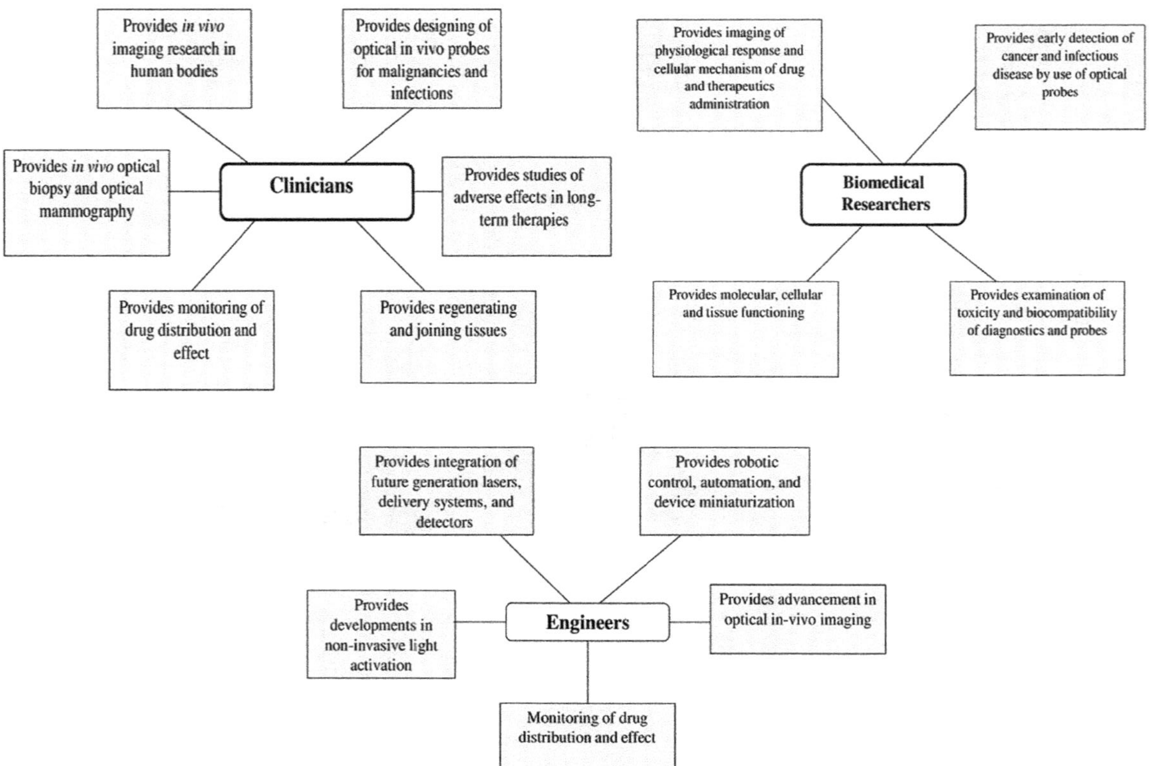

FIGURE 5.2 Biomedical imaging offers multidisciplinary and symbiotic expertise to clinicians, biomedical researchers, engineers, physicists, and computer scientists, depending on multivariate approaches of technologies.

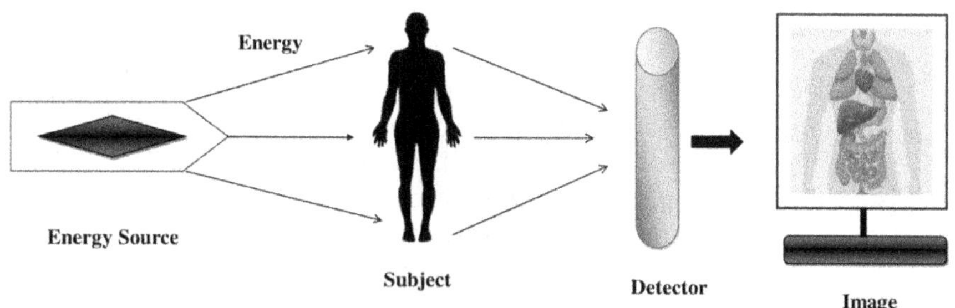

FIGURE 5.3 Principle of Imaging: Energy or signal travels through the human body, organs, tissues which is picked up by the detectors and manipulated to produce an image.

X-Ray Radiography Limitation and Risk [30, 31]

- X-ray imaging is not suitable to examine soft tissue and organ.
- Prolonged and constant radiation exposure raises the risk of developing cancer.
- High levels of radiation exposure may cause skin reddening, hair loss, or cataracts.

5.3.1.2 Computed Tomography

CT is a great advancement in imaging technology. Transverse axial imaging of the inside of the body exposed to external X-ray radiation was proposed by Allan M. Cormack (1924–1998) in the United States [32]. In 1972, Godfrey N. Hounsfield (1919–2004) of the UK developed and commercialized the X-ray computed tomography technology. The development of CT has significantly altered the medical industry due to its innovative non-invasive imaging technique [33]. Computed tomography (CT) is a diagnostic technique that creates images of cross sections of the human body using X-ray equipment, a computer, and a cathode ray tube display. A detector that measures the X-ray profile has been used in place of radiographic film. Each profile is converted by the computer into a 2D representation of the scanned slice. Spiral CT can be used to obtain 3D CT images since it collects a large amount of data while keeping the patient's anatomy in one place [34]. To produce three-dimensional (3D) photographs of intricate structures, this volume data set can then be computer-reconstructed. The generated 3D CT pictures make the tumor masses more easily visible in three dimensions [35]. Four-dimensional (4D) CT has recently been developed to address challenges brought on by respiratory motions [36]. Both geographical and temporal data on organ movements are produced by 4D CT. For improved anatomy delineation while using CT, radiocontrast agents are frequently employed. Continuous CT technological advancements, such as faster scan times and better resolution, have significantly increased the accuracy and usefulness of CT scanning, which has led to an increase in its use for medical diagnosis [37].

CT Medical Applications [36-38]

- In order to diagnose several urgent and emergent illnesses, such as cerebral hemorrhage, pulmonary embolism (clots in the arteries of the lungs), aortic dissection (tearing of the aortic wall), appendicitis, and obstructive kidney stones, CT scanning has emerged as the test of choice.
- It is used in monitoring the effectiveness of drug therapies.

CT Limitation and Risk [40]

Prolonged and constant radiation exposure can damage body tissues and cells, raises the risk of developing cancer.

5.3.1.3 Magnetic Resonance Imaging (MRI)

In the USA in 1973, Paul Christian Lauterbur (1929–2007) proposed the MRI tenet [42]. In contrast to CT and PET, MRI uses only three types of electromagnetic fields: radio frequency (RF) electromagnetic fields that operate at the magnetic resonant frequencies, time-varying pulsed magnetic fields, and static magnetic fields [43]. The positional variation of the proton resonance in the tissues provides us with the spatial information needed to perform the Fourier Transformation [44]. The use of MRI produces high-quality pictures of soft tissue and anatomical structures with outstanding contrast resolution without subjecting the patient or person to ionizing radiation (X-rays) [45]. Due to its wide range of possible uses, MRI has been well developed in a variety of applications, including MR angiography, diffusion tensor imaging (DTI), functional magnetic resonance imaging (fMRI), impedance and current imaging based on MRI, and molecular imaging [46]. The finest soft tissue contrast is provided by MRI scans among all imaging modalities. With improvements to computer 3D algorithms, scanning speed, and spatial resolution, MRI has vastly grown into a tool in neuroradiology and musculoskeletal radiology [47].

MRI Medical Applications [48-52]

- Examining injuries or abnormalities of the joints.
- Checking tumors, cysts, and other abnormalities in various sections of the body.
- Studying and examining abnormalities of the brain and spinal cord.
- Examining the liver and other abdominal organs.
- Imaging of brain, spine, and musculoskeletal system.
- Potentially advanced fields include cardiovascular MRI, functional imaging, and MR image-guided therapy.

MRI Risks and Limitations [53, 54]

- Prolonged post-processing and scan times.
- A large number of probes may be required.
- Fairly poor sensitivity.
- Can make some people feel claustrophobic.

- Unable to identify intraluminal anomalies.
- If a young child has trouble staying still, sedation may be necessary.
- Relatively costly.

5.3.1.4 Ultrasonography

Ultrasonography is a diagnostic technique that creates medical images by using high frequency, wideband sound waves in the megahertz range that reflect differently off tissue. The patient's skin is touched by the ultrasonic transducer close to the area of interest [55]. High frequency sound waves are emitted by the transducer, entering the body and bouncing off the inside organs. The transducer picks up sound waves as they return from the internal organ structures. These sound waves are reflected differently by various tissues, creating a signature that can be quantified and converted into an image [56].

Ultrasonography Medical Applications [57-63]

- Contrary to radiography, CT scans, and nuclear medicine imaging procedures, ultrasound does not employ ionizing radiation; hence it is considered safer.
- Imaging the greater part of the head and neck structures, such as the salivary, lymphatic, and thyroid glands.
- Examining the pancreas, aorta, inferior vena cava, liver, gall bladder, bile ducts, kidneys, and spleen, and other abdominal solid organs.
- Echocardiography is used to diagnose cardiac conditions and to examine the ventricles and valves of the heart.
- Directing the injection of needles when local anesthetic solutions are placed close to nerves.
- For image-guided procedures like biopsies and drainages like thoracentesis, ultrasound is helpful.

Ultrasonography Limitations [59,63]

- Affected by one's level of hydration.
- Blinding techniques are difficult.

5.3.1.5 Positron Emission Tomography

David Kuhl and Roy Edwards established the idea of emission and transmission tomography in the late 1950s. Later additional tomographic machines were designed at the University of Pennsylvania. Gordon Brownell, Charles Burnham, and team's work in 1950, made a key contribution to the advancement of PET technology and featured the first instance of annihilation radiation used for medical imaging. A PET scan is a sort of imaging test that searches for diseases in the body using a radioactive material called a tracer for example, ^{11}C, ^{13}N, ^{15}O or

^{18}F, a positron-emitting radionuclide [64]. The tracer circulates in the blood and accumulates in tissues and organs. When a positron interacts with an ordinary electron to form signals, back-to-back annihilation photons are produced, which are detected by PET detectors. The impulses are converted into 3D images by a computer [65].

Positron Emission Tomography Medical Application [66, 67]

- Non-invasive in-vivo imaging methods that show protein-protein interactions and gene expression.
- Visualization and monitoring of molecular events.
- Cellular targeting.
- *In-vivo* evaluation of the molecular and physiological mechanisms at early stages of disease progression.

Positron Emission Tomography Limitation

- Short imaging time.
- Ionizing radiation side effects.
- The tracer substance may cause allergic reactions in certain people.

5.3.1.6 Single-Photon Emission Computerized Tomography

A SPECT scan is a form of nuclear imaging test, which produces 3D images using a radioactive material and a specialized camera. Different organs and/or tissue types can absorb radiopharmaceuticals, depending on their biodistribution characteristics. The majority of radiopharmaceuticals used in SPECT and nuclear medicine are marked with radionuclides that release γ-ray photons. Many different isotopes are utilized for SPECT imaging (e.g., ^{99m}Tc, ^{13}1I, ^{123}I and ^{111}In) [68]. A CT scan and a radioactive substance are combined in a SPECT scan to study the human body. The scanner can pick up on the gamma rays that the tracer releases. Gamma ray information is gathered by the computer and converted into two-dimensional cross-sections. These cross-sections are modulated to form 3D images. The major advantage of SPECT is that it may be used to monitor physiological and metabolic processes, as well as the dimensions and volume of in the body [69].

SPECT Medical Applications [70, 71]

- Diagnosis of brain disorders: Determining brain blood artery blockages that could cause stroke and transient ischemic attacks.
- Diagnosis of heart problems: Blockage in the coronary arteries: This condition causes impaired blood flow and may block portions of the heart muscle that the coronary artery supplies.
- Diagnosis and tracking bone disorders: Complicated bone fractures and bone disorders.

SPECT Limitations

- Ionizing radiation has side effects like skin rashes, redness, burns, hair loss.

5.4 ADVANCEMENTS IN BIOIMAGING: MOLECULAR AND CELLULAR IMAGING

Medical imaging has developed from plain radiography (radioisotope imaging) through X-ray imaging, computer-assisted tomography (CAT scans), ultrasound imaging, and magnetic resonance imaging (MRI), which has resulted in dramatic changes in the quality of healthcare that is currently offered. However, the majority of these methods are centered on tissue- or organ-level structural and anatomical imaging [72]. There is an obvious need to extend imaging to the cellular and molecular biology levels in order to develop novel imaging approaches for the early detection, screening, diagnosis, and image-guided therapy of life-threatening disorders like cancer. Only knowledge at the molecular and cellular levels can enable the early detection of molecular changes during intervention or therapy, as well as the onset of disease or cancer [73].

There are a variety of limitations to the medical imaging modalities currently in use, including X-rays, radiography, CAT scans, ultrasound imaging, and MRI as discussed can be summarized as:

- Ionizing radiation side effects from CAT scan and X-ray imaging.
- The incapacity of X-ray imaging to differentiate between benign and malignant tumors.
- Harmful radioactivity in radioisotope imaging
- MRI cannot deliver precise chemical information or any dynamic information.
- Inability of ultrasonography to differentiate between a benign and malignant tumor as well as to offer resolution lower than millimeters.

A number of these shortcomings are addressed by optical imaging. It deals with how light interacts with objects ranging in size from viruses at about 100 nm to macro-objects (live organisms). Whether it is a cell, a tissue, an organ, or a complete live thing, optical imaging makes use of the spatial variation in the optical properties of a biospecies [74]. Molecular imaging shows how cells or molecular processes operate inside the body. The noninvasive viewing, characterization, and frequent quantification of structures and biologic processes at the cellular and molecular levels are made possible by optical methods for molecular imaging [75]. The ability to distinguish between anatomical and molecular structures is particularly important for both early diagnosis and therapeutic intervention, and one of the goals of molecular imaging is to apply laboratory tools and methods to an in-vivo setting. Researchers and clinicians can study intact organisms over time using nontoxic and noninvasive imaging techniques without having to remove or otherwise change the tissues being studied. This allows them to study cells and tissue in their natural environments without affecting how they interact with the microenvironment [76]. This makes optical imaging a highly effective method for both basic research and clinical diagnostics. Additionally, with real-time imaging, dynamic studies of the target biology can impart an informative picture. In order to do optical imaging, a probe with the ability to emit a detectable and focused signal must typically be injected. A successful imaging probe needs to have a good pharmacokinetic profile, be able to cross biological membranes and barriers separating the target from the bloodstream, and approach a target molecule with high affinity [74]. The probe also needs to be promptly, accurately, and with high sensitivity detected. For clinical translation, molecular imaging technologies may also need to fit into current environments and workflows (e.g., be the right size, weight, and comfort) depending on the applications [77]. Optical imaging technique is the oldest bioimaging tool, which is well recognised for using the visible light spectrum to take images of objects. Innovations in optical imaging techniques for molecular level have geared novel biomedical applications for functional imaging and has targeted towards research and clinical use [78]. The various techniques involving optical imaging summarized in Figure 5.4 are bioluminescence imaging (BLI), fluorescence microscopy, photoacoustic microscopy/tomography, total internal reflection fluorescence microscopy (TIRF), differential interference contrast microscopy (DIC), and fluorescence reflectance imaging (FRI) [79].

Currently, a variety of diagnostic imaging methods are employed for clinical imaging and illness diagnosis. Based on the type of energy used to generate the visual information (such as X-rays, sound waves, photons or positrons), the level of spatial resolution (mesoscopic or microscopic), and the amount of information that can be gathered, these techniques can be compared at physiological, anatomical or molecular

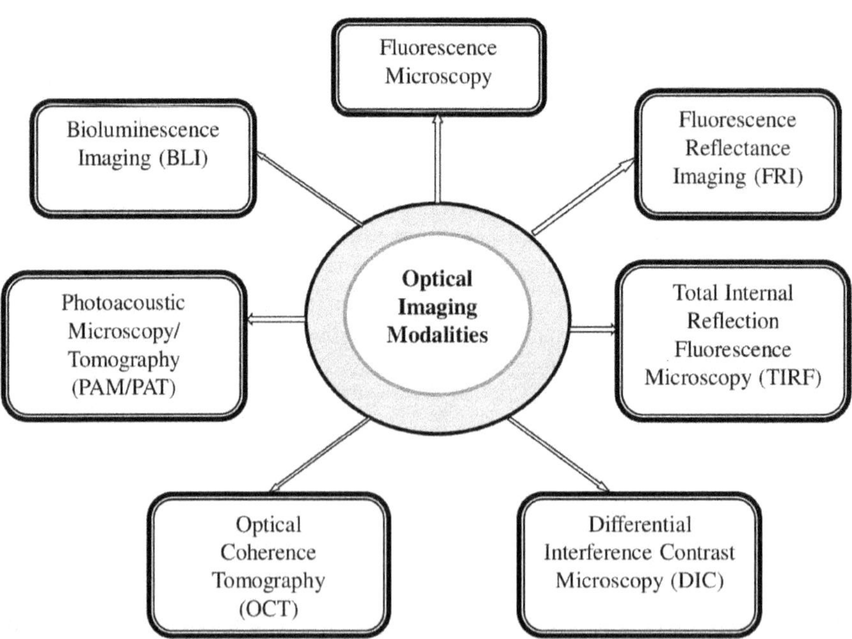

FIGURE 5.4 Optical imaging modalities.

TABLE 5.1
Comparison between Bioimaging Modalities

SN	Imaging Modalities	Spectrum	Type	Penetration Depth	Resolution	Penetration Level
1	CT Scan	X-ray	Ionizing	No limit	50μm	Organ -tissue
2	MRI	Magnetic field variation	Non-Ionizing	No limit	25-100 μm	Organ-tissue
3	Ultrasound imaging	Sound Waves	Non-Ionizing	mm to cm	50μm	Organ-tissue
4	PET	γ-rays	Non-Ionizing	1-2mm	1-2 mm	Tissue-cellular-molecular
5	SPECT	γ-rays	Non-Ionizing	1-2mm	1-2 mm	Tissue-cellular
6	Fluorescence Microscopy	Light	Non-Ionizing	<1 cm	<1 cm	Cellular-molecular

level. Additionally, combining two or more imaging modalities into a single imaging unit has received a lot of study interest [80]. Nonionizing imaging techniques use safe low-energy input radiations that are safer to photograph the targeted subjects than ionizing medical imaging modalities. For image clarification and verification, they also support repeated imaging sessions with higher energy dosage levels. MRI, ultrasound, and optical imaging techniques are the most often utilised non-ionizing imaging modalities [80]. Table 5.1 discusses the features of imaging modalities with comparison to other existing bioimaging techniques.

5.5 EVOLUTION OF MOLECULAR PROBES AND QUANTUM DOTS IN BIOIMAGING

Bioimaging is a visual representation of biological structures and processes, which uses several methods, techniques and imaging modalities, each with their own advantages been developed to meet clinical and preclinical examinations. The imaging can be broadly classified according to examining process involved and can be on the basis of alteration in anatomical changes of diseases such as X ray, CT Scan, ultrasound and second category in which measures biological processes such as single photon emission computed tomography (SPECT), optical, molecular magnetic resonance imaging (mMRI) and magnetic resonance spectroscopy (MRS) [81]. When using imaging techniques like PET, SPECT, and optical imaging, molecular probes are required to inject into the test subject in order to capture the imaging signal, while others, like optical imaging and mMRI, can track the disease using either molecular exogenous probe or endogenous molecules. To increase visibility and make it possible to collect more precise structural and functional data, imaging probes that can mark target molecules or organs are used [82]. As a result, imaging probes are increasingly necessary for biological study and disease detection. Advancement in the imaging probe has led imaging to subcellular or molecular level. Imaging probes also called as contrast agent compromises of inorganic molecules or metal–organic compounds, protein molecules, polymers and now recent development includes quantum dots and nanoparticles providing advantage of increasing binding specificity and ability [83]. The present developments involve the evolution of magnetic, optical, and electrical functional nanoparticles attached to biological molecules like peptides, proteins, and nucleic acids. Additionally, a number of techniques for nanoparticle surface modification have been created to offer improved biocompatibility and functions such as treatment, targeted imaging, and stimuli-responsiveness [84].

5.5.1 Magnetic Resonance Imaging Contrast Agents

MRI, a noninvasive technique of bioimaging modalities which is based on Nuclear Magnetic Resonance principle (NMR) provides sufficient penetration depth and strong contrast of soft tissues to visualize the image of human body. Though MRI machines provide anatomical, physiological and molecular imaging but with low resolution, use of exogenous contrast agents improves the resolution. Imaging probes used are iron oxide [IO], gadolinium [Gd]) which increases MRI signal intensities [85,86].

5.5.2 CT Contrast Agents

Computed Tomography (CT) relies on how an X-ray interacts with a body. Because of its great spatial resolution and speedy image capture, CT is one of the most frequently utilized whole body imaging procedures. As a result, it is widely used to depict a variety of anatomical structures, including disorders of the brain, lungs, circulatory system, and abdomen. For the majority of applications, CT's inherent sensitivity is insufficient; hence contrast agents are frequently needed to pick up small changes in soft tissues [87]. Clinical applications of contrast agents based on barium and iodine have been used as a contrast agent. Also, gold nanoparticles, meanwhile, have attracted interest as a CT contrast agent [88,89].

5.5.3 Positron Emission Tomography and Single-Photon Emitted Tomography

A nuclear medicine imaging technique called PET creates 3-D functional pictures of the body. Without external excitation, the positron signals (γ-rays) produced by radioactive isotopes or tracers are recognized. With the help of this great sensitivity, PET can be used for ongoing disease monitoring and early detection [90]. The drawbacks of PET are radiation to the subject, a very low spatial resolution, and a quick acquisition time caused by radioisotopes with short half-lives. PET captures γ -rays coming from the body. In order to achieve that, a radioisotope with a brief half-life must be administered into the body before the scan. Various positron-emitting isotopes have been used, including ^{18}F, ^{15}O, ^{13}N, ^{11}C, ^{14}O, ^{64}Gu, ^{62}Gu, ^{12}4I, ^{76}Br, ^{82}Rb and ^{68}Ga. One such similar method is single-photon emission computed tomography (SPECT). Many different isotopes are utilised for SPECT imaging (e.g., ^{99m}Tc, ^{13}1I, ^{123}I and ^{111}In) [91]. Although SPECT has a lower sensitivity than PET, it can simultaneously image multiple radioisotopes because each radioisotope emits a different set of γ -rays with a different energy [82].

5.5.4 Optical Imaging: Fluorescence Microscopy and Quantum Dots

One of the most in-depth and simple methods in fundamental research and medical diagnostic imaging has long been the visualization of cells and tissues using light. Bright field, dark field, and differential interference contrast are all effective optical microscopy techniques that offer a considerable depth of resolution and three-dimensional optical effects The development of fluorescence microscopy has transformed the study of cellular biology over the past 20 years [92]. In-vivo non-invasive fluorescence imaging of preclinical animal models is a subject that is continuously evolving and has recently seen the development of novel methods [93]. Fluorescence imaging in the NIR-I (700-900 nm) window has been thoroughly studied by scientists over the past few decades, ranging from basic research to preclinical and clinical applications [94]. Tissue autofluorescence, fluorescence quenching, photobleaching of the fluorescence photons, and inadequate tissue penetration are the key challenges in fluorescence imaging. However, Fluorescence imaging in the NIR-II (1000-1400nm) window has developed recently as a novel platform that offers greater tissue penetration [95]. Though few numbers of fluorophores, are suitable for NIR-II fluorescence imaging, given these characteristics, quantum dots (QDs), a subgroup of nanoparticles, can be utilised instead of organic dyes, fluorescent proteins (e.g., GFP), as fluorescent NIR probes [96, 97].

Quantum Dots are a special type of fluorescent probe inorganic nanoparticles semiconductor crystals with a range of sizes (1-100 nm) and a small number of atoms (a few hundred to a few thousand). Atoms from groups II and VI, III and V, or IV and VI of the periodic table are frequently used to make the highest-quality QDs [99]. Infrared QDs are generally III–V (InAs, InSb), II–VI (HgCdTe, HgSe, HgTe), I–VI (Ag_2S, Ag_2Se), and ternary I–III–VI ($CuInS_2$, $CuInSe_2$, $AgBiS_2$, $AgInSe_2$) semiconductors, their alloys, core/shell. QDs used for biological applications have a semiconductor core (CdSe, CdS, and CdTe) that is covered in a semiconductor shell with a wider bandgap than the core material (ZnS, CdS), in order to significantly increase the quantum efficiency [100, 101]. QDs are a novel biological labelling material that, when compared to organic fluorophores, offer a number of significant advantages. Conventional organic dyes have a number of chemical and photophysical limitations, including pH dependence, photobleaching susceptibility, narrow wavelength absorption windows, and asymmetric emission spectra [102]. QDs have unique electrical and optical features, such as high quantum yield and wide absorption, in contrast to organic dyes and fluorescent proteins, excellent resistance to photo-bleaching, as well as to chemical and photo-degradation [103]. Due to their electrical and optical properties, which can be changed by altering the dimension, shape, and composition of NPs, QDs represent the best contenders for optical bioanalysis. Size regulated emission and absorbance permits the manufacture of multiplexed probes with various optical characteristics from the same material [104]. QDs are successfully employed for multicolor imaging because of their distinctive spectral characteristics (wide excitation spectra and strong Stockes shifts), necessitating only a single light source for the excitation of QDs of various colors [105]. A multimodal probe QDs with a high surface area-to-volume ratio enables imaging of disease conditions at various scale lengths and tissue depths providing dynamic imaging [106]. Also, longer excited state may facilitate the creation of probes for time-delayed imaging to prevent the autofluorescence of cells. High brightness due to large excitation cross section and high photostability results in high resolution images for extended time as required in image-based surgery [106]. These exclusive characteristics of QD probes discussed above can be studied through Figure 5.5.

QDs distinctive optical properties make them useful in vivo and in vitro for a variety of biological and therapeutic applications, including cell labelling, cell tracking, in-vivo imaging, and DNA detection [107, 108]. QDs offer a lot of potential for use in fields including drug delivery, sensors, and bioimaging. Theranostic platforms are continuously being created by mixing QDs with different kinds of nanoparticles and/or biologically active compound [109, 110].

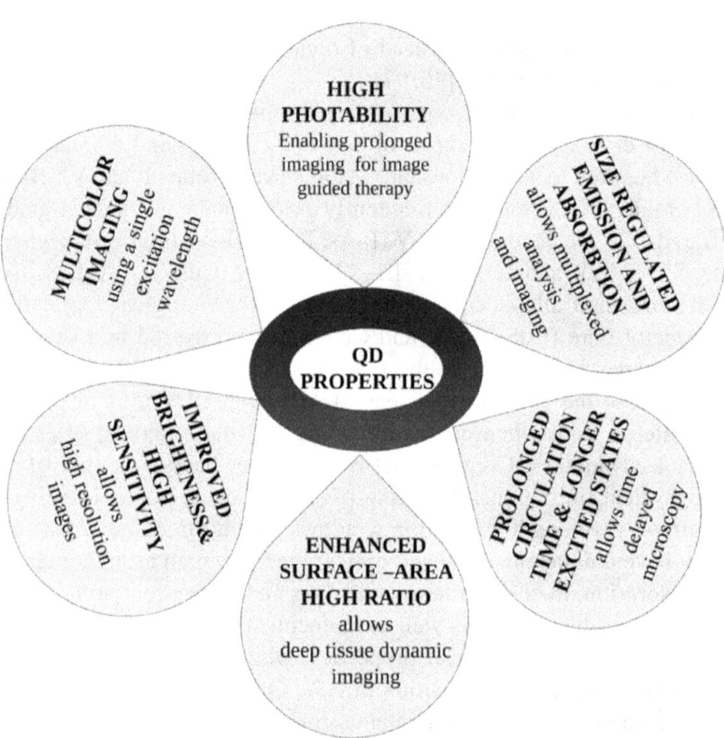

FIGURE 5.5 Properties of Quantum Dots: Unique properties of QD probe making them widely applicable and favorable in biological imaging.

5.6 CONCLUSION

Bioimaging is evolved from visualization and magnification of cell through microscopy to revolution brought by discovery of early X-ray radiography used for the diagnosis of fracture and numerous advancement in imaging techniques emerged are digital radiography, computed tomography (CT), ultrasound sonography, magnetic resonance imaging (MRI), positron emission tomography (PET), optical microscopy for examining the internal organs, disorders or diseases and obtaining biological and physiological processes noninvasively.

The application of imaging probes or contrast agents and multiple or combination of imaging modalities can extend imaging to penetration depth of millimeters to micron or even subcellular level for clinical diagnostic and therapeutic paradigm and increase the sensitivity and specificity *in-vivo* 3D quantitative imaging. As the understanding of molecular basis of disease grows by the image driven modalities will provide customized therapies at each clinical research.

REFERENCES

1. Vadivambal R, Jayas DS (2015). Bio-Imaging: Principles, Techniques, and Applications (1st ed.). Boca Raton: CRC Press.
2. Vo-Dinh T (Ed.) (2003). Biomedical Photonics Handbook (1st ed.). Boca Raton: CRC Press.
3. Masters, Barry R (2008). History of the Optical Microscope in Cell Biology and Medicine. In: Encyclopedia of Life Sciences (ELS). Chichester: John Wiley & Sons, Ltd.
4. Ford BJ (1985). Single Lens, the Story of the Simple Microscope. New York: Harper & Row, Publishers.
5. Ober RJ, Sally Ward E, Chao J (2021). Quantitative Bioimaging: An Introduction to Biology, Instrumentation, Experiments, and Data Analysis for Scientists and Engineers. First Edition. Boca Raton: CRC Press.
6. Evens RG (1995). Rontgen retrospective: one hundred years of a revolutionary technology. J Am Med Assoc; 274(11):912–917.
7. Bradley WG (2008). History of medical imaging, Proceedings of the American Philosophical Society. 152(3):349–361.
8. Macovski A (1983). Medical Imaging Systems. New Jersey: Prentice-Hall.
9. Weissleder R, Mahmood U (2001). Molecular imaging. Radiology. 219(2):316–333.
10. Shung KK, Smith MB, Tsui B (1992). Principles of Medical Imaging. New York: Academic Press.
11. Cho ZH, Jones JP, Singh M (1993). Foundations of Medical Imaging. New York: Wiley.
12. Arridge SR (1999). Optical tomography in medical imaging. Inverse Problems. 15, R41–R93.
13. Webb AG (2003). Introduction to Biomedical Imaging. New York: IEEE.
14. Fessler (2009). Introduction to Medical Imaging Systems. Available from: http://web.eecs.umich.edu/~fessler/course/516/l/c1-intro.pdf
15. Robb, RA (2006). Biomedical imaging: past, present and predictions. Med. Imaging Technology. 24(1): 25–37.
16. Dormann D, Weijer CJ (2006). Imaging of cell migration. EMBO J. 25(15):3480–3493.
17. Nichols L, Pike VW, Cai L, Innis RB (2006). Imaging and *in vivo* quantitation of beta amyloid: an exemplary biomarker for Alzheimer's disease? Biol Psychiatry. 59(10):940–947.

18. Sorensen AG (2006). Magnetic resonance as a cancer imaging biomarker. J Clin Oncol. 24(20):3274–3281.
19. Workman P (2003). The opportunities and challenges of personalized genome-based molecular therapies for cancer: targets, technologies, and molecular chaperones. Cancer Chemother Pharmacol. 52(1):S45–S56.
20. Prasad PN (2003). Introduction to Biophotonics. Hoboken, New Jersey: John Wiley & Sons, 4–6
21. Barentsz J, Takahashi S, Oyen W, et al. (2006). Commonly used imaging techniques for diagnosis and staging. J Clin Oncol. 24(20):3234–3244.
22. Hendee WR, Gazelle GS. (2006). Biomedical imaging research opportunities workshop III: a white paper. Ann Biomed Eng. 34: 188–198.
23. Angenent S, Pichon E, Tannenbaum A (2006). Mathematical methods in medical image processing. Bulletin of the American Mathematical Society. 43:365–396.
24. Kasban H, El-Bendary MAM, Salama DH (2015). A comparative study of medical imaging techniques. International Journal of Information Science and Intelligent System. 4(2):37–58
25. Seibert JA, Boone JM (2005). X-ray imaging physics for nuclear medicine technologists. Part 2: X-ray interactions and image formation. J Nucl Med Technol. 33(1):3–18.
26. Ritman EL (2006). Medical x-ray imaging, current status and some future challenges. JCPDS. International Centre for Diffraction Data.
27. Dobbins JT, Godfrey DJ (2003). Digital x-ray tomosynthesis: current state of the art and clinical potential. Phys Med Biol. 48(19):R65–R106.
28. Noel A, Thibault F (2004). Digital detectors for mammography: the technical challenges. Eur Radiol. 14:1990–1998.
29. Xu T, Ducote JL, Wong JT, Molloi S (2006). Feasibility of real time dual-energy imaging based on a flat panel detector for coronary artery calcium quantification. Med Phys. 33(6):1612–1622.
30. Yuasa T, Sugiyama H, Zhong Z, et al. (2005). Highpass-filtered diffraction microtomography by coherent hard X rays for cell imaging: theoretical and numerical studies of the imaging and reconstruction principles. J Opt Soc Am A Opt Image Sci Vis. 22(12):2622–2634.
31. Snigireva I, Snigirev A (2006). A. X-Ray microanalytical techniques based on synchrotron radiation. J Environ Monit. 8(1):33–42.
32. Hiriyannaiah HP, Cupertino CA (1997). X-ray computed tomography for medical imaging. IEEE Signal Processing Magazine. 14(2):42–59.
33. Xu J, Tsui BMW (2014). Quantifying the importance of the statistical assumption in statistical x-ray CT image reconstruction. IEEE Transactions on Medical Imaging. 33(1):61–73.
34. Rubin GD (2003). 3-D imaging with MDCT. Eur J Radiol. 45(1):S37–41.
35. Wolbarst AB, Hendee WR (2006). Evolving and experimental technologies in medical imaging. Radiology. 238:16–39.
36. Shikhaliev PM, Xu T, Molloi S (2005). Photon counting computed tomography: concept and initial results. Med Phys. 32:427–436.
37. Siewerdsen JH, Moseley DJ, Bakhtiar B, Richard S, Jaffray DA (2004). The influence of antiscatter grids on soft-tissue detectability in cone-beam computed tomography with flat-panel detectors. Med Phys. 31:3506–3520.
38. Rubin GD (2003). MDCT imaging of the aorta and peripheral vessels. Eur J Radiol. 45(1):S42–S49.

39. Rietzel E, Pan T, Chen GT. (2005). Four-dimensional computed tomography: image formation and clinical protocol. Med Phys. 32:874–889.
40. Kalra MK, Maher MM, Toth TL, et al. (2004). Strategies for CT radiation dose optimization. Radiology. 230(3):619–628.
41. Prince MR, Meaney JF (2006). Expanding role of MR angiography in clinical practice. Eur Radiol. 16(2): B3–B8.
42. Damadian RV (1971). Tumor detection by nuclear magnetic resonance. Science. 171(3976):1151–1153.
43. Lauterbur PC (1973). Image formation by induced local interactions: examples of employing nuclear magnetic resonance. Nature. 242(5394):190–191.
44. Mansfield P, Maudsley AA (1997). Medical imaging by NMR. Br J Radiol. 50(591):188–194.
45. Hennig J, Nauerth A, Friedburg H (1986). RARE imaging: A fast imaging method for clinical MR. Magn Reson Med. 3(6):823–833.
46. Carr JC, Simonetti OP, Bundy J, Li D, Pereles S, Finn JP (2001). Cine MR angiography of the heart with segmented true fast imaging with steady-state precession. Radiology. 219 (3):828–834.
47. Seshamani S, Cheng X, Fogtmann M, Thomason ME, Studholme C (2014). A method for handling intensity inhomogeneities in fMRI sequences of moving anatomy of the early developing brain. Medical Image Analysis. 18(2):285–300.
48. Kwong KK, Belliveau JW, Chesler DA, et al. (1992) Dynamic magnetic resonance imaging of human brain activity during primary sensory stimulation. Proc Natl Acad Sci U S A. 89(12):5675–5679.
49. Wehrli FW, Leonard MB, Saha PK, Gromberg BR (2004). Quantitative high-resolution magnetic resonance imaging reveals structural implications of renal osteodystrophy on trabecular and cortical boneeasonagn Reson Imaging. 20:83–89.
50. Munley MT, Marks LB, Hardenbergh PH, Bentel GC (2001). Functional imaging of normal tissues with nuclear medicine: applications in radiotherapy. Semin Radiat Oncol. 11, 28–6.
51. Jolesz FA, McDannold N (2008). Current status and future potential of MRI-guided focused ultrasound surgeryeasonagn Reson Imaging. 27:391–399.
52. Rouvière O, Yin M, Dresner MA, et al. (2006) MR elastography of the liver: preliminary results. Radiology. 240(2):440–448.
53. Frates MC, Kumar AJ, Benson CB, Ward VL, Tempany CM (2004). Fetal anomalies: comparison of MR imaging and US for diagnosis. Radiology; 232(2):398–404.
54. Binks DA, Hodgson RJ, Ries ME, et al. (2013) Quantitative parametric MRI of articular cartilage: A review of progress and open challenges. Br J Radiol. 86(1023):20120163.
55. Deserno T, Burtseva L, Secrieru I, Popcova O (2009). CASAD – Computer aided sonography of abdominal diseases – the concept of joint technique impact. Computer Science Journal of Moldova. 17, 3, 278–297.
56. Szabo TL (2004). Diagnostic Ultrasound Imaging: Inside Out. Elsevier Academic Press.
57. Ovland R (2012). Coherent Plane-Wave Compounding in Medical Ultrasound Imaging. Master thesis. Norwegian University of Science and Technology.
58. Sahuquillo P, Tembl JI, Parkhutik V, Vázquez JF, Sastre I, Lago A (2013). The study of deep brain structures by transcranial duplex sonography and imaging resonance correlation. Ultrasound in Medicine and Biology. 39(2):226–232.
59. Wells PN (2006). Ultrasound imaging. Phys Med Biol. 51(13):R83–R98.
60. Murtagh J, Foerster V (2006). Transient elastography (FibroScan) for non-invasive assessment of liver fibrosis. Issues Emerge Health Technol. 90:1–4.

61. Klibanov AL (2006). Microbubble contrast agents: targeted ultrasound imaging and ultrasound-assisted drug-delivery applications. Invest Radiol. 41(3):354–362.
62. Dwivedi G, Ahmed-Hayat S, Janardhanan R, Senior R (2006). Myocardial contrast echocardiography: role in clinical cardiology. Curr Vasc Pharmacol. 4:229–237.
63. National Research Council (US) and Institute of Medicine (US) Committee on the Mathematics and Physics of Emerging Dynamic Biomedical Imaging (1996). Mathematics and Physics of Emerging Biomedical Imaging. Washington (DC): National Academies Press (US). Chapter 6, Positron Emission Tomography.
64. Bazañez-Borgert M (2006). Basics of SPECT, PET and PET/CT Imaging. JASS.
65. Gropler RJ, Soto P (2004). Recent advances in cardiac positron emission tomography in the clinical management of the cardiac patient. Curr Cardiol Rep. 6(1):20–26.
66. Ding YS, Gatley SJ (2006). Positron radiopharmaceuticals and their chemistry. In: Henkel RE, Bova D, Dillehay GL, Karesh SM, Halama JR, Wagner RH, editors. Nuclear medicine. second ed. Philadelphia, PA: Elsevier; 2006. pp. 439–56.
67. RE Bova D, Dillehay G, et al, eds (2006). Nuclear medicine. Volume 2. 2nd ed. Philadelphia, PA: Mosby Elsevier.
68. National Research Council (US) and Institute of Medicine (US) Committee on the Mathematics and Physics of Emerging Dynamic Biomedical Imaging (1996). Mathematics and Physics of Emerging Biomedical Imaging. Washington (DC): National Academies Press (US). Chapter 5, Single Photon Emission Computed Tomography.
69. McVeigh ER (2006). Emerging imaging techniques. Circ Res. 98, 879–886.
70. Fletcher JW (2006). Positron emission tomography: applications in oncology. In: Henkin RE, Bova D, Dillehay G, et al, eds. Nuclear Medicine. 2nd ed. Philadelphia, PA: Mosby Elsevier.
71. Park MA, Moore SC, Kijewski MF (2005). Brain SPECT with short focal-length cone-beam collimation. Med Phys. 32:2236–2244.
72. Kherlopian AR, Song T, Duan Q, Neimark MA, Po MJ, Gohagan JK, Laine AF (2008). A review of imaging techniques for systems biology. BMC Syst. Biol. 2, 74.
73. Garini Y, Vermolen BJ, Young IT (2005). From micro to nano: recent advances in high resolution microscopy. Curr Opin Biotechnol. 16(1):3–12.
74. Wilkinson JM, Kuok MH, Adamson G. (2004). Biomedical applications of optical imaging. Med Device Technol. 15(10):22–24.
75. Wagner HN Jr. (2006). From molecular imaging to molecular medicine. J Nucl Med. 47(8):13N–39N.
76. Weissleder R (1999). Molecular imaging: exploring the next frontier. Radiology. 212:609–661.
77. Massoud TF, Gambhir SS. (2003). Molecular imaging in living subjects: seeing fundamental biological processes in a new light. Genes Dev. 17(5):545–580.
78. Margolis DJ, Hoffman JM, Herfkens RJ, Jeffrey RB, Quon A, Gambhir SS (2007). Molecular imaging techniques in body imaging. Radiology. 245:333–356.
79. Xu M, Wang LV (2006). Photoacoustic imaging in biomedicine (Review Article). Review of Scientific Instruments. 77:041101.
80. Dhawan AP, D'Alessandro B, Fu X (2010). Optical imaging modalities for biomedical applications. IEEE Reviews in Biomedical Engineering. 3.
81. Ntziachristos V, Leroy-Willig A, Tavitian B (eds) (2007). Text book of *in-vivo* Imaging in Vertebrates. West Sussex: John Wiley & Sons.
82. Massoud TF, Gambhir SS (2003). Molecular imaging in living subjects: seeing fundamental biological processes in a new light. Genes Dev. 17(5):545–580.
83. Mu Q, Jiang G, Chen L et al. (2014). Chemical basis of interactions between engineered nanoparticles and biological systems. Chem Rev: 114:7740–7781.

84. Kunjachan S, Ehling J, Storm G, Kiessling F, Lammers T (2015). Noninvasive imaging of nanomedicines and nanotheranostics: Principles, progress, and prospects. Chem Rev. 115:10907–10937.
85. Lee N, Kim H, Choi SH, et al. (2011). Magnetosome-like ferrimagnetic iron oxide nanocubes for highly sensitive MRI of single cells and transplanted pancreatic islets. Proc Natl Acad Sci USA. 108(7):2662–2667.
86. Adam G, Neuerburg J, Spuntrup E, Mühler A, Scherer K, Günther RW (1994). Gd-DTPA-cascade-polymer: potential blood pool contrast agent for MR imagingeasonagn Reson Imaging. 4(3):462–466.
87. Pietsch H. CT contrast agents. In: Kiessling F, Pichler BJ, eds. (2011). Small animal imaging—basics and practical guide. Heidelberg: Springer; 141–149.
88. Aviv H, Bartling S, Kieslling F, Margel S (2009). Radiopaque iodinated copolymeric nanoparticles for X-ray imaging applications. Biomaterials. 30(29):5610–5616.
89. Rabin O, Manuel Perez J, Grimm J, Wojtkiewicz G, Weissleder R (2006). An X-ray computed tomography imaging agent based on long-circulating bismuth sulphide nanoparticles. Nat Mater. 5(2):118–122.
90. Rosenthal MS, Cullom J, Hawkins W, Moore SC, Tsui BMW, Yester M (1995). Quantitative SPECT imaging: a review and recommendations by the Focus Committee of the Society of Nuclear Medicine Computer and Instrumentation Council. J Nucl Med. 36(8):1489–1513
91. Hahn M, Singh A, Sharma P, Brown S, Moudgil B (2011). Nanoparticles as contrast agents for *in vivo* bioimaging: current status and future perspectives. Anal Bioanal Chem. 399(1):3–27.
92. Lichtman JW, Conchello J (2005) Fluorescence microscopy. Nat Methods. 2:910–919.
93. Salzer R (2012). Biomedical Imaging – Principles and Applications. Hoboken, NJ: John Wiley & Sons.
94. Reineck P, Gibson BC (2017). Near-infrared fluorescent nanomaterials for bioimaging and sensing. Adv. Optical Mater. 5:1600446.
95. Guo Z, Park S, Yoon J, Shin I (2016). Recent progress in the development of near-infrared fluorescent probes for bioimaging applications. Chem Soc Rev. 43, 16–29.
96. Smith AM, Mancini MC, Nie S (2009). Second window for *in vivo* imaging. Nat. Nanotechnol. 4, 710–711.
97. Sevick-Muraca EM, Houston JP, Gurfinkel M (2002). Fluorescence-enhanced, near infrared diagnostic imaging with contrast agents. Curr Opin Chem Biol. 6:642–650.
98. Lim YT, Kim S, Nakayama A, Stott NE, Bawendi MG, Frangioni JV (2003). Selection of quantum dot wavelengths for biomedical assays and imaging. Mol Imaging. 2:50–64.
99. Wang Y, Tang Z, Kotov NA (2005). Bioapplication of nanosemiconductors. Materials Today. 8(5):20–31.
100. Ji X, Pen F, Zhong Y, Su Y, He Y (2014). Fluorescent quantum dots: synthesis, biomedical optical imaging, and biosafety assessment. Colloids Surf B Biointerfaces. 124:132–139.
101. Reiss P, Protiere M, Li L (2009). Core/shell semiconductor nanocrystals. Small. 5:154–68.
102. Marukhyan SS, Gasparyan VK. (2017). Fluorometric immunoassay for human serum albumin based on its inhibitory effect on the immunoaggregation of quantum dots with silver nanoparticles. Spectrochim Acta A Mol Biomol Spectrosc. 173:34–38.

103. Dabbousi BO et al. (1997). (CdSe)ZnS core-shell qds: synthesis and characterization of a size series of highly luminescent nanocrystallites. J Phys Chem. 101, 9463–9475.
104. Zahavy E, Freeman E, Lustig S, Keysary A, Yitzhaki S (2005). Double labeling and simultaneous detection of B- and T cells using fluorescent nano-crystal (q-dots) in paraffin-embedded tissues. J Fluoresc. 15(5):661–665.
105. Yu GT, Luo MY, Li H, Chen S, Huang B, Sun ZJ, Cui R, Zhang M (2019). Molecular targeting nanoprobes with non-overlap emission in the second near-infrared window for in vivo two-color colocalization of immune cells. ACS Nano. 13, 12830–12839.
106. Pons T, Bouccara S, Loriette V, Lequeux N, Pezet S, Fragola A (2019). *In vivo* imaging of single tumor cells in fast-flowing bloodstream using near-infrared quantum dots and timegated imaging. ACS Nano. 13, 3125–3131.
107. Sitbon G, Bouccara S, Tasso M, Francois A, Bezdetnaya L et al. (2013). Multimodal Mn-doped I-III-VI quantum dots for near infrared fluorescence and magnetic resonance imaging: from synthesis to *in vivo* application. Nanoscale. RSC Publishing, 1–3.
108. Liu Q, Deng R, Ji X, Pan D (2012). Alloyed Mn–Cu–In–S nanocrystals: a new type of diluted magnetic semiconductor quantum dots. Nanotechnology. 23, 255706.
109. Kelkar SS, Reineke TM (2011). Theranostics: combining imaging and therapy Bioconjugate Chem. 22(10): 1879–1903.
110. Blanco-Canosa JB, Wu M, Susumu K, Petryayeva E, Jennings TL, Dawson PE, Algar WR, Medintz IL (2014). Recent progress in the bioconjugation of quantum dots. Coord. Chem. Rev. 263–264, 101–137.

6 Fluorescent Magnetic Quantum Dots in Bioimaging

Dipti Rawat and Ragini Raj Singh
Nanotechnology Laboratory, Department of Physics and Materials Science, Jaypee University of Information Technology, Waknaghat, Solan, India

CONTENTS

6.1 INTRODUCTION

Magneto luminescent core-shell nanostructures, a fast-evolving class of materials with numerous applications, have drawn much interest. These magneto fluorescent core-shell nanostructures' magnetic and optical properties can be retrieved simultaneously by a single entity, with the magnetic and optical units intact [1-4]. The luminous shell completely encircles the magnetic core. Core-shell nanostructures are a novel class of hybrid nanostructures with spectacular results. In biological imaging, fluorescent compounds are essential. They've increased our understanding

DOI: 10.1201/9781003319870-6

of basic biology, offered cell separation and characterization procedures, and been used to detect and evaluate disease. Fluorescent molecules, on the other hand, have several drawbacks, including a photostability falling-off (i.e., bleaching) and wide-ranging emission spectra, which make multiplexing difficult [2, 5]. Because of broad excitation spectra, limited emission bandwidth, and increased photostability, QDs are an attractive alternative to standard fluorescent dyes. QDs are semiconductor nanocrystals with a diameter of 1-10 nm that exhibit quantum confinement [6]. Size-dependent fluorescence is the result of this feature. Furthermore, this characteristic leads to fluorescence that is dependent on size. QD fluorescent signals can be adjusted as a consequence. Moreover, the emission bandwidth is much narrower than conventional fluorescent dyes because fluorescence emission is based partly on exciton-hole recombination [6]. With QDs advancement, more people are interested in mixing them with other materials to generate multimodal nanocomposites [7-8]. Theranostics, in particular, could benefit from using QDs and magnetic materials in imaging, cell and molecule separations, and theranostics. Though many different magnetic materials exist, iron oxides, specifically SPIONs, are the most commonly studied. There has been a lot of attention in combining the advantages of QDs with those of other materials to create composites with multifunctional capabilities as the potential of QDs for biological imaging is realized. A magnetic substance in the form of ions or nanoparticles is among the most often used materials to mix with QDs. The

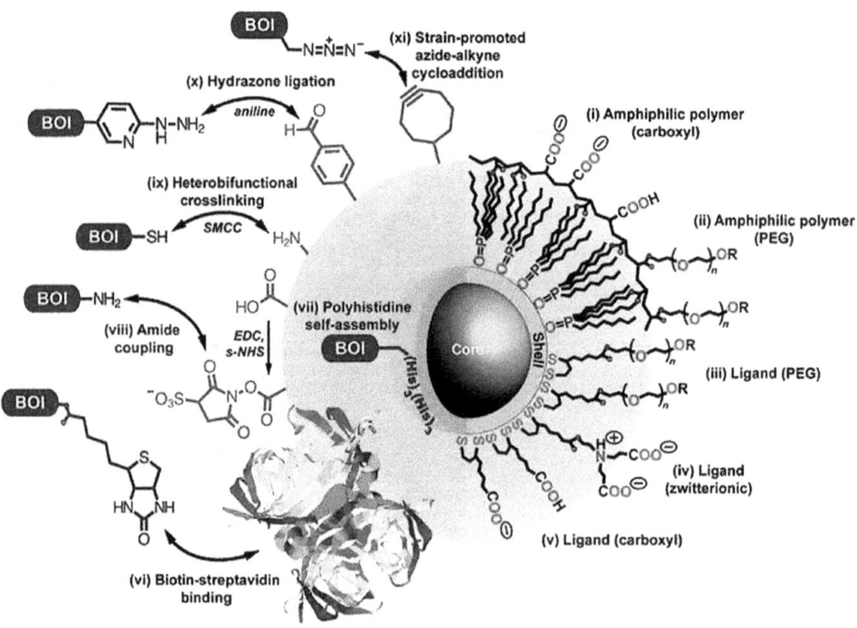

FIGURE 6.1 Overview of different bioconjugation (left side, BOI = biomolecule of interest) and surface coating (right side) strategies for QDs. Two surface coating strategies are presented: encapsulation with amphiphilic polymers (I, ii) and cap exchange with hydrophilic ligands exploiting the thiol-affinity of the ZnS shell of the QD (iii–v) [10].

QDs' fluorescent properties allow for vision, while the composite's magnetic properties allow for imaging, magnetic separation, and maybe therapeutic effect. This work describes magnetofluorescent nanoparticles and magnetic QDs and their uses in theranostics, imaging, and separation. QDs have the potential to have a significant influence on biological imaging, diagnosis, and treatment as the characteristics of these materials improve. Magnetic quantum dots [9] with a size smaller than that of the magnetic crystal domain. It means that the nanoparticles have paramagnetic activity in solution, which reduces their aggregation potential, but inducible magnetic behavior when exposed to a magnetic field. During magnetic cycling, SPIONs have low or no hysteresis losses. SPIONs have been employed in several biomedical applications, including cell separation, single-molecule manipulation, and magnetic resonance imaging (MR imaging). This chapter looks at some of the most prevalent synthesis methods for making magnetic and other nanocomposites and their biological applications. Nanoparticle-based magnetofluorescent composites are brand-new, but molecular-based fluorescent-magnetic composites have been known for a while. Figure 6.1 displays various surface coating (right side) and bioconjugation (left side, BOI = biomolecule of interest) methods for QDs. Encapsulation with amphiphilic polymers (i-ii) and cap exchange with hydrophilic ligands, both of which take advantage of the thiol-affinity of the ZnS shell of the QD, are described as two surface coating techniques (iii–v).

6.2 SYNTHESIS OF FLUORESCENT MAGNETIC QUANTUM DOTS

Making "fluorescent-magnetic nanoparticles" is extensively known [11]. Although the contents and morphologies of these nanocomposites vary, most synthesis techniques fall into one of four groups: doping, high-temperature precursor breakdown, crosslinking, or encapsulation. In prime technique, a particular kind of nanocrystal is initially produced, and then a second material grows on top of the first to form a composite. In another technique, doping transition metals into bulk nanocrystal is done to offer magnetofluorescent properties. Using extra linkers or electrostatic forces, magnetic and fluorescent species are manufactured independently and combined in a third way. In the fourth technique, fluorescent and magnetic components are enclosed in a matrix often formed of silica or polymer materials.

6.2.1 High-Temperature Decomposition

"Organometallic precursors are heated during the organic phase, which causes decomposition" [12–13]. High-temperature decomposition results in particle nucleation and growth. This process creates dual samples successively, starting with a core material (such as FePt,Fe_3O_4, Fe_2O_3, FeP, or Co) that assists as a model for the development of the subsequent material. This differs from doping or encapsulation methods, which assemble pre-fabricated magnetic and fluorescent components (e.g., CdSe, CdTe, CdS). This method can be used to make two types of fluorescent-magnetic nanoparticles. The first is a core/shell heterodimer, which is made up of two different materials [14-15]. The second kind starts as a heterodimer, but interfacial instabilities cause constituent atoms to migrate, producing dumbbell-shaped particles with different domains. Crystal lattice mismatch and reaction circumstances are the key determinants of the final shape of the nanocomposites generated (e.g., temperature, reactant ratios). The initial nanocrystal can be created under ideal conditions; nevertheless, the initial crystal can be subjected to a range of temperature fluctuations that influence its structure during the synthesis of the succeeding material. “Also, it is possible for unintended interfacial doping to occur between the two materials” [16]. “Disorder in the magnetic crystal structure, which happens due to sub-optimal synthesis conditions, doping, or interfacial instabilities in the two crystal lattices, is thought to be the cause of magnetic property loss” [16].

6.2.2 Doping

The employment of dopants to insert transition elements into the nanocrystal lattice is the second main technique for creating fluorescent-magnetic nanostructures. Contrary to the first method, nanocomposites have many functions because they have two independent material domains; in doped materials, the energy states brought about by the dopant ions provide the materials with their fluorescence and magnetic properties. Manganese-doped nanoparticles are one of the most well-studied of these materials, with applications in magnetic resonance imaging, spintronics, and photoluminescence imaging [17-18]. Mn^{2+} can be doped in the core or shell of both II–VI and III–V nanocrystals. Rare earth minerals doped in the host lattice are another dopant material [19–22]. Doped materials have an advantage over materials made through high-temperature precursor breakdown in that interactions between the two types of materials are desirable and drive the fluorescent/magnetic features. On the other hand, the range of available dopants and the difficulty of creating doped structures limit the use of doped materials.

6.2.3 Crosslinking

Researchers have studied technologies that allow for the distinct creation of nanoparticles, which are then merged by another approach to optimizing the attributes of constituent materials (e.g., crosslinking, encapsulation)[23]. SPIONs are used as the magnetic material in fluorescent-magnetic nanoparticles made by crosslinking, while QDs, or fluorescent proteins, are used as the fluorescent material [24–26]. Functional groups binding both entities, such as siloxane groups, thiol, or carboxyl, are commonly found in coupling ligands [27], although proteins have also been utilized to arbitrate these interactions [28]. On the other hand, some crosslinking techniques do not use a chemical linker and instead rely on charge attraction to mediate connections. Static electric forces, for example, were used to bind positively charged Fe_3O_4 nanoparticles and negatively charged carboxy-CdTe QDs together [29]. The disadvantages of sequential synthesis or doping are reduced considerably by crosslinking; nonetheless, the structures produced can be colossal, and the constancy of the binding ligands and crosslinkers varies with the environment.

6.2.4 Encapsulation

In encapsulation-based assembly processes, a matrix contains already-formed magnetic and fluorescent entities. Because the synthesis of fluorescent and MNPs is done separately, encapsulation approaches avoid the inherent challenges of controlled crystallization and doping. Furthermore, the need for stable nanoparticle-ligand-linker interactions is eliminated because coupling agents are not needed. Top-down or bottom-up procedures can be used to create the encapsulation matrix [30]."The use of polymers (e.g., poly (lactic-co-glycolic acid), PLGA) to co-encapsulate QDs and SPIONs via an adaption of the single emulsion technique used in producing particles for drug administration [31] is an example of top-down encapsulation." Alternatively, nanoprecipitation techniques have been used [32]. The particles formed by top-down encapsulation methods are quite bulky, which is detrimental in many applications, like diffusion into deep tumors. For example, synthesizing disaggregated PLGA particles with dimensions less than 200 nm is challenging. It is due to adding energy to top-down production methods to build the necessary structure, which might limit particle amount and size distribution. By relying on thermodynamic forces rather than energy input to propel the final structures' self-assembly, bottom-up encapsulation avoids several of these problems. According to theory, the matrix might be any self-assembling structure with a hydrophobic zone more significant than the total volume of the encapsulants (e.g., micelles, liposomes [33], and silica [34]). Figure 6.2. illustrate the different synthesis methods for the Fluorescent magnetic quantum dots.

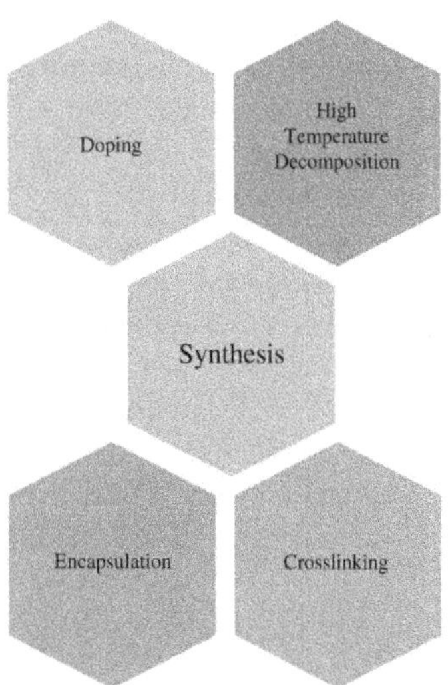

FIGURE 6.2 Different synthesis methods for the fluorescent magnetic quantum dots.

6.3 TYPES OF MAGNETIC QUANTUM DOTS

6.3.1 Core/Shell and Heterostructures (Type I)

In Type I structure, core/shell or heterostructure is created by fusing MNPs and QDs. Despite the enormous lattice mismatch between magnetic and semiconductor nanocrystals, two can be combined within one nanocrystal. However, the attachment method is yet unknown [35]. The four samples below were all created via a wet chemical technique, where the magnetic core was made before the semiconductor material was attached (i.e., the QD).

Two groups investigated the coupling of FePtMNPs with CdS or CdSe QDs, demonstrating that both core/shell and heterostructures may be manufactured. "The production of FePt cores with a diameter of 3 nm, enclosed by a 3–5 nm CdSe shell was reported" [36-37]. In addition to CdSe, the authors show that a CdS or CdTe shell can be used to coat the FePt cores. The fluorescence emission of FePt/CdSe "core/shell" nanocrystals was about 465 nm, which was unexpected given the massive volume of the CdSe shell. The fluorescence QE of the FePt/CdSe nanocrystals was 7–10%, which was lower than that of CdSe QDs alone. The light quenching was attributed to a QD-metallic core interaction, but this was not further studied. "The FePt cores' superparamagnetic characteristics were preserved thanks to the low blocking temperature of 14 K established for the core/shell particles. Gu and colleagues demonstrated the synthesis of FePt–CdS heterodimers similarly" [37]. They believe that when the FePt/CdS core/shell complex was heated, it was converted into a heterodimer. The heterodimers displayed a QE of 3%, and the fluorescence emission peak at 438 nm was consistent with CdS QDs of comparable size in this sample.

"Kim et al. described a core/shell structure with a superparamagnetic cobalt core surrounding by a CdSe shell" [38]. The cobalt particles were first precipitated and redispersed in a coordinating solvent before adding the CdSe precursor materials, in contrast to the one-pot synthesis mentioned above. "The cobalt cores of the final particles have a diameter of 8 nm and a CdSe shell of 2 nm. In comparison to conventional CdSeQDs, the fluorescence emission peak of the Co/CdSe core/shell particles was about 580 nm with a Stokes shift of 40–50 nm." Surprisingly, when the temperature dropped, and the rate of radiative decay increased, the QE of these particles (2–3%) increased. When a persistent magnetic field was applied to the particle suspension, all of the particles were drawn to the magnet and a clear solution was left in their place. The Fe2O3 core had an 8–10 nm diameter, and by varying the reaction period, the size of the QD could be changed from 2 to 5 nm. Thus, the emission color of the QDs could be changed between 550 and 600nm. Comparing the final heterodimers to the preceding samples, the QE was high at 13–18%.

6.3.2 Doped QDs (Type II)

Doping paramagnetic luminous ions into QDs allows fluorescence and magnetic characteristics to be combined in a single nanoparticle. Doping paramagnetic ions into semiconductors has a long history [39]. The paramagnetic ion itself or recombination with paramagnetic dopant bandgap states could be the light source. "After an early report in 1994, which showed a high luminous quantum yield and a remarkable shortening of the luminescence life time for ZnS: Mn^{2+} nanocrystals, research on QDs doped with (paramagnetic) transition-metal ions exploded" [40]. Despite the fact that later research revealed that the claimed lifetime shortening was incorrect [41], a whole new sector had opened up. Many unique and fascinating occurrences and uses for doped QDs have been described since then, including investigations on the use of doped QDs as multifunctional bioimaging probes.

Integrating the dopant ions into the semiconductor core is a crucial feature. Signals for the dopant presence inside the semiconductor particle and not rivetted on the exterior is difficult to come by. To synthesize doped QDs, four different approaches are used. The simplest approach is growing semiconducting QDs in a solution containing precursors for the semiconducting QDs, the dopant ion, and a passivating ligand. The synthesis of the broadly premeditated "ZnS: Mn^{2+} and ZnO: Co^{2+} QDs" is an example of successful doped QD preparation utilizing this method. The development of QDs inside water droplets from precursors dissolved in the aqueous phase is accomplished using an alternate method called an inverse micelle solution [42]. Recently, it has been discovered that QDs doped with paramagnetic ions can function as multimodal (magnetic and bright) imaging probes. The majority of prior research focused on the magnetic characteristics of spintronics and doped QDs in light-emitting devices, with ferromagnetism in dilute magnetic semiconductor (DMS) nanoparticles being of special interest. The inverse micelle technique was used to create 3.1-nm particles, which were then coated with a thin layer of silica and changed with APS to add amine surface groups for bioconjugation. The QDs had a robust magnetic response and released a vivid yellow Mn^{2+}. "Louie's group demonstrated bimodal imaging using CdSe/Zn1–x MnxS nanoparticles. The extremely luminous (undoped) CdSe core was overrun with a ZnS shell doped with Mn^{2+} to introduce the paramagnetic functionality in this core/shell particle" [43]. The fluorescence of this bimodal particle, which had a QE of more than 20%, originated from the CdSe core.

The presence of paramagnetic Mn2+ ions in the outer shell allows for good water interaction and works well as an MRI contrast agent. The poly(acrylic acid)-a modified octyl-amine coating made the nanoparticles (5 nm) water-soluble. Large-scale Cd-free QD development is taking place for biological imaging applications. A recent study on silicon nanoparticles doped with Mn^{2+} was published. While maintaining a good quantum yield, adding Mn2+ caused the QD emission to shift from 430 to 520 nm (up to 16 percent). The EPR and transient absorption spectra show that the Mn2+ is incorporated into the nanocrystal and firmly bonded to the Si surface. These results are a step in the right direction for creating doped Si QDs as multimodal imaging labels. Future integrated fluorescence and MRI probes may be made from luminescent and ferromagnetic doped QDs.

6.3.3 Composite Particles Containing Semiconducting Nanoparticles and MNPs (Type III)

Another method for giving QDs magnetic characteristics is to combine them with semiconductor nanocrystals using a carrier material to form a composite particle. The various NP types may be incorporated into the carrier material, adhered to the exterior of the carrier material, or both. For instance, the MNPs may be attached to the carrier particle's surface, and the QD may be inserted into the carrier particle. As discussed further below, silica and polymer matrices can also be employed as a carrier material, resulting in a construct typically larger than Type I or Type II particles. This section will be divided into two parts. The first will examine composite particles employing silica as a carrier material, which will include polymer capsules.

Fe_2O_3 MNPs and CdSe/ZnS QDs were combined to form composite particles matching Type IIIa particles in 50 nm silica spheres. The composite particles were created by separately manufacturing MNPs (12 nm) and QDs (3.5 nm) and combining them in a conventional microemulsion system [44]. It led to the formation of monodisperse silica particles rich in MNPs and QDs. "After being incorporated into silica, the QE of the CdSe/ZnS QDs was reduced from 15% to 5%, and the QDs had an emission peak at 554 nm" [45]. Although the reason for the drop in QE was not specified, it might have something to do with interactions with the magnetic particles or be a result of the QDs being directly incorporated into a silica matrix using the microemulsion approach [43]. The superparamagnetic properties of the MNPs contained within the silica spheres were corroborated by the blocking temperature (Tb) of 165 K. The presence of the QDs did not affect the magnetism at saturation. A well-known layer-by-layer (LbL) deposition technique was then used to place negatively charged CdTe QDs onto the silica spheres after they had been coated with positively charged polymers. The final composite particles had a diameter of 220 nm after the magnetic core and fluorescent shell silica particles were covered with a 20 nm silica shell. The scientists assert that the relatively large distance between the MNPs and the QDs prevented the quenching of the QD emission, despite the fact that light scattering of the enormous silica particles made it difficult to measure the QE of the CdTe QDs. A "Superconducting quantum interference device (SQUID)," which offered the blocking temperature (Tb) of 150 K, coercivity of 175 Oe (at 5 K), and saturation magnetization of 1.34 emu/g, was used to measure the magnetic characteristics of the composite particles. The last example of silica serving as a carrier material for MNPs and QDs that we'll look at here was released by the Hyeon group. In this example, 100-nm silicon spheres were coated with both kinds of nanocrystals on the outside [46]. The silica spheres were created with Stober, and then APS was used to add amine groups on the surface. "The covalent attachment of 2-bromo2-methylpropane acid (BMPA)-functionalized Fe3O4 particles to the silica surface was achieved by a nucleophilic substitution reaction between the terminal Br groups of the ligands and the amine groups of the silica spheres" [47].

6.3.4 QDs with a Paramagnetic Coating of Gd-Chelates (Type IV)

The last kind of particles that will be looked at here are QDs covered with organic complexes (chelates) that contain paramagnetic ions. The lanthanide ion Gd^{3+} is the most popular paramagnetic ion for MRI contrast agents due to its substantial magnetic moment and symmetric electronic ground state. The Gd^{3+} ions are complex in organic chelates that coordinate to the paramagnetic ions by an ionic contact to reduce toxicity and boost stability.

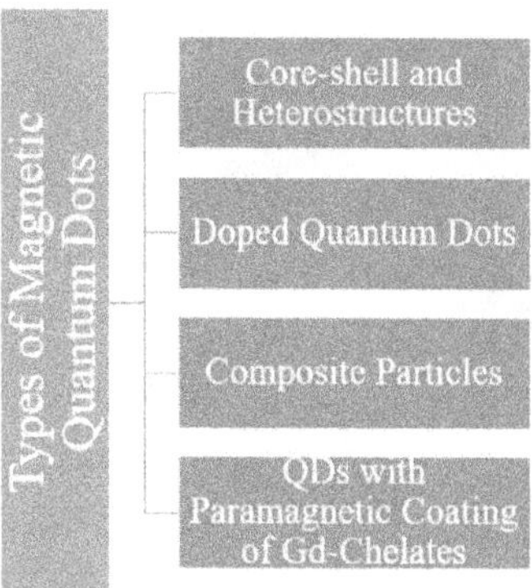

FIGURE 6.3 Different types of fluorescent magnetic quantum dots.

Gd-DTPA (diethylenetriaminepentaacidic acid) is the extensively used MRI contrast agent in both experimental and clinical settings. However, a wide range of alternative chemicals is being investigated to improve molar relaxivity and expand the spectrum of possible biological uses. Organic dye molecules have been covalently [59, 60] and noncovalently linked to the paramagnetic chelates to give them fluorescent characteristics [48-49]. Attaching paramagnetic chelates to QDs is another way to obtain bimodality (MRI and fluorescence imaging) [50].

"Prinzen et al. described an additional method for obtaining QDs covered with a paramagnetic covering of Gd-chelates" [51]. Instead of lipidic coating surrounding the QDs, the authors used biotinylated "Gd-DTPA and annexin A5 proteins," which were coupled to streptavidin in coated QDs. To increase MRI detection sensitivity, a novel dendrimer-like chelate (Gd-wedge) with eight Gd-DTPA complexes per wedge was created. Using fluorescence imaging and MRI, it was determined that the QD-Gd wedge particles marked Jurkat cells undergoing apoptosis significantly in vitro, but the specific binding was weak. Furthermore, an excised murine carotid artery that had been physically wounded was subjected to target-specific tagging, resulting in an overexpression of PS (phosphatidylserine, binds to annexin A5). The target specificity was validated by the absence of labelling in an intact control artery. Ex vivo T1-weighted MRI revealed that the injured artery had a brighter signal than the control artery, supporting the bimodal nanoparticles' value. Figure 6.3 shows the schematic for the different types of Fluorescent magnetic quantum dots.

6.4 APPLICATIONS OF FLUORESCENT-MAGNETIC NANOPARTICLES

Owing to the multiple functionalities, magnetic quantum dots have found their applications in varied fields. Some of the applications of fluorescent-magnetic nanoparticles are illustrated in Figure 6.4 and are discussed in detail in the next section.

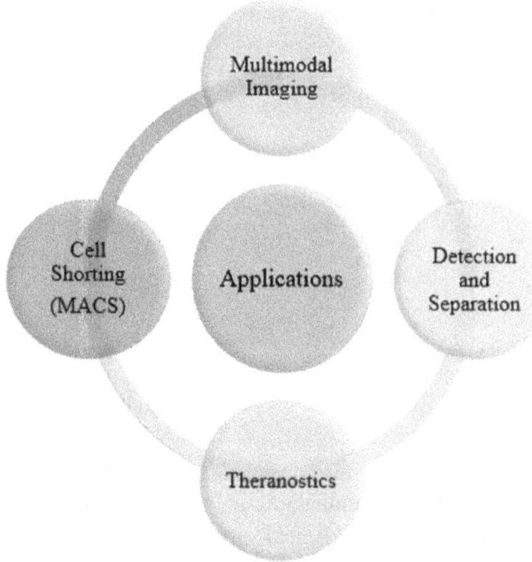

FIGURE 6.4 Some of the applications of fluorescent magnetic quantum dots.

6.4.1 Multimodal Imaging

"Since Nie and Alivisatos first exhibited QDs for in-vitro imaging in 1998, these semiconductor materials have been widely used in biomedical imaging [2-3]." QDs have been coupled with magnetic components like magnetic ions and SPIONs to create two-fold functional nanocomposites that can be used with MRI and fluorescence microscopy for more potent and complete imaging capabilities. Using hydrogen protons, which are abundant in fat and water, MRI uses signals produced by these protons to build images of internal human tissues. Two distinct relaxing methods are required to restore excited hydrogen nuclei to their initial low-energy state. Although tissues' natural characteristics can produce significant MRI signals, contrast chemicals that reduce T1 and T2 relaxation durations are employed in 40% of all MRI studies [52]. Numerous multimodal nanocomposites with optical and MR imaging capabilities were used by the researchers. The initial investigations [17–18] mainly concentrated on doping manganese ions into QDs. Using QDs doped with paramagnetic ions, such as when they were employed for arterial labelling, it was shown that multimodal QDs can cross the blood-brain barrier [18]. It implies that it may be possible to deliver therapeutic drugs to the brain via a QD-coupled delivery system. Encapsulation techniques have also been utilized to combine QDs with paramagnetic ligands for in-vitro integrin imaging and fluorescence/MR detection of apoptosis.

There have been a few attempts to associate QDs with SPIONs as the magnetic component [54]. For in-vitro imaging, thin silica-coated CdSe-Fe_2O_3 heterodimers have been used, whereas micelle-encapsulated QDs rods and SPIONs have been used for combined fluorescence/MR imaging [55]. Yet, as an MR agent, magnetic

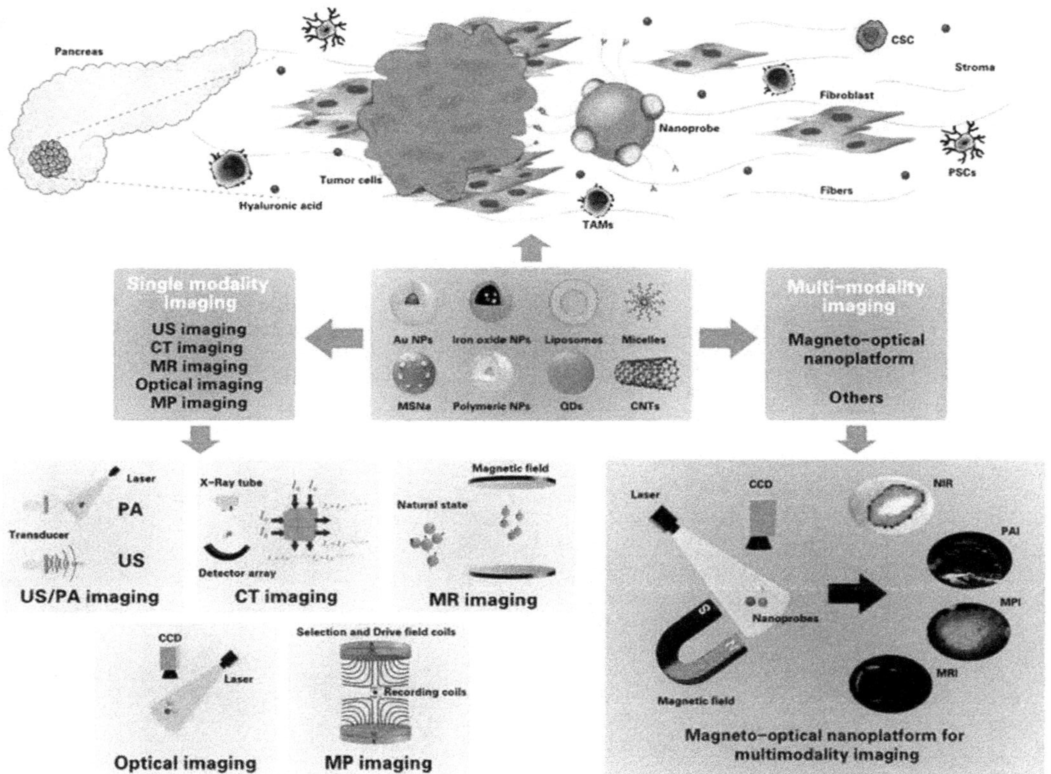

FIGURE 6.5 An example of several imaging agents that may be simply added to or modified nanoparticles for PDAC imaging and their application in different imaging modalities for the diagnosis of DPAC [56].

ions have drawn much more interest than SPIONs. This might be because the signals produced are different or because these materials are relatively new; their fabrication was first proven in 2005 [53].

Potential toxicity is one of the main issues with multimodal treatments based on QDs, including magnetic ions or SPIONs. However, the range of materials accessible for in-vivo applications will also increase as several groups work on non-toxic QDs.

6.4.2 Detection and Separation

Molecular recognition and parting is another key application of "fluorescent-magnetic nanoparticles." Detecting signals from low concentration solutions or distinguishing between minor changes in biomolecular structures is impossible with bulk measurement techniques like absorbance. These minute structural alterations have a major impact on biomolecules' overall activity. Hence precise exposure is critical for proper diagnosis. Because nanoparticle-based detection approaches are similar in size to biomarkers, they have a distinct advantage. Nanoparticle-based detection technologies provide great specificity and accuracy due to size similarities and nanoscale phenomena. "Fluorescent-magnetic nanoparticles" have been utilized to detect and separate DNAs, RNAs, and proteins in diagnostics, gene profiling, and the detection of microbial contamination in the environment [57-58]. "The ability to discriminate DNA/RNA targets with a single base pair mismatch has been proven [58-59], and detection levels on the order of femtomolar concentrations have been reached." "Fluorescent-magnetic nanoparticles" have also been utilized to separate uncommon cells for illness diagnosis, in-vitro disease progression models, and medication screening. Fluorescence functionality added to magnetic cell separation enables visualization of the separated cells and further analysis by immune cytochemistry or flow cytometry, providing details on the disease's genetic basis at the gene or protein level.

6.4.3 Theranostics

When used with imaging agents, nanoparticles have several advantages as drug delivery systems; they can be used for treatment and diagnosis, a process known as theranostics. The ideal theranostic probe would be small enough to circulate in the circulation, non-immunogenic, have recognition motifs to recognize the target location (for example, a tumor), penetrate the target site, and be able to detect and cure disease in response to an external signal. A good theranostic should also be able to be eliminated by the body after usage or be biodegradable. These goals have been greatly advanced thanks to nanotechnology. Because of their small size, sub-micron polymeric nanoparticle carriers are simple to make and can be tailored to enter and aggregate in specific anatomical locations. Nanoparticle accumulation in tumors, for example, is boosted for particles with diameters between 10-100 nm, a size that enhances delivery via the tumor microenvironments [60]. Surfaces of nanoparticles can be tailored to reduce interactions with the immune system, improving circulation and uptake. Nanoparticle surfaces can be coupled with targeting molecules like peptides and antibodies.

Multifunctional particles, especially magneto-fluorescent nanoparticles, have several advantages. The fluorescent and/or magnetic components, most often QDs and organic dye molecules or SPIONs, can be used to introduce fluorescence or MR imaging capabilities. Because of their biocompatibility, iron oxide nanoparticles are perfect for imparting magnetic functionality. Transferrin, ferritin, and hemosiderin receptors allow cells to absorb iron oxide, which is then destroyed by natural mechanisms [8]. The bulk of fluorescent-magnetic nanoparticles are made by encapsulating them in polymers, which improves circulation and uptake. "Poly(caprolactone) [59], poly (lactic-co-glycolic acid) [61], chitosan [62], polystyrene [63], and poly (allyl – amine hydrochloride) are all commonly used materials. These encapsulating polymers provide bioconjugation sites and the potential for stimuli-dependent releases, such as chitosan's pH responsiveness [62]." Gene therapy has drawn a lot of attention because it uses DNA, RNA, and small interfering ribonucleic acid (siRNA) to modulate the expression of genes and proteins to produce therapeutic effects. An external magnetic field can be used to focus and distribute therapeutics associated with SPIONs to a particular area, which can enhance cellular absorption. This technique, which creates particle aggregation around a tumor location using an externally controlled magnetic micromesh, has already been commercialized [64]. This intriguing class of therapeutics needs effective and focused transfection with few adverse effects to be effective. Because of their safety, simplicity in synthesis, large loading capacities, and ease of surface modification, which minimize immune system involvement, nanoparticles are replacing viral vectors as nonviral delivery systems. Because targeted therapies, which are predominantly mediated by antibodies, might cause uptake via the reticuloendothelial system (RES), decreasing efficacy and increased particle accumulation by non-targeted techniques is crucial [65]. It has become one of the most crucial elements in the administration of RNA therapies because RNA treatments degrade rapidly in the bloodstream in the absence of a targeting mechanism.

Nanoparticles can be employed to induce therapeutic effects in addition to releasing medicines. QDs, for example, can be utilized in photodynamic treatment to activate photosensitizers locally and produce hazardous species. Magnetic nanoparticles can also cause hyperthermia, which is caused by the SPION reaction to an oscillating magnetic field.

6.4.4 Magnetic-Activated Cell Sorting (MACS)

Magnetic nanoparticles processed with fluorescent QDs can find their application in cell shorting. Magnetic-activated cell sorting (MACS) is a technique for dividing different cell populations according to the antigens on their surfaces. Through an antibody interaction with cell surface indicators of the intended cells, MACS binds magnetic particles to cells, often called immunomagnetic cell separation. The targeted cells are then separated from the remainder of the biological sample using magnets. Figure 6.6 shows the schematic for magnetic cell separation. Here the magnetic material helps attach the antibody related to the respective antigen, and fluorescent QDs over the magnetic material help track the path of the cell.

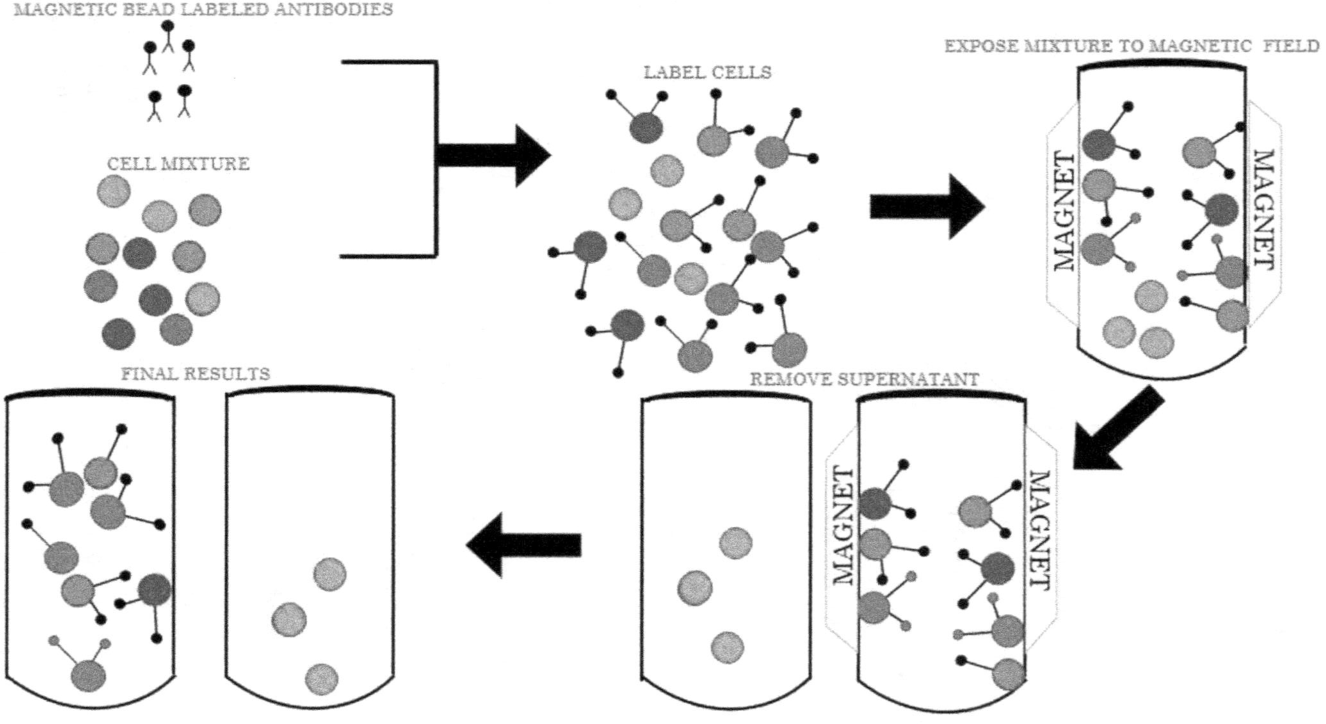

FIGURE 6.6 Shows the schematic for the magnetic cell sorting.

6.5 CONCLUSION

Despite these recent breakthroughs, the practice of "fluorescent-magnetic nanoparticles" is still in its early stages, with significant difficulties to overcome before this promising technology can be widely applied in medicine. The poisonousness of QDs is a major concern, which has sparked a lot of interest in creating non-toxic fluorescence probes and QDs. However, the development of alternative probes has been hampered by a lack of understanding of nanoparticle poisonousness processes and established methodologies for toxicity assessment. In diseases like cancer, the efficacy of passive versus aggressive targeted techniques is also contested [65]. Active delivery, particularly when using antibody-conjugated probes, can lead to higher RES absorption even while passive distribution is inconsistent. Aptamers, a newer class of targeting probes, may reduce this interaction and enable more precise delivery [66-67].

Even if these biological concerns are handled, high-volume manufacturing procedures will be necessary to produce adequate quantities and quality of particles for regulatory approval. Self-assembly procedures, rather than traditional top-down production processes, may offer a solution. Despite these drawbacks, "fluorescent-magnetic nanocomposites, particularly magnetic QDs," offer a plethora of fascinating biomedical applications. Particles can be made in a variety of ways, but the encapsulation approaches have shown to be the most successful since they allow for a separate synthesis of different nanospecies, which are then merged by encapsulation in a matrix. Not only does the matrix improve water solubility and biocompatibility, but it can also be used to deliver therapeutics. These particles can be used for a variety of purposes. The initial focus was focused on multimodal imaging, such as in-vitro and in-vivo combined fluorescence/MR imaging.

Furthermore, magnetic properties can be employed to capture, separate, and manipulate certain species, and fluorescent-magnetic nanocomposites can be used to isolate rare cells and molecules from heterogeneous solutions. Also, "fluorescent-magnetic nanoparticles" can be used to create theranostics, which uses either precise drug release or the exceptional properties of nanoparticles to diagnose and treat disease. As "fluorescent-magnetic nanoparticles" develop, further advantages will become apparent, maybe resulting in a novel class of diagnostic and therapeutic materials.

REFERENCES

[1] Cho, N. H., Cheong, T. C., Min, J. H., Wu, J. H., Lee, S. J., Kim, D., ... & Seong, S. Y. (2011). A multifunctional core–shell nanoparticle for dendritic cell-based cancer immunotherapy. *Nature Nanotechnology*, *6*(10), 675–682.

[2] Chan, W. C., & Nie, S. (1998). Quantum dot bioconjugates for ultrasensitive nonisotopic detection. *Science*, *281*(5385), 2016–2018.

[3] Bruchez Jr, M., Moronne, M., Gin, P., Weiss, S., & Alivisatos, A. P. (1998). Semiconductor nanocrystals as fluorescent biological labels. *Science*, *281*(5385), 2013–2016.

[4] Dubertret, B., Skourides, P., Norris, D. J., Noireaux, V., Brivanlou, A. H., & Libchaber, A. (2002). In vivo imaging of quantum dots encapsulated in phospholipid micelles. *Science, 298*(5599), 1759–1762.

[5] Bruchez Jr, M., Moronne, M., Gin, P., Weiss, S., & Alivisatos, A. P. (1998). Semiconductor nanocrystals as fluorescent biological labels. *Science, 281*(5385), 2013–2016.

[6] Alivisatos, A. P. (1996). Perspectives on the physical chemistry of semiconductor nanocrystals. *The Journal of Physical Chemistry, 100*(31), 13226–13239.

[7] Pellegrino, T., Kudera, S., Liedl, T., Muñoz Javier, A., Manna, L., & Parak, W. J. (2005). On the development of colloidal nanoparticles towards multifunctional structures and their possible use for biological applications. *Small, 1*(1), 48–63.

[8] Quarta, A., Di Corato, R., Manna, L., Ragusa, A., & Pellegrino, T. (2007). Fluorescent-magnetic hybrid nanostructures: preparation, properties, and applications in biology. *IEEE Transactions on Nanobioscience, 6*(4), 298–308.

[9] Gupta, A. K., & Gupta, M. (2005). Synthesis and surface engineering of iron oxide nanoparticles for biomedical applications. *Biomaterials, 26*(18), 3995–4021.

[10] Wegner, K. D., & Hildebrandt, N. (2015). Quantum dots: bright and versatile in vitro and in vivo fluorescence imaging biosensors. *Chemical Society Reviews, 44*(14), 4792–4834.

[11] Wang, G., & Su, X. (2011). The synthesis and bio-applications of magnetic and fluorescent bifunctional composite nanoparticles. *Analyst, 136*(9), 1783–1798.

[12] Murray, C., Norris, D. J., & Bawendi, M. G. (1993). Synthesis and characterization of nearly monodisperse CdE (E= sulfur, selenium, tellurium) semiconductor nanocrystallites. *Journal of the American Chemical Society, 115*(19), 8706–8715.

[13] Katari, J. B., Colvin, V. L., & Alivisatos, A. P. (1994). X-ray photoelectron spectroscopy of CdSe nanocrystals with applications to studies of the nanocrystal surface. *The Journal of Physical Chemistry, 98*(15), 4109–4117.

[14] Kim, H., Achermann, M., Balet, L. P., Hollingsworth, J. A., & Klimov, V. I. (2005). Synthesis and characterization of Co/CdSe core/shell nanocomposites: bifunctional magnetic-optical nanocrystals. *Journal of the American Chemical Society, 127*(2), 544–546.

[15] Du, G. H., Lu, Q. H., Xia, X., Jia, L. H., Yao, K. L., Chu, Q., & Zhang, S. M. (2006). Fe3O4/CdSe/ZnS magnetic fluorescent bifunctional nanocomposites. *Nanotechnology, 17*(12), 2850.

[16] Deng, S., Ruan, G., Han, N., & Winter, J. O. (2010). Interactions in fluorescent-magnetic heterodimer nanocomposites. *Nanotechnology, 21*(14), 145605.

[17] Wang, S., Jarrett, B. R., Kauzlarich, S. M., & Louie, A. Y. (2007). Core/shell quantum dots with high relaxivity and photoluminescence for multimodality imaging. *Journal of the American Chemical Society, 129*(13), 3848–3856.

[18] Santra, S., Yang, H., Holloway, P. H., Stanley, J. T., & Mericle, R. A. (2005). Synthesis of water-dispersible fluorescent, radio-opaque, and paramagnetic CdS: Mn/ZnS quantum dots: a multifunctional probe for bioimaging. *Journal of the American Chemical Society, 127*(6), 1656–1657.

[19] Wang, F., & Liu, X. (2009). Recent advances in the chemistry of lanthanide-doped upconversion nanocrystals. *Chemical Society Reviews, 38*(4), 976–989.

[20] Bünzli, J. C. G. (2010). Lanthanide luminescence for biomedical analyses and imaging. *Chemical reviews, 110*(5), 2729–2755.

[21] Petoral Jr, R. M., Soderlind, F., Klasson, A., Suska, A., Fortin, M. A., Abrikossova, N., ... & Uvdal, K. (2009). Synthesis and characterization of Tb3+-doped Gd2O3

nanocrystals: a bifunctional material with combined fluorescent labeling and MRI contrast agent properties. *The Journal of Physical Chemistry C*, *113*(17), 6913–6920.

[22] Das, G. K., Zhang, Y., D'Silva, L., Padmanabhan, P., Heng, B. C., Chye Loo, J. S., ... & Yang Tan, T. T. (2011). Single-phase Dy2O3: Tb3+ nanocrystals as dual-modal contrast agent for high field magnetic resonance and optical imaging. *Chemistry of Materials*, *23*(9), 2439–2446.

[23] Wang, D., He, J., Rosenzweig, N., & Rosenzweig, Z. (2004). Superparamagnetic Fe2O3 beads– CdSe/ZnS quantum dots core– shell nanocomposite particles for cell separation. *Nano Letters*, *4*(3), 409–413.

[24] Veiseh, O., Sun, C., Gunn, J., Kohler, N., Gabikian, P., Lee, D., ... & Zhang, M. (2005). Optical and MRI multifunctional nanoprobe for targeting gliomas. *Nano Letters*, *5*(6), 1003–1008.

[25] Huh, Y. M., Jun, Y. W., Song, H. T., Kim, S., Choi, J. S., Lee, J. H., ... & Cheon, J. (2005). In vivo magnetic resonance detection of cancer by using multifunctional magnetic nanocrystals. *Journal of the American Chemical Society*, *127*(35), 12387–12391.

[26] Lee, J. H., Jun, Y. W., Yeon, S. I., Shin, J. S., & Cheon, J. (2006). Dual-mode nanoparticle probes for high-performance magnetic resonance and fluorescence imaging of neuroblastoma. *Angewandte Chemie International Edition*, *45*(48), 8160–8162.

[27] Thakur, D., Deng, S., Baldet, T., & Winter, J. O. (2009). pH sensitive CdS–iron oxide fluorescent–magnetic nanocomposites. *Nanotechnology*, *20*(48), 485601.

[28] Nikitin, M. P., Zdobnova, T. A., Lukash, S. V., Stremovskiy, O. A., & Deyev, S. M. (2010). Protein-assisted self-assembly of multifunctional nanoparticles. *Proceedings of the National Academy of Sciences*, *107*(13), 5827–5832.

[29] You, X., He, R., Gao, F., Shao, J., Pan, B., & Cui, D. (2007). Hydrophilic high-luminescent magnetic nanocomposites. *Nanotechnology*, *18*(3), 035701.

[30] Ruan, G., Vieira, G., Henighan, T., Chen, A., Thakur, D., Sooryakumar, R., & Winter, J. O. (2010). Simultaneous magnetic manipulation and fluorescent tracking of multiple individual hybrid nanostructures. *Nano Letters*, *10*(6), 2220–2224.

[31] Astete, C. E., & Sabliov, C. M. (2006). Synthesis and characterization of PLGA nanoparticles. *Journal of Biomaterials Science, Polymer Edition*, *17*(3), 247–289.

[32] Tan, Y. F., Chandrasekharan, P., Maity, D., Yong, C. X., Chuang, K. H., Zhao, Y., ... & Feng, S. S. (2011). Multimodal tumor imaging by iron oxides and quantum dots formulated in poly (lactic acid)-d-alpha-tocopheryl polyethylene glycol 1000 succinate nanoparticles. *Biomaterials*, *32*(11), 2969–2978.

[33] Beaune, G., Dubertret, B., Clément, O., Vayssettes, C., Cabuil, V., & Ménager, C. (2007). Giant vesicles containing magnetic nanoparticles and quantum dots: feasibility and tracking by fiber confocal fluorescence microscopy. *AngewandteChemie International Edition*, *46*(28), 5421–5424.

[34] Kim, J., Lee, J. E., Lee, J., Yu, J. H., Kim, B. C., An, K., ... & Hyeon, T. (2006). Magnetic fluorescent delivery vehicle using uniform mesoporous silica spheres embedded with monodisperse magnetic and semiconductor nanocrystals. *Journal of the American Chemical Society*, *128*(3), 688–689.

[35] Kwon, K. W., & Shim, M. (2005). Γ-Fe2O3/II– VI sulfide nanocrystal heterojunctions. *Journal of the American Chemical Society*, *127*(29), 10269–10275.

[36] Gao, J., Zhang, B., Gao, Y., Pan, Y., Zhang, X., & Xu, B. (2007). Fluorescent magnetic nanocrystals by sequential addition of reagents in a one-pot reaction: a simple preparation for multifunctional nanostructures. *Journal of the American Chemical Society*, *129*(39), 11928–11935.

[37] Gu, H., Zheng, R., Zhang, X., & Xu, B. (2004). Facile one-pot synthesis of bifunctional heterodimers of nanoparticles: a conjugate of quantum dot and magnetic nanoparticles. *Journal of the American Chemical Society*, *126*(18), 5664–5665.

[38] Kim, H., Achermann, M., Balet, L. P., Hollingsworth, J. A., & Klimov, V. I. (2005). Synthesis and characterization of Co/CdSe core/shell nanocomposites: bifunctional magnetic-optical nanocrystals. *Journal of the American Chemical Society*, *127*(2), 544–546.

[39] Klasens, H. A. (1953). On the nature of fluorescent centers and traps in zinc sulfide. *Journal of The Electrochemical Society*, *100*(2), 72.

[40] Bhargava, R. N., Gallagher, D., Hong, X., & Nurmikko, A. J. P. R. L. (1994). Optical properties of manganese-doped nanocrystals of ZnS. *Physical Review Letters*, *72*(3), 416.

[41] Murase, N., Jagannathan, R., Kanematsu, Y., Watanabe, M., Kurita, A., Hirata, K., ... & Kushida, T. (1999). Fluorescence and EPR characteristics of Mn2+-doped ZnS nanocrystals prepared by aqueous colloidal method. *The Journal of Physical Chemistry B*, *103*(5), 754–760.

[42] Fainblat, R., Muckel, F., Barrows, C. J., Vlaskin, V. A., Gamelin, D. R., & Bacher, G. (2014). Valence-band mixing effects in the upper-excited-state magneto-optical responses of colloidal Mn2+-doped CdSe quantum dots. *ACS Nano*, *8*(12), 12669–12675.

[43] Wang, S., Jarrett, B. R., Kauzlarich, S. M., & Louie, A. Y. (2007). Core/shell quantum dots with high relaxivity and photoluminescence for multimodality imaging. *Journal of the American Chemical Society*, *129*(13), 3848–3856.

[44] Yi, D. K., Selvan, S. T., Lee, S. S., Papaefthymiou, G. C., Kundaliya, D., & Ying, J. Y. (2005). Silica-coated nanocomposites of magnetic nanoparticles and quantum dots. *Journal of the American Chemical Society*, *127*(14), 4990–4991.

[45] Osseo-Asare, K., & Arriagada, F. J. (1990). Preparation of SiO2 nanoparticles in a non-ionic reverse micellar system. *Colloids and Surfaces*, *50*, 321–339.

[46] Kim, J., Lee, J. E., Lee, J., Jang, Y., Kim, S. W., An, K., ... & Hyeon, T. (2006). Generalized fabrication of multifunctional nanoparticle assemblies on silica spheres. *Angewandte Chemie International Edition*, *45*(29), 4789–4793.

[47] Kim, J., Lee, J. E., Lee, J., Jang, Y., Kim, S. W., An, K., ... & Hyeon, T. (2006). Generalized fabrication of multifunctional nanoparticle assemblies on silica spheres. *Angewandte Chemie International Edition*, *45*(29), 4789–4793.

[48] Kim, J., Lee, J. E., Lee, J., Jang, Y., Kim, S. W., An, K., ... & Hyeon, T. (2006). Generalized fabrication of multifunctional nanoparticle assemblies on silica spheres. *Angewandte Chemie International Edition*, *45*(29), 4789–4793.

[49] Kim, J., Lee, J. E., Lee, J., Jang, Y., Kim, S. W., An, K., ... & Hyeon, T. (2006). Generalized fabrication of multifunctional nanoparticle assemblies on silica spheres. *Angewandte Chemie International Edition*, *45*(29), 4789–4793.

[50] Mulder, W. J., Strijkers, G. J., van Tilborg, G. A., Griffioen, A. W., & Nicolay, K. (2006). Lipid-based nanoparticles for contrast-enhanced MRI and molecular imaging. *NMR in Biomedicine: An International Journal Devoted to the Development and Application of Magnetic Resonance In vivo*, *19*(1), 142–164.

[51] Prinzen, L., Miserus, R. J. J., Dirksen, A., Hackeng, T. M., Deckers, N., Bitsch, N. J., ... & Reutelingsperger, C. P. (2007). Optical and magnetic resonance imaging of cell death and platelet activation using annexin A5-functionalized quantum dots. *Nano Letters*, *7*(1), 93–100.

[52] Weinstein, J. S., Varallyay, C. G., Dosa, E., Gahramanov, S., Hamilton, B., Rooney, W. D., ... & Neuwelt, E. A. (2010). Superparamagnetic iron oxide nanoparticles: diagnostic magnetic resonance imaging and potential therapeutic applications in neurooncology and central nervous system inflammatory pathologies, a review. *Journal of Cerebral Blood Flow & Metabolism*, *30*(1), 15–35.

[53] Prinzen, L., Miserus, R. J. J., Dirksen, A., Hackeng, T. M., Deckers, N., Bitsch, N. J., ... & Reutelingsperger, C. P. (2007). Optical and magnetic resonance imaging of cell death and platelet activation using annexin A5-functionalized quantum dots. *Nano Letters*, *7*(1), 93–100.

[54] Yi, D. K., Selvan, S. T., Lee, S. S., Papaefthymiou, G. C., Kundaliya, D., & Ying, J. Y. (2005). Silica-coated nanocomposites of magnetic nanoparticles and quantum dots. *Journal of the American Chemical Society*, *127*(14), 4990–4991.

[55] Park, J. H., von Maltzahn, G., Ruoslahti, E., Bhatia, S. N., & Sailor, M. J. (2008). Micellar hybrid nanoparticles for simultaneous magnetofluorescent imaging and drug delivery. *Angewandte Chemie*, *120*(38), 7394–7398.

[56] Zhang, X., Zeng, Z., Liu, H., Xu, L., Sun, X., Xu, J., & Song, G. (2022). Recent development of a magneto-optical nanoplatform for multimodality imaging of pancreatic ductal adenocarcinoma. *Nanoscale*, *14*(9), 3306–3323.

[57] Rusling, J. F., Kumar, C. V., Gutkind, J. S., & Patel, V. (2010). Measurement of biomarker proteins for point-of-care early detection and monitoring of cancer. *Analyst*, *135*(10), 2496–2511.

[58] Son, A., Dosev, D., Nichkova, M., Ma, Z., Kennedy, I. M., Scow, K. M., & Hristova, K. R. (2007). Quantitative DNA hybridization in solution using magnetic/luminescent core–shell nanoparticles. *Analytical Biochemistry*, *370*(2), 186–194.

[59] Song, E. Q., Wang, G. P., Xie, H. Y., Zhang, Z. L., Hu, J., Peng, J., ... & Pang, D. W. (2007). Visual recognition and efficient isolation of apoptotic cells with fluorescent-magnetic-biotargeting multifunctional nanospheres. *Clinical Chemistry*, *53*(12), 2177–2185.

[60] Maeda, H., Sawa, T., & Konno, T. (2001). Mechanism of tumor-targeted delivery of macromolecular drugs, including the EPR effect in solid tumor and clinical overview of the prototype polymeric drug SMANCS. *Journal of Controlled Release*, *74*(1-3), 47–61.

[61] Kim, J. E. L. J., Lee, J. E., Lee, S. H., Yu, J. H., Lee, J. H., Park, T. G., & Hyeon, T. (2008). Designed fabrication of a multifunctional polymer nanomedical platform for simultaneous cancer-targeted imaging and magnetically guided drug delivery. *Advanced Materials*, *20*(3), 478–483.

[62] Li, L., Chen, D., Zhang, Y., Deng, Z., Ren, X., Meng, X., ... & Zhang, L. (2007). Magnetic and fluorescent multifunctional chitosan nanoparticles as a smart drug delivery system. *Nanotechnology*, *18*(40), 405102.

[63] Cho, H. S., Dong, Z., Pauletti, G. M., Zhang, J., Xu, H., Gu, H., ... & Shi, D. (2010). Fluorescent, superparamagnetic nanospheres for drug storage, targeting, and imaging: a multifunctional nanocarrier system for cancer diagnosis and treatment. *ACS Nano*, *4*(9), 5398–5404.

[64] Cho, H. S., Dong, Z., Pauletti, G. M., Zhang, J., Xu, H., Gu, H., ... & Shi, D. (2010). Fluorescent, superparamagnetic nanospheres for drug storage, targeting, and imaging: a multifunctional nanocarrier system for cancer diagnosis and treatment. *ACS Nano*, *4*(9), 5398–5404.

[65] Huang, X., Peng, X., Wang, Y., Wang, Y., Shin, D. M., El-Sayed, M. A., &Nie, S. (2010). A reexamination of active and passive tumor targeting by using rod-shaped gold nanocrystals and covalently conjugated peptide ligands. *ACS Nano*, *4*(10), 5887–5896.

[66] Smith, J. E., Medley, C. D., Tang, Z., Shangguan, D., Lofton, C., & Tan, W. (2007). Aptamer-conjugated nanoparticles for the collection and detection of multiple cancer cells. *Analytical Chemistry*, *79*(8), 3075–3082.

[67] Zheng, D., Seferos, D. S., Giljohann, D. A., Patel, P. C., &Mirkin, C. A. (2009). Aptamer nano-flares for molecular detection in living cells. *Nano Letters*, *9*(9), 3258–3261.

7 Magnetic Quantum Dots for Magnetic Resonance Imaging (MRI) and Biomedical Applications

Pranita Rananaware and Varsha Brahmkhatri
Centre for Nano and Material Sciences, Jain University, Jain Global Campus, Bengaluru, Karnataka, India

CONTENTS

7.1 INTRODUCTION

The nanosized semiconductor nanocrystals with optoelectronic properties are named Quantum Dots (QDs) [1]. Due to the advantages of QDs, their use in the biomedical field have been increased tremendously. In another way, they also can be stated as nanocrystals composed of the core of semiconductor material, enclosed within a shell of another semiconductor having a larger spectral band gap. QDs are composed of groups II-VI or III-V elements [2, 3]. QDs show several exceptional optical properties like broad absorption with narrow symmetric photoluminescence spectra, a high quantum yield, and high resistance to photobleaching and chemical degradation. Additionally, surface modification can be achieved easily on QDs for pharmacokinetics and bio applicability [4, 5].

QDs are inorganic fluorophores that have size-tunable emission, strong light absorbance, bright fluorescence, narrow symmetric emission bands, and high photostability. Having a size from 1nm to 10nm, it provides adaptive photoluminescence (PL) as a result of quantum confinement effects that are associated with the control of nanocrystal size and composition [6]. QDs have characteristics that make them special for applications in spectroscopy which embrace the narrow, symmetrical emission

DOI: 10.1201/9781003319870-7

bands, and stock shifts that are greater than 200 nm. The brightness of QDs provides high sensitivity in the development of labels and assays. Collectively all these significant features make it viable to track the multiplexed analyses and imaging mythologies [7, 8].

The sustainability of photobleaching and bright emission makes QDs significantly important for in-vivo applications. Gratifyingly QDs have better properties compared to the fluorophores such as quantum yield, photostability, blinking, size scale, bioconjugation, and chemical properties. Subsequently, due to its unique properties, its usage has been demonstrated in various fields such as bioimaging, biosensing, and drug delivery [9]. Additionally, the QDs can be fabricated in such a way that they can show the (super)paramagnetic properties and can be used for magnetic resonance imaging (MRI) [10]. Using MRI, the 3D images of soft tissues can be observed along with the functional and metabolic parameters [11]. QDs were firstly discovered by Alexei Ekimov in the 1980s [12]. For MRI enhancement, IA labelled with Gd-chelates was synthesized and studied as an intravascular, blood pool-enhancing agent [13]. Combinatory probes, which have both fluorescence and (super)paramagnetic capabilities, offer considerable advantages due to the complementary qualities of optical techniques and MRI. Therefore, various research groups have recently reported (super)paramagnetic QDs [14]. Magnetic materials are one of the most common materials combined with QDs like ions (e.g., gadolinium (Gd) [15], and nanoparticles (e.g., superparamagnetic iron oxide nanoparticles known as SPIONs [16, 17]. The fluorescent property of the QDs permits visualization, whereas the magnetic property of the composite enables imaging, and magnetic separation,

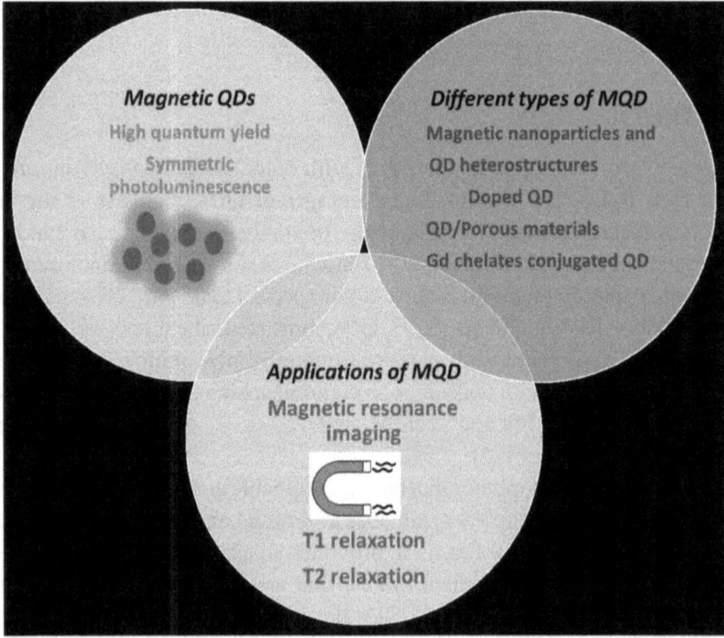

FIGURE 7.1 Different types of MQDs and their applications in MRI.

and may even have therapeutic benefits [18]. The various MQDs and synthesis strategies and applications in MRI are summarized in Figure 7.1 which gives the overall insides of this chapter.

7.2 DIFFERENT TYPES OF MQDs

The synthesis and application of contrast agent for MRI using magnetic properties of metal, metal oxides, and metal ions combined with the luminescent properties of QD may give rise to the highly efficient and attractive approach to the synthesis of the MQDs. These MQDs can be used as a contrast agent for MRI and fluorescence imaging very effectively. MCQDs are synthesized using different approaches: (1) Magnetic nanoparticle (MNP)/QDs heterostructures, where different types of magnetic nanoparticles like iron oxide are used as a core and semiconductor QDs decorated on the core to form a core shell structure; (2) doped QD is a method where QDs are doped with the magnetic transition metal ions, ions doped onto the surface or inside the QDs structure. Doping can create defects due to which the activity may increase; (3) QDs and magnetic nanoparticles with porous materials, in this type of highly porous material, are used to synthesize a heterostructure combining the magnetic entities and semiconductor QDs, the purpose is to use the heterostructure as a theranostic material; (4) This synthesis strategy, QDs coated or encapsulated with organic complexes like chelated gadolinium (Gd chelates).

7.3 MNP/QDS HETEROSTRUCTURES FOR MRI

Due to the long emission wavelength of QDs which could extend to the NIR region, they can be a promising material and diagnostic tool in bioimaging. This section covers magnetic nanoparticles (MNP) and QD combined to form either a core/shell or heterostructure. Magnetic nanoparticles like Fe_3O_4 are either capped with the fluorescent QD or the magnetic nanoparticle/QDs heterostructures formed by the one step in situ synthesis method [19, 20]. The various attempts for synthesizing the same are described here.

In 2010 Fan et al. synthesized the magnetic fluorescent nanoprobe using quantum dot (CdSe/ZnS) capped magnetite nanoring (NRs-Fe_3O_4 NRs) with high luminescence and magnetic vortex core. Where they observed that synthesized nanoprobe shows a much stronger magnetic resonance (MR) T2* effect where the r2* relaxivity and r2*/r1 ratios are 4 times and 110 times respectively larger than those of commercial superparamagnetic iron oxide [21].

In the year 2014 Ye et al. synthesized PLGA (poly lactic-co-glycolic acid) vesicles fabricated using encapsulation of superparamagnetic iron oxide nanoparticles (SPION), manganese-doped zinc sulfide (Mn: ZnS) quantum dots (QDs) and the anticancer drug busulfan. Here the authors aimed to study the in-vivo and in-vitro imaging to study the cellular uptake as well as biodistribution of polymeric nanoparticles by magnetic resonance imaging (MRI) as shown in Figure 7.2A. In-vitro studies observed the transverse relaxivity (r2) is obtained as the slope of linear fitting for the relaxation rate at different iron concentrations, calculated as 523 $mM^{-1}S^{-1}$ Fe for PLGAeSPIONeMn: ZnS vesicles [22].

In 2018 Li et al. synthesized the Fe_3O_4@chitosan graphene QDs (Fe_3O_4@CS-GQD) for fluorescence and magnetic resonance imaging. The T2 weighted MRI of Fe_3O_4@CS-GQD was taken at a 3T clinical MRI scanner [23].

Seleci et al. synthesized the transferrin (Tf)-decorated niosomes integrated with magnetic iron oxide nanoparticles (MIONs), and QDs termed as (PEGNIO/QDs/MIONs/Tf). PEGNIO/QDs/MIONs/Tf nanocomposite was used for the

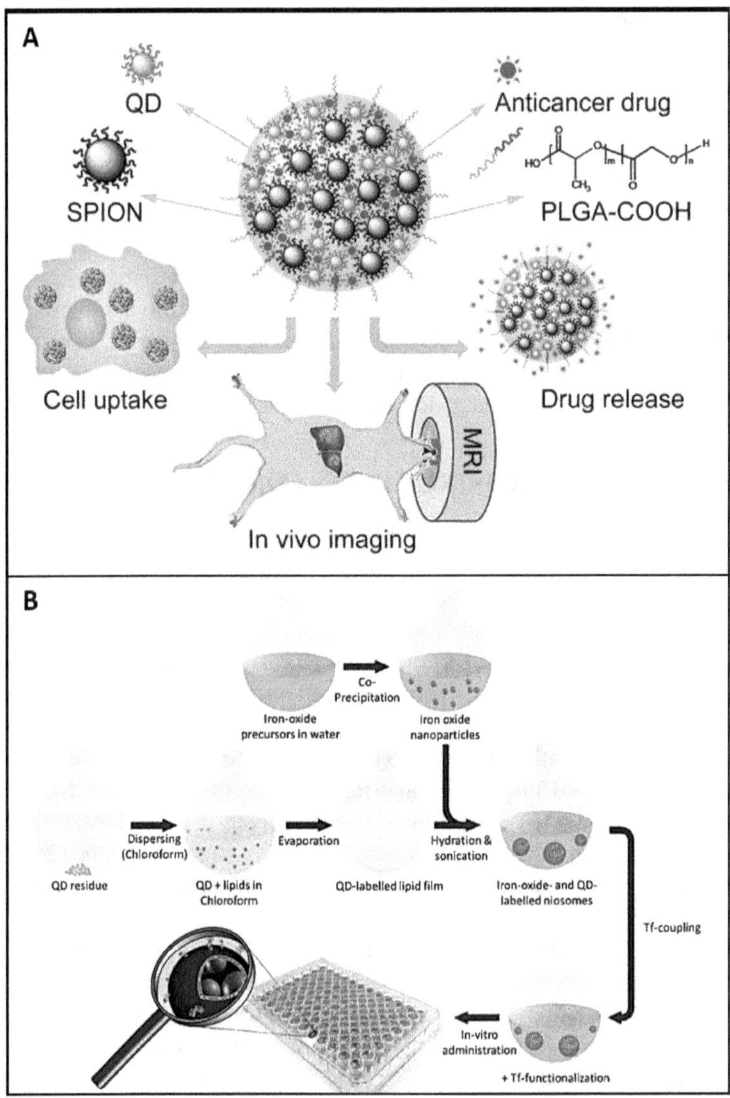

FIGURE 7.2 A) Schematic illustration of the composition of PLGAeSPIONeMn: ZnS nanoparticles and their multiple applications [22]. B) Schematic representation of the PEGNIO/QDs/MIONs/Tf synthesis and application [24].

in-vitro dual imaging of glioma by Magnetic resonance imaging and fluorescence imaging as shown in Figure 7.2B. In the synthesis protocol, the magnetic iron oxide nanoparticles (MIONs) and InP/ZnS QDs were synthesized first and then integrated into the niosomes, and this is the first report to do so. The results suggested that the PEGNIO/QDs/MIONs/Tf shows less cytotoxicity and produces the negative contrast enhancement effect on glioma cells by magnetic resonance imaging and improved fluorescence intensity under fluorescence microscopy [24].

In 2015 Li et al. synthesized Fe_3O_4-Carbon Dots (Fe_3O_4-CDs) Hybrid NPs by one-step hydrothermal synthesis method and studied its application in dual model imaging like magnetic resonance imaging (MRI) and fluorescent imaging. The prepared (Fe_3O_4-CDs) show excellent superparamagnetic properties (Ms = 56.8 emu g^{-1}) for MRI, and also excitation-independent photoluminescent for down-conversion and up-conversion at 445 nm for fluorescent imaging. Here MRI tests were performed on a 7.0 T MRI scanner, Fe_3O_4 -CDs NPs were dispersed in deionized water with various concentrations for T2 -weighted images to calculate R2 relaxation. The r2 value of the Fe_3O_4 -CDs NPs was r2 = 0.02588 mL $\mu g^{-1} s^{-1}$ and suggest that as-prepared Fe_3O_4 -CDs NPs could be used as a T2 contrast agent in MRI [25].

Recently in 2022 Hassan et al. synthesized the heterostructure chitosan-coated magnetic iron oxide/graphene quantum dots (MGC) and loaded them with 5-fluorouracil. This synthesized MGC-FU was used for the bimodal magnetic resonance/fluorescence imaging. The results shows that for MRI a high value of transverse relaxivity was observed (r2 = 690 mM^{-1} s^{-1}) and strong fluorescence emission at 493 nm wherein loading of FU is useful for cancer treatment [26].

Similarly, Chen et al. synthesized aptamer-modified graphene quantum dots (GQDs) loaded onto the chitosan-coated Fe_3O_4 (DOX-Fe_3O_4@CS@GQD-Apt (DOX-Fe_3O_4@CGA), and loaded with the photothermal and chemotherapy. They used the 3.0 T clinical MR scanner to study the MRI of Fe_3O_4@CGA, and observed relaxation rate (r2) of Fe_3O_4@CGA was calculated to be 16.70 m/M/S [27].

Magnetic nanoparticle/QDs heterostructures show both strong magnetic as well as fluorescent properties and can be used for dual-modality bioimaging. Although MQDs can be used as both positive as well as negative MRI contrast agents, the capping or conjugating with QDs can reduce the magnetic property of magnetic nanoparticles which is a matter of concern. Hence, these aspects need to be very well considered while synthesizing new MQDs by integrating them with magnetic nanoparticles.

7.4 DOPED QDs FOR MRI

In this section, we have explained the MQDs synthesized by doping paramagnetic luminescent ions into QDs. Doping of the semiconductor QDs with paramagnetic transition metal ions is a yet attractive approach toward fluorescent/paramagnetic dual-modal probes. This method is useful gives a synthesis method where fluorescence and magnetic properties can be incorporated into a single nanoparticle. Doped QDs having luminescent and ferromagnetic can be used as a combined fluorescence and MRI probe [28, 29].

In 2013 Saha et al. synthesized Fe doped CdTeS MQDs by a simple solvothermal method and then capped them with N-Acetyl-Cysteine (NAC) ligands, containing thiol and carboxylic acid functional groups to give stable aqueous dispersion. The synthesized nanoparticle shows a size range of around 3-6 nm and is water-dispersible. T2 weighted MR images of Fe dope were d MQDs studied at concentrations of 6.8, 3.4, 1.7, 0.8, 0.4, 0.2- and 0-mM shows that higher Fe concentration produced larger signal intensity decrements, and relaxivity of the Fe doped MQDs (738 nm) at 4.7 T was determined to be 3.6 mM^{-1} S^{-1}. Therefore, MRI results suggest that magnetic MQDs can be used as effective T2 contrast agents and also can be used for cancer diagnosis and therapy [30].

In 2020 Huang et al. synthesized a bimodal imaging nanoprobe (Gd-CQDs@N-Fe_3O_4) comprising fluorescence ability of carbon quantum dots (CQDs), T1 and T2 contrast-enhancing functionality owing to Gd (III) ions, and Fe_3O_4 nanoparticles as shown in Figure 7.3A. In-vitro longitudinal relaxation times (T1) and transverse relaxation time (T2) studies were performed for the different dilutions of Gd-CQDs@N- Fe_3O_4 in a 1% aqueous solution of agarose and were measured on a 0.5 T MRI instrument. XPS data revealed the high binding energy from Gd 4d5/2 and Gd 4d3/2 lines at 142.6 and 148.1 eV and the XPS of the Gd^3d spectrum displayed the Gd 3d5/2 line at 1187.0 eV which may contribute in MRI. Gd-CQDs@N- Fe_3O_4 nanoparticles exhibit superparamagnetic nature and display properties as contrast agents with r1 and r2 relaxivities of 5.16 and 115.6 $mM^{-1}s^{-1}$ respectively [31].

Sun et al. synthesized Manganese doped carbon dots denoted as (Mn-CDs) using a one-step solvothermal method. Uniform-sized Mn-CDs having a size of around 5 nm shows great properties like redshifted orange emission and enhanced longitudinal relaxation, and are applied for fluorescence and MR imaging. The T1-weighted MR imaging of Mn-CDs was studied at different concentrations and observed MR signal got brighter by increasing concentration., The Mn-CDs exhibit a longitudinal MR relaxation rate of 12.69 mM^{-1} S^{-1}. The high relaxation rate could be observed due to the movement limit of Mn^{2+} in Mn-CDs. The Mn^{2+} ions could restrict their escape into the blood and also slow their relaxation speed. In the in-vivo MR imaging, after injecting Mn-CDs into the tumor, the MR signal became bright and obvious fluorescence was observed. The results demonstrated the prepared Mn-CDs could be safe and efficient dual-modal nanoprobes for in-vivo MR and fluorescence imaging [32].

In 2019 Wang et al. synthesized an efficient MRI contrast agent using a simple microwave-assisted thermal decomposition method based on a stable Fe (III) complex of fluorine and nitrogen co-doped carbon dots (F, N-CDs), prepared from glucose and levofloxacin. They introduced heteroatoms such as fluorine (F) and nitrogen (N) into CDs simultaneously to enhance the fluorescence properties of CDs. The reason behind this is as F and N atoms with lone pair of electrons facilitate coordination bonds with Fe^{3+} cations as shown in Figure 7.3B. The MRI experiments were carried out using HeLa cells, in vitro. The comparison studies shows that Fe^{3+}@F,N-CD complex exhibits higher longitudinal relaxivity (r1 = 5.79 mM^{-1}·S^{-1}) than that of the control samples of the Fe^{3+}@CD complex (r1 = 4.23 mM^{-1}·S^{-1}) and free Fe^{3+} (r1 = 1.59 mM^{-1}·S^{-1})[33].

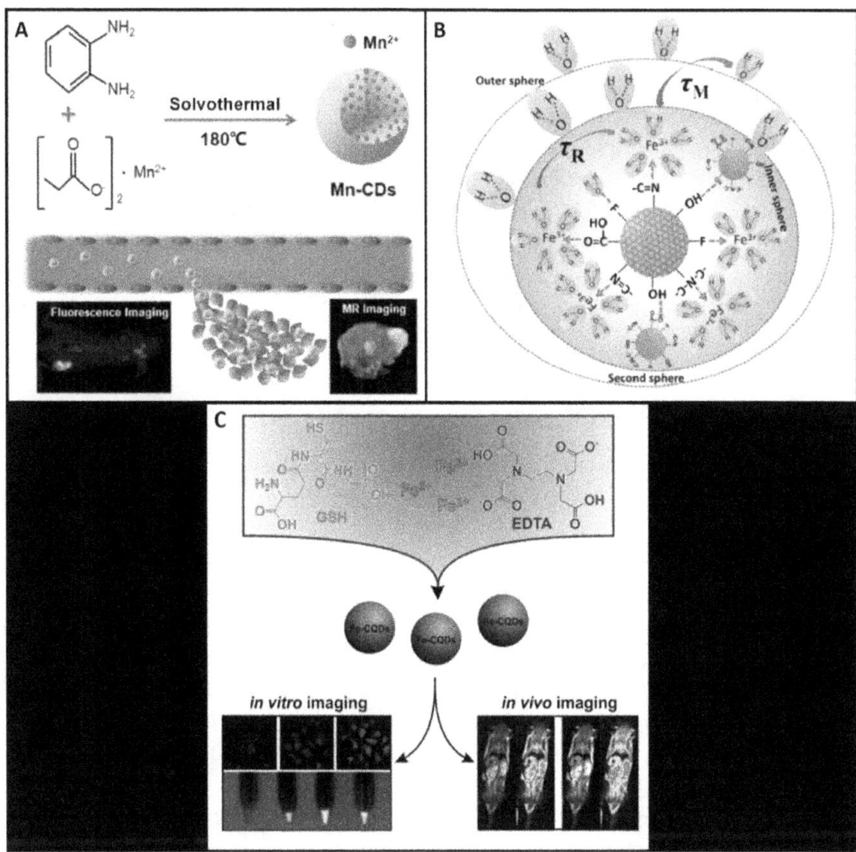

FIGURE 7.3 A) Schematic Diagram of Mn-CDs as Dual-Modal Nanoprobes for Fluorescence and MR Imaging [32]. B) The possible mechanism of the Fe3+@F, NCD-enhanced MRI effect [33]. C) synthesis of Fe-CQDs and applications to magnetic resonance and fluorescent dual-mode imaging [34].

In 2019 Huang et al. synthesized imaging nanoprobes consisting of iron-doped carbon quantum dots (Fe-CQDs) for use in dual-mode imaging, magnetic resonance (MR), and fluorescent imaging as shown in Figure 7.3C. In-vitro analysis of the T1 and T2 relaxivity was carried out using a clinical 3 T MR system, using different concentrations of Fe-CQD (0, 62.5, 125, and 250 $\mu g{\cdot}mL^{-1}$) in the normal water for the analysis in A549 cells. T1-weighted MR images of Fe-CQDs produced a positive contrast signal which got brighter with the increase of iron content. T2-weighted MR images became darker with the increase of iron content. Fe-CQDs produced a high longitudinal relaxivity of 3.92 $mM^{-1}{\cdot}s^{-1}$ and a low transverse relaxivity of 4.99 $mM^{-1}{\cdot}s^{-1}$. In-vivo MRI experiments were carried out using six-week-old female Balb/c mice which were initially anesthetized with 5% isoflurane. The concentration of Fe-CQDs solution used was 5 mg $Fe{\cdot}kg^{-1}$and injected via tail vein. From *in-vivo* studies, it was observed that Fe-CQDs can be effectively used as a T1-weighted contrast agent. Subsequently synthesized Fe-CQDs exhibit advantages like high contrast efficiency, biocompatibility, and good optical stability [34].

The design and synthesis strategies of various magnetically doped MQDs demonstrate that factors like dopant location, doping level, and concentration of the dopants per QD are very important parameters. These parameters affect the photoluminescence and magnetic properties of MQDs. Magnetically doped QD shows a very small size range and tenable and strong emission, and T1 contrast enhancement. The small size of magnetic QD exhibits fast excretion of intravenously administered probes, which results in reducing possible side effects, as the small size of MQDs can reduce the nonspecific retention.

7.5 QDs WITH POROUS MATERIAL NANOPARTICLES FOR MRI

An additional approach is the use of a carrier material to create a composite particle, where magnetic and semiconductor nanocrystals can be integrated onto the highly porous material. The semiconductor QDs and MNPs (Fe_3O_4) are attached or incorporated into the highly porous carrier particle such as silica and polymer nanoparticles [35, 36].

In 2018 Zhou et al. synthesized Mn-doped QDs, especially Mn-doped ZnS (ZnSe) QDs, and decorated them onto the mesoporous silica nanoparticles (MSNs) for Fluorescence/MRI dual-modal bioimaging as shown in Figure 7.4A. The amine-functionalized large pore MSNs were biocompatible and have large porous volumes attesting as a suitable carrier for QDs. The carboxyl-functionalized Mn-doped QDs were loaded into the pores of MSNs via EDC/NHS coupling. Mn-doped QDs can enhance the Mn^{2+} concentration which leads to improved MRI sensitivity. The in-vitro MRI images for longitudinal relaxation time (T1) were measured at different nanoparticle concentrations. In-vivo MR imaging of QDs and MSN@QD (31.2 mg mL^{-1}) studied on the Balb/c mice Mn-doped ZnSe QDs exhibited enhanced fluorescence brightness and magnetic signal for studying both in-vitro and in-vivo imaging as compared with the single QDs [37].

Su et al. prepared a nanoprobe using novel graphene quantum dot (GQD)@ Fe_3O_4@SiO_2 carboxyl-terminated GQD (C-GQD) was firstly conjugated with Fe_3O_4@ SiO_2 and then functionalized with cancer-targeting molecule folic acid (FA). The synthesized nanoprobe was studied for the targeted drug delivery, sensing, dual-modal imaging, and therapy. They have studied in-vitro, MRI, and fluorescence imaging of living Hela cells. T2 weighted MRI images were obtained in a clinical MR scanner at different concentrations of Fe 0.00, 0.04, 0.08, 0.12, and 0.16 mM, and for the cell experiments. So the results suggested that relaxation rate and Fe concentrations with a high relaxation value of 62.8 mM^{-1} S^{-1}, indicating the as-prepared Fe_3O_4@SiO_2@ GQD-FA can be used as efficient T2-weighted MRI [38].

In 2018 Zhang et al. synthesized multifunctional SWCNTs-PEG-Fe_3O_4@CQDs nanocarriers where PEG 2000N modified Fe_3O_4@carbon quantum dots (CQDs) coated with the single-walled carbon nanotubes (SWNTs). These nanocarriers are used as dual-modal targeted imaging and chemo/photodynamic/photothermal triple-modal therapeutic agents. In the properties best for magnetic resonance, imaging SWCNTs-PEG-Fe_3O_4@CQDs showed the saturation magnetization of 13.2 emu/g and from the hysteresis loop, it is confirmed that SWCNTs-PEG- Fe_3O_4@CQDs showed ferromagnetic structure. The synthesized nanocarrier was used as the T2 contrast agent using deionized water as a medium using different concentrations of the material. As the concentration of SWCNTs-PEG- Fe_3O_4@CQD increases increased contrast enhancement of MR signal was observed. The T2 relaxation time of the SWCNTs-PEG- Fe_3O_4 @CQDs from MRI from the slope of the relaxation time was determined as 16.419 mM^{-1} S^{-1} [39].

Recently in 2022 Teng and his coworkers synthesized a novel multifunctional approach comprising the graphene quantum dots (GQDs) with cobalt ferrite ($CoFe_2O_4$) loaded on the surface of porous SiO_2 and this nanocomposite is functionalized with folic acid and loaded with anticancer drug DOX demonstrated as DOX/GQD-CFO@SiO_2/FA. The synthesized nanocomposite were investigated for drug delivery and dual model imaging like fluorescence/MRI. The advantage of this system is DOX/GQD-CFO@SiO_2/FA monitoring the drug delivery system by detecting the FRET signals. GQD-CFO@SiO_2/FA was used as the biomarker to track the targeted tumor [40].

7.6 GD CHELATES CONJUGATED QD FOR MRI

Gd-loaded nanoparticles (GdNPs) are considered as one of the most promising theranostic agents, used for diagnostics (molecular imaging) as well therapeutics (molecular therapy) functions in a single platform. Also, Gd chelates, such as Gd-diethylenetriaminepentaacetic acid (Gd-DTPA), are studied to prevent unwanted toxicity induced by free Gd ions [41, 42].

In 2019 Pereira et al. developed a bimodal nano system comprising hydrophilic (QDs) directly conjugated to Gd (III)-DO3A monoamide chelates, Gd (III)-DO3A is the derivative of Gd (III)-DOTA. The advantage of this particulate chelate towards MRI is it improves the local concentration of paramagnetic chelates which leads to the improvement in the contrast, and high relativities. Here the relaxometry properties of QDs-Gd (III) chelates were calculated by measuring longitudinal (T1) and/or transverse (T2) relaxation times on the Bruker Minispec mq60 (1.5T, 37 °C) and mq20 (0.49T, at 25 and 37 °C) relaxometers. In comparison with the reported studies, this nanosystems shows interesting advantages, such as a direct conjugation approach with stable Gd(III)-DOTA derivative chelates on as-prepared hydrophilic CdTe QD surfaces, very intense fluorescence, and effective longitudinal relaxivity per QD and per paramagnetic ion, at least five times [per Gd(III)] and 100 times (per QD) higher than the r1 for Gd(III)-DOTA chelates [43].

Yang et al. synthesized a bimodal contrast nano agent by chelating Gd ion to the DTDTPA modified CuInS2/ZnS QDs as shown in Figure 7.4B. QDs@DTDTPA-Gd shows the longitudinal relaxivity r1 value of 9.91 $mM^{-1}s^{-1}$, which was 2.5 times as high as that of clinically-approved Gd-DTPA (3.9 $mM^{-1}s^{-1}$)[44].

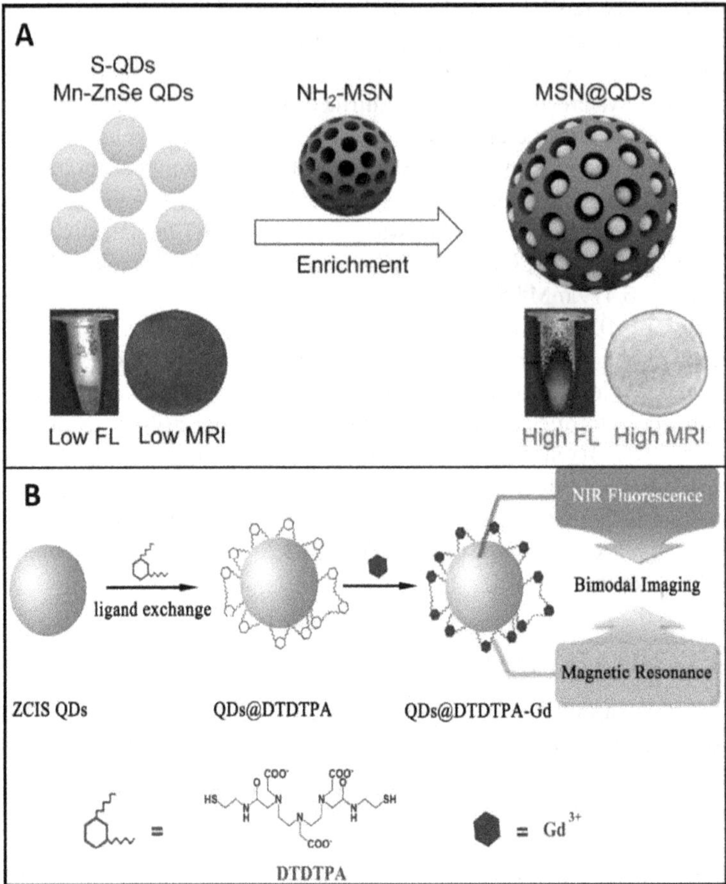

FIGURE 7.4 A) Mn-doped ZnSe QDs onto Mesoporous silica nanoparticles dual-modal bio-imaging [37]. B) Fabrication procedure and functional description of QDs@DTDTPA-Gd NPs [44].

Though the Gd chelates exhibit numerous advantages they also show some disadvantages. These problems can be minimized by using low molecular masses (around 500Da) Gd chelates, that can be rapidly cleared from the intravascular space after injection. Gd chelates show non-specific biodistribution and rapid renal clearance which results in short diagnostic time, which can be remedied by using Gd chelates with effective MQDs. The use of porous material with MQDs can enhance the applicability of nanocomposites which can be used as theranostic nanocomposites for diagnosis as well as drug delivery applications. The various magnetic QD systems showing potential applications in MRI are summarized in Table 7.1. MQDs can be used for MRI as well as fluorescent imaging. As most of the QDs are used for dual-modal biomedical imaging due to the magnetic and fluorescent properties, magnetic QD used for T1 or T2 imaging and in vivo or in vitro are depicted in Table 7.1.

TABLE 7.1
Magnetic Quantum Dots and Their Applications in MRI

Sr No	MQDs	Application	MRI	Reference
1	Fe_3O_4-Carbon Dots	Dual modal MRI and fluorescent imaging T2 imaging In vitro	In-vitro T2 imaging	[25]
2	PLGAeSPIONeMn: ZnS	MRI	In-vitro T2 imaging	[22]
3	(GQD)@Fe_3O_4@SiO_2	MRI and fluorescence imaging	In-vitro T2-weighted MRI	[38]
4	SWCNTs-PEG-Fe_3O_4@ CQDs	Dual modal MRI and fluorescent imaging	In-vitro T2-weighted MRI	[39]
5	CdSe/ZnS@ Fe_3O_4 nanoring's	Fluorescence and Magnetic Resonance Imaging	In-vitro T1, T2-weighted MRI	[21]
6	Fe_3O_4@chitosan graphene QDs (Fe_3O_4@CS-GQD)	Fluorescence and Magnetic Resonance Imaging	In-vitro	[23]
7	PEGNIO/QDs/ MIONs/Tf	Fluorescence and Magnetic Resonance Imaging	In-vitro T2-weighted MRI	[45]
8	Paramagnetic graphene quantum dots (PGQD)	tumor-targeted Theranostics Fluorescence and Magnetic Resonance Imaging	in-vitro T1 weighted MRI	[46]
9	$CuInS_2$ /ZnS Quantum Dots Conjugating Gd (III) Chelate	Fluorescence and Magnetic Resonance Bimodal Imaging	in-vivo and in-vitro T1 weighted MRI	[44]
10	Manganese-doped carbon QDs	fluorometric and magnetic resonance (dual mode) bioimaging and biosensing	in-vitro T1 weighted MRI	[47]
11	Iron (II)-doped carbon dots	magnetic resonance (MR) and fluorescent (dual-mode) imaging nanoprobes	In-vivo and in-vitro T1 weighted MRI	[34]
12	Paramagnetic, Silicon QDs	Magnetic Resonance and Two-Photon Imaging	In-vitro T1 weighted MRI	[48]
13	Magnetically engineered Cd-free QDs	fluorescence/magnetic resonance imaging of tumors	In-vivo and in-vitro T1 weighted MRI	[49]
14	Fe Doped CdTeS MQDs	Bioimaging	In-vitro T2 weighted MRI	[50]
15	Fluorine and Nitrogen Co-Doped Carbon Dot Complexation with Fe(III)	Magnetic Resonance Imaging	In-vitro T1 weighted MRI	[33]

7.7 CONCLUSION AND FUTURE PERSPECTIVES

MQDs have attracted tremendous interest over the years for different biomedical imaging techniques and cancer treatments. Consequently, magnetic QD with various functionalities and with reduced toxicity can meet various requirements necessitated for biomedical applications. In most magnetic quantum dot-based imaging systems, dual-modal imaging has been performed including fluorescence imaging and magnetic resonance imaging. Eventually the enhanced structural and chemical properties of QDs enhance the contrast imaging in MRI for better resolution.

In the future perspective, the reduction of cytotoxic effects caused due to the semiconductor QD as well as magnetic entities need to be reduced, and modification in the heterostructure by surface modification. As new theranostic approaches are being studied by encapsulating/loading targeting agent/anticancer drugs the magnetization properties of magnetic nanoparticles are getting affected. Hence, for MRI the magnetization effect of MQD is the most important aspect so after loading of anticancer drug and modification of surface the final nanocomposite should not lose its magnetization. Hence, futuristic studies o magnetic QDs should be focused on the synthesis design and strategies as well as on the toxicity effects.

Acknowledgements: We are thankful to Jain University, Bangalore, India, for providing facilities. VB also acknowledges TARE-SERB File NO: TAR/2018/000547.

REFERENCES

1. Chen, F., et al., *Graphene quantum dots in biomedical applications: Recent advances and future challenges.* Frontiers in Laboratory Medicine, 2017. 1(4): p. 192–199.
2. Wagner, A.M., et al., *Quantum dots in biomedical applications.* Acta Biomaterialia, 2019. 94 : 44–63.
3. Azzazy, H.M.E., M.M.H. Mansour, and S.C. Kazmierczak, *From diagnostics to therapy: Prospects of quantum dots.* Clinical Biochemistry, 2007. 40(13): 917–927.
4. Chan, W.C., and S. Nie, *Quantum dot bioconjugates for ultrasensitive nonisotopic detection.* Science, 1998. 281(5385): 2016–8.
5. Wang, C., X. Gao, and X. Su, *In vitro and in vivo imaging with quantum dots.* Analytical and Bioanalytical Chemistry, 2010. 397(4): 1397–1415.
6. Zhao, P., et al., *Near infrared quantum dots in biomedical applications: current status and future perspective.* WIREs Nanomedicine and Nanobiotechnology, 2018. 10(3): e1483.
7. Medintz, I.L., H. Mattoussi, and A.R. Clapp, *Potential clinical applications of quantum dots.* International Journal of Nanomedicine, 2008. 3(2): 151–167.
8. Gil, H.M., et al., *NIR-quantum dots in biomedical imaging and their future.* iScience, 2021. 24(3): 102189.
9. Bajwa, N., et al., *Pharmaceutical and biomedical applications of quantum dots.* Artificial Cells, Nanomedicine, and Biotechnology, 2016. 44(3): 758–768.
10. Tavares, A.J., et al., *Quantum dots as contrast agents for in vivo tumor imaging: progress and issues.* Anal Bioanal Chem, 2011. 399(7): 2331–42.
11. Jing, L., et al., *Magnetically Engineered Semiconductor Quantum Dots as Multimodal Imaging Probes.* Advanced Materials, 2014. 26(37): 6367–6386.

12. Pohanka, M., *Quantum dots in the therapy: current trends and perspectives.* Mini Rev Med Chem, 2017. 17(8): 650–656.
13. Schmiedl, U., et al., *Albumin labeled with Gd-DTPA as an intravascular, blood pool-enhancing agent for MR imaging: biodistribution and imaging studies.* Radiology, 1987. 162(1): 205–210.
14. Gupta, A.K., et al., *Recent advances on surface engineering of magnetic iron oxide nanoparticles and their biomedical applications.* Nanomedicine, 2007. 2(1): p. 23–39.
15. Liu, Y. and N. Zhang, *Gadolinium loaded nanoparticles in theranostic magnetic resonance imaging.* Biomaterials, 2012. 33(21): 5363–75.
16. Avasthi, A., et al., *Magnetic Nanoparticles as MRI Contrast Agents.* Top Curr Chem (Cham), 2020. 378(3): 40.
17. Rümenapp, C., B. Gleich, and A. Haase, *Magnetic nanoparticles in magnetic resonance imaging and diagnostics.* Pharm Res, 2012. 29(5): 1165–79.
18. Mahajan, K.D., et al., *Magnetic quantum dots in biotechnogy—synthesis and applications.* Biotechnol J, 2013. 8(12): 1424–34.
19. Gao, J., et al., *Fluorescent magnetic nanocrystals by sequential addition of reagents in a one-pot reaction: A simple preparation for multifunctional nanostructures.* Journal of the American Chemical Society, 2007. 129(39): 11928–11935.
20. Gu, H., et al., *Facile one-pot synthesis of bifunctional heterodimers of nanoparticles: A conjugate of quantum dot and magnetic nanoparticles.* Journal of the American Chemical Society, 2004. 126(18): 5664–5665.
21. Fan, H.M., et al., *Quantum dot capped magnetite nanorings as high performance nanoprobe for multiphoton fluorescence and magnetic resonance imaging.* J Am Chem Soc, 2010. 132(42): 14803–11.
22. Ye, F., et al., *Biodegradable polymeric vesicles containing magnetic nanoparticles, quantum dots and anticancer drugs for drug delivery and imaging.* Biomaterials, 2014. 35(12): 3885–3894.
23. Li, Y., et al. *Magnetic-fluorescent Fe3O4@ chitosan-graphene quantum dots nanocomposites for dual-modal nanoprobes of fluorescence and magnetic resonance imaging.* in 2018 IEEE International Magnetics Conference (INTERMAG). 2018.
24. Ag Seleci, D., et al., *Transferrin-decorated niosomes with integrated inp/zns quantum dots and magnetic iron oxide nanoparticles: Dual targeting and imaging of glioma.* International Journal of Molecular Sciences, 2021. 22(9): 4556.
25. Li, B., et al., *One-pot synthesis of polyamines improved magnetism and fluorescence Fe3O4–carbon dots hybrid NPs for dual modal imaging.* Dalton Transactions, 2016. 45(13): 5484–5491.
26. Hassani, S., N. Gharehaghaji, and B. Divband, *Chitosan-coated iron oxide/graphene quantum dots as a potential multifunctional nanohybrid for bimodal magnetic resonance/fluorescence imaging and 5-fluorouracil delivery.* Materials Today Communications, 2022. 31: 103589.
27. Chen, L., et al., *Graphene quantum dots mediated magnetic chitosan drug delivery nanosystems for targeting synergistic photothermal-chemotherapy of hepatocellular carcinoma.* Cancer Biology & Therapy, 2022. 23(1): 281–293.
28. Klasens, H.A., *On the nature of fluorescent centers and traps in zinc sulfide.* Journal of The Electrochemical Society, 1953. 100(2): 72.
29. Bhargava, R.N., et al., *Optical properties of manganese-doped nanocrystals of ZnS.* Physical Review Letters, 1994. 72(3): 416–419.
30. Saha, A.K., et al., *Fe doped cdtes magnetic quantum dots for bioimaging.* Journal of Materials Chemistry. B, 2013. 1(45): 6312–6320.

31. Huang, Y., et al., *Gadolinium-doped carbon quantum dots loaded magnetite nanoparticles as a bimodal nanoprobe for both fluorescence and magnetic resonance imaging.* Magnetic Resonance Imaging, 2020. 68: 113–120.
32. Sun, S., et al., *Manganese-doped carbon dots with redshifted orange emission for enhanced fluorescence and magnetic resonance imaging.* ACS Appl Bio Mater, 2021. 4 (2): 1969–1975.
33. Wang, J., et al., *Fluorine and nitrogen co-doped carbon dot complexation with Fe(III) as a T1 contrast agent for magnetic resonance imaging.* ACS Applied Materials & Interfaces, 2019. 11(20): 18203–18212.
34. Huang, Q., et al., *Biocompatible iron(II)-doped carbon dots as T1-weighted magnetic resonance contrast agents and fluorescence imaging probes.* Mikrochim Acta. 2019. 186: 1–10.
35. Kim, J., et al., *Magnetic fluorescent delivery vehicle using uniform mesoporous silica spheres embedded with monodisperse magnetic and semiconductor nanocrystals.* Journal of the American Chemical Society, 2006. 128(3): 688–689.
36. Salgueiriño-Maceira, V., et al., *Composite silica spheres with magnetic and luminescent functionalities.* Advanced Functional Materials, 2006. 16(4): 509–514.
37. Zhou, R., et al., *Enriching mn-doped ZnSe quantum dots onto mesoporous silica nanoparticles for enhanced fluorescence/magnetic resonance imaging dual-modal bio-imaging.* ACS Appl Mater Interfaces, 2018. 10(40): 34060–34067.
38. Su, X., et al., *A graphene quantum dot@Fe3O4@SiO2 based nanoprobe for drug delivery sensing and dual-modal fluorescence and MRI imaging in cancer cells.* Biosensors and Bioelectronics, 2017. 92: 489–495.
39. Zhang, M., et al., *Magnetofluorescent Fe3O4/carbon quantum dots coated single-walled carbon nanotubes as dual-modal targeted imaging and chemo/photodynamic/photothermal triple-modal therapeutic agents.* Chemical Engineering Journal, 2018. 338: 526–538.
40. Teng, Y., et al., *A multifunctional nanoplatform based on graphene quantum dots-cobalt ferrite for monitoring of drug delivery and fluorescence/magnetic resonance bimodal cellular imaging.* Advanced NanoBiomed Research, 2022. 2200044.
41. Hüber, M.M., et al., *Fluorescently detectable magnetic resonance imaging agents.* Bioconjugate Chemistry, 1998. 9(2): 242–249.
42. Pierre, V.C., M. Botta, and K.N. Raymond, *Dendrimeric gadolinium chelate with fast water exchange and high relaxivity at high magnetic field strength.* Journal of the American Chemical Society, 2005. 127(2): 504–505.
43. Pereira, M.I.A., et al., *Hydrophilic quantum dots functionalized with Gd(III)-DO3A monoamide chelates as bright and effective T1-weighted bimodal nanoprobes.* Scientific Reports, 2019. 9(1): 2341.
44. Yang, Y., et al., *CuInS(2)/ZnS quantum dots conjugating Gd(III) chelates for near-infrared fluorescence and magnetic resonance bimodal imaging.* ACS Appl Mater Interfaces, 2017. 9(28): 23450–23457.
45. Ag Seleci, D., et al., *Transferrin-decorated niosomes with integrated InP/ZnS quantum dots and magnetic iron oxide nanoparticles: Dual targeting and imaging of glioma.* Int J Mol Sci, 2021. 22(9).
46. Yang, Y., et al., *Engineered paramagnetic graphene quantum dots with enhanced relaxivity for tumor imaging.* Nano Lett, 2019. 19(1): 441–448.
47. Yue, L., et al., *Manganese-doped carbon quantum dots for fluorometric and magnetic resonance (dual mode) bioimaging and biosensing.* Mikrochim Acta, 2019. 186(5): 315.

48. Tu, C., et al., *Paramagnetic, silicon quantum dots for magnetic resonance and two-photon imaging of macrophages.* Journal of the American Chemical Society, 2010. 132(6): 2016–2023.
49. Ding, K., et al., *Magnetically engineered Cd-free quantum dots as dual-modality probes for fluorescence/magnetic resonance imaging of tumors.* Biomaterials, 2014. 35(5): 1608–1617.
50. Saha, A.K., et al., *Fe doped CdTeS magnetic quantum dots for bioimaging.* Journal of Materials Chemistry B, 2013. 1(45): 6312–6320.

8 A Siege Cancer Phototherapies by Magnetic Quantum Dots

An Overview, Challenges, and Recent Advancements

Vimal Patel[1], *Vivek Mewada*[1], *Jigar Shah*[1], *and Hiral Shah*[2]
[1]Department of Pharmaceutics, Institute of Pharmacy, Nirma University, Ahmedabad, India
[2]Department of Pharmaceutics, Arihant School of Pharmacy & BRI, Gandhinagar, India

CONTENTS

DOI: 10.1201/9781003319870-8

8.1 CANCER PHOTOTHERAPIES

Cancer phototherapy is a part of the clinical treatment that uses radiance to eliminate tumor cells and peripheral infections, and to also stimulate the body and reduce stress. With technological developments over the last four decades, use of NIR lasers as light sources and nanoparticles (NPs) as photosensitizers (PS), it has achieved new heights (1). The two different types of phototherapies employed for cancer treatment so far are photothermal therapy (PTT) and photodynamic therapy (PDT) (2). In PTT, a photothermal (PT) compound is employed photophysical mechanism to heat aberrant tissues or cells selectively, in PDT the treatment is received by a sequence of photochemical processes activated by photoactivated compounds or molecules known as PS agents. It also creates an in-situ tumor vaccine by destructing established tumor cells via releasing tumor antigens. Phototherapy has also performs by chronically releasing and immunologically delivering therapeutic and diagnostic substances, as well as eliciting tissue immunological responses and harmonizing immunotherapy (1,2,3). Combining phototherapies has the potential to improve clinical outcomes because of their synergistic effects. With a greater awareness of the induction of antitumor response, we predict major breakthroughs in the field of phototherapies in cancer treatment (4).

8.2 PHOTODYNAMIC THERAPY (PDT) IN CANCER TREATMENTS

8.2.1 Mechanism

A cascade of photochemical reactions induced by a photosensitive PS drug is the primary mechanism underlying cancer PDT. After a PS accumulates in cancer cells at a faster rate than in nonmalignant tissue, irradiation with near-infrared (NIR) laser causes death of cell. The PS is activated by a certain specific electromagnetic spectrum. Intersystem crossing causes a PS drug to be triggered to the excited singlet (S1) and then to the triplet (T1) state after irradiation at a specific wavelength. The T1 state has a longer duration than the S1 state, allowing for more protracted interactions between the PS drug and the surrounding molecules. At the T1 stage, there are different kinds of mechanisms for such interactions: Type I and Type II. The generation of free radicals is caused either by absorption of a hydrogen atom or the exchange of an electron in between excited PS and substrate (5). The transfer of energy in between molecular oxygen and with photoactivated PS in ground state, also known as triplet oxygen, occurs through the Type II mechanism ($^3O_2$). The development of a sequence of reactive oxygen intermediates (ROI) such as singlet oxygen ($^1O_2$), the hydroxyl radical, superoxide, and hydrogen peroxide will occur from this energy transfer (6). PDT involves photophysical and photochemical activities like these. Reactive oxygen species (ROS) are formed as an effect of photochemical reactions with oxygen. Superoxide O_{2-} is produced when ROS and oxygen combine. PS interacts with oxygen $^3O_2$, resulting in oxygen transformation in toxic singlet oxygen $^1O_2$ (7, 8). As shown in Figure 8.1, $^1O_2$ and O_{2-} cause cell death through two mechanisms: direct (apoptosis and necrosis) and indirect (antitumor immune responses and microvascular damage) (6). Several PS were designed, but Photofrin° being the initial to be used. Diamond and colleagues presented the first study of PDT's efficiency in treating glioma cells in culture and subcutaneous glioma in rats in the Lancet in 1972 (9). Kelly and colleagues presented the first preclinical outcomes with PDT in the treatment of different types of malignancies in 1975, including bladder superficial transitional cell carcinoma (10). Dermatology first saw the clinical PDT outcomes, particularly in the treatment of mycosis fungoides, psoriasis and skin cancer. The most common side effect of this treatment was skin irritation, which varied depending on the doses and duration of treatment (11).

8.2.2 Limitations of PDT

The PDT has such a tolerable adverse reaction profile, which can increase anti-cancer immune reaction, and has noninvasive properties, according to the findings. PDT is not widely used in clinical setups for the management of tumor malignancies, which is unsatisfactory. Figure 8.2 depicts several difficulties that continue to obstruct the development of PDT, including innate photosensitizers, tissue oxygen, and light density.

8.2.2.1 Inherent Photosensitizers

PSs are a key component of PDT activities which are accumulated in tumor areas and stimulated by irradiation of a specific intensity and wavelength. PSs that are little interact with oxygen and conduct PDT. PSs' therapeutic efficacies are dependent on key factors such as pharmacokinetics, amphiphilicity, dosimetry, purity, and single oxygen yield capabilities. PSs' solubilities are the most well-known drawbacks (12). Lower soluble PSs aggregate in aqueous conditions, making them photodynamically inactive and rendering them useless in in-vivo activities. The solubility, absorption, and emission of PSs in DMSO have all been studied recently (13). Functionalization with moieties such as polysaccharides and polyhydroxylates could lead to non-ionic water-soluble PSs. PSs can also be synthesized with hydrophilic substitutions, such as ionic substitutions, to increase singlet oxygen production in watery environments. In order to use in aqueous mediums, insoluble PSs were previously associated into emulsions, NPs, and liposomes (14).

8.2.2.2 Tissue Oxygen

PDT's therapeutic efficacy is determined by the amount of molecular oxygen located inside tumor, singlet oxygen effective produced by the $^3O_2$; so, tissue oxygenation seems to be extremely significant for the PDT efficiency. On either side, tumor tissues are well recognized to be oxygen-depleted as a product of its fast growth, poor angiogenesis, and higher oxygen diffusion rates. Owing to the fact that oxygen is a fundamental component of PDT, hypoxic conditions, which show oxygen concentrations of

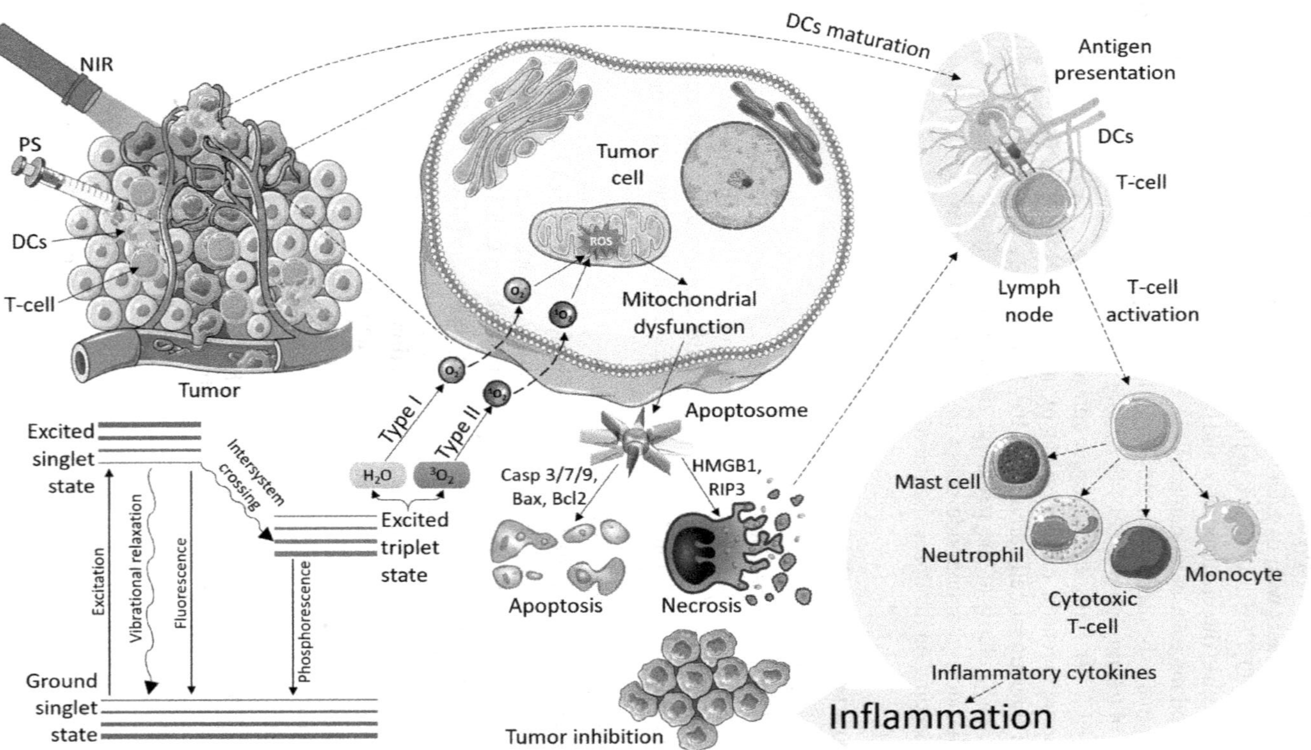

FIGURE 8.1 PDT's effects on tumor inhibition and associated immunological response.

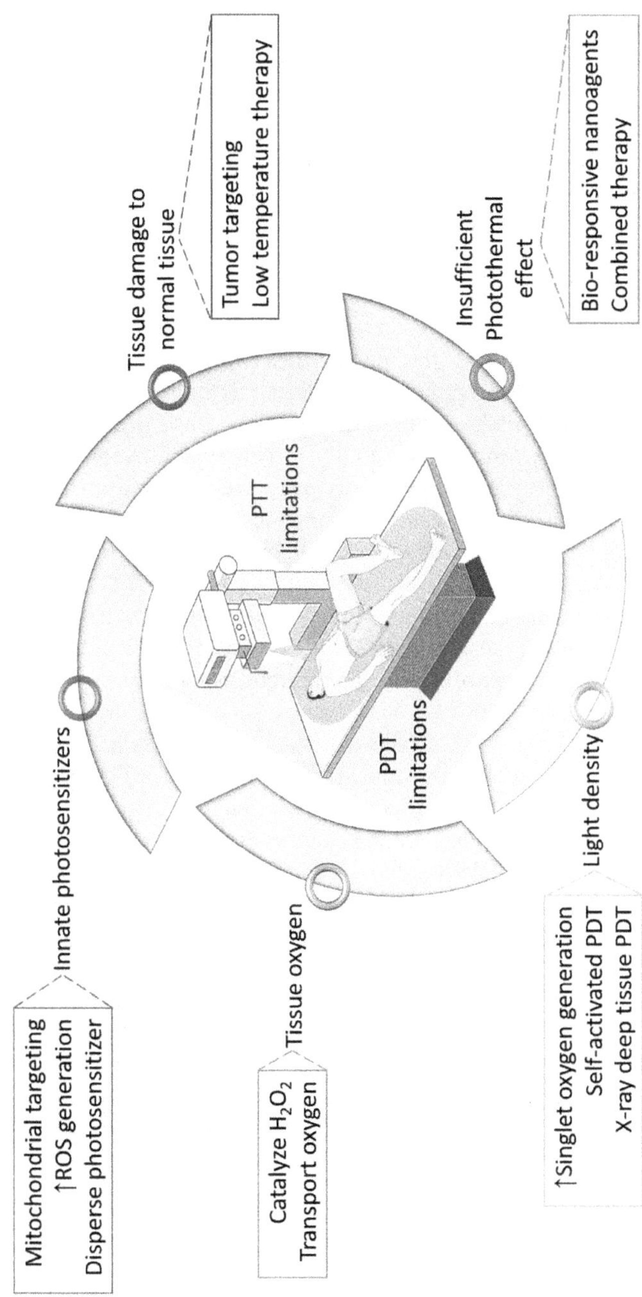

FIGURE 8.2 Limitations and strategies for improving cancer phototherapy outcomes.

typically less than 1 μM, are still a significant issue (15). In reality, both PDT's clinical applications and tumor models have shown that photosensitization rapidly drains oxygen level in the cells. As a result, the light dose must be managed appropriately, and the light should preferably be delivered in fractions (pulses). As a reason, PDT is seen to be a self-limiting modality that hinders itself. Certain chemotherapies and radiotherapies are extremely resistant to hypoxic zones of solid tumors (16). Drugs like carboplatin, cyclophosphamide, and Doxorubicin (DOX), as per the latest research, exhibit hypoxia-induced chemotherapy resistance. Likewise, when a major vessel of the tumor tissue is disrupted, the efficiency of PDT is dramatically diminished. Hence, Hypoxic tissues reflect the optimum dose of the drug utilized for PDT (17).

8.2.2.3 Density of Light

PDT is an irradiation therapy, and its therapeutic efficacy is determined by the quality of the radiation used to stimulate the specific chromophore. The need of laser irradiation has both advantages and drawbacks. Unless it is stimulated, the PS would not be active, and it will not produce cytotoxic singlet oxygen (18). The procedure gains intrinsic selectivity as a result of the ability to choose the timing and location of irradiation. However, because light can only penetrate a few millimeters into tissue, the treatment efficacy is limited to surface malignancies. Light penetration is affected by the optical property of tissues as well as the light wavelength. Between tissues, and even within a tissue, there is heterogeneity. Light scattering, reflecting, transmitting, and absorption are all caused by inhomogeneity sites (such as nuclei and membranes) (19). Furthermore, water absorbs a photon of longer wavelengths, limiting the depth of light penetration into tissue. Furthermore, natural dyes like melanin and hemoglobin absorb light at shorter wavelengths. As a result, they have an impact on light penetration. These findings showed that spectral range of light (also known as the phototherapeutic window) is critical for PDT. Between 600 and 1,300 nm is the phototherapeutic window (20). Light with wavelengths between 620 and 850 nm, on the other hand, has the best penetrating capacity for achieving maximum skin permeability. The light does not deliver enough energy to activate the PS at wavelengths above 850 nm, hence the PS cannot provide enough energy transfer to its triplet state to produce singlet oxygen. Finally, NIR with a wavelength range of 620–850 nm produces the best tissue penetration for PDT applications (21). Incandescent light and lasers are two effective devices for delivering light. The source of light used for PDT has various distinct properties. As a result, a universal source of light will not be capable of covering all PDT applications. First and foremost, the type of disease indicates the sort of light to be used (tumor location, type, and size). Second, a light source must be acceptable for the PS's spectral properties (e.g., absorption spectrum) (21).

8.2.3 Solution to the Drawbacks of PDT

The strategies associated to PS selectivity are focused on enhancing the sensitivity of PDT to tumor tissues, hence boosting the treatment's specificity. PSs can also be customized to target specific subcellular compartments or organelles, in addition to tumor tissue. Such approaches have the probable to strengthen PDT's effectiveness.

It depends largely on targeting groups (such as triphenyl phosphonium) to direct PS delivery to the most crucial subcellular organelle in terms of singlet oxygen-mediated apoptosis. PDT's efficacy and selectivity could be improved with this method (4).

Because $^1O_2$ is very reactive, it has long been recognized and have a very short lifetime. Despite the fact that $^1O_2$ has a 3 μs lifetime in water, its lifetime in cells is believed to be around 200 ns due to its high reactivity to biological substances (22). The electronic to vibrational energy transfer facilitates quenching in water, resulting in deactivation via interactions with the vibrational states of O–H bonds. Because $^1O_2$ has such a short lifetime, it has a small diffusion range in cells, estimated to be around 45 nm. Considering that most human cells have a diameter of 10–100 μm, the site of $^1O_2$ synthesis defines which subcellular targets it will target, as well as limiting photodamage to the regions where the PS is situated (23). Despite that several organelles are around 1–5 μm (e.g., mitochondria are 1–3 μm), the sizes of most cellular organelles are substantially larger in comparison to such a limited diffusion range. As a result, it appears that the PS's subcellular position has a significant influence on the performance of PDT. As a result, various research looked at the effects of using organelle-targeted PSs, such as those that are accumulated in the nucleus, mitochondria, or lysosomes (24). Such strategies could, in fact, enhance the effectiveness of PDT. As shown in the studies, nucleus-targeted peptides and mitochondria-targeting groups (e.g., triphenyl phosphonium) target the most crucial organelles, nucleus, and mitochondria, respectively, to efficiently induce singlet oxygen mediated apoptosis (25).

In-vivo, light irradiation can be employed as an external "ON-OFF" switch to control PDT for tumor tissues. PDT's ability to penetrate deep tissues is significantly limited by light penetration. In PDT, it's critical to overwhelmed the short penetration depth of light. As a result, multiple studies in PDT have attempted to construct various devices, light resources, and targeted delivery techniques. X-ray radiation, NIR, bioluminescence methods, radio-luminescent NPs, and quantum dots can all be used to excite PS (26). The PS are designed to have significant absorption visible spectrum's range between red to NIR. This spectrum area reduces light scattering due to tissue heterogeneity, achieves penetration depths of several mms, and enhances the signal to noise ratio due to the low background emission in NIR. The purpose of adopting two-photon excitation sensitive of PSs is improving the depth of therapeutic penetration using NIR light (27). To improve therapeutic efficacy, these PSs can be conjugated with NPs. A BODIPY derivative sensitizer that absorbs NIR was produced by Akkaya et al. in their study. In the case of excessive intracellular glutathione (GSH) levels inside the tumor, lethal singlet oxygen was produced by BODIPY-based PS (28). X-ray radiation, in addition to NIR light, is a promising candidate for use as an indirect activation source of energy in the therapeutic use of PDT in depth tumors (29). However, there is a noticeable energy discrepancy between the therapeutic X-ray (keV–MeV photon energy) and the PSs (eV single–triplet energy gap). Like a consequence, this strategy enables a usage of stimulated NPs radioluminescence (scintillation) or Cherenkov radiation, which emits light in response to X-ray stimulation and can activate adjacent PSs. The auto-PDT technique has been used, which aims to improve PDT efficiency except using an additional light source (30). As a PDT agent, self-illuminating NPs were developed that can be triggered by enzyme-mediated bioluminescence methods. Another possible way for generating cytotoxic singlet oxygen in PDT is persistent luminescence, which could provide an internal source of light. PSs can be excited for a long period of time through NPs that emit NIR light with long luminescence lifespan. As a result, PDT may be free of the need for external stimulations (31).

Hyperbaric oxygen (HBO_2) therapy was integrated with PDT to resolve the issues related with low oxygen level in tumors. As a result, when used during hyperoxygenation, PDT efficacy could be significantly improved. Hetzel et al. investigated the hyperoxygenation associated PDT-Photofrin in order to reduce hypoxia. The xenografted breast carcinoma cells bearing mice preserved at HBO_2 with three atp (atmospheric pressure). When combined with hyperoxygenation, PDT proved to be more effective in treating hypoxic cancers (32). Ultimately, molecular oxygen was found to be a critical component of PDT-induced cytotoxicity. PDT induces cytotoxic singlet oxygen production; photochemical oxygen depletion is related to light fluence rate. In the Colon 26 tumor model, rate of fluence was enhanced the PDT tumor response investigated by Snyder et al. (33). Tumor control was improved by using light with a lower fluence. Higher fluence light, on either side, rapidly reduced the tumor oxygen concentrations. Furthermore, both before and after PDT, the levels of tumor oxygenation were evaluated. The findings revealed that tumor oxygenation during PDT has a significant impact upon post-treatment. PDT cytotoxicity and tumor vasculature damage could be the source of these side effects. Ultimately, in order to get excellent outcomes in PDT, it is indeed essential to pick the best combinations PS dosage, light sources, and treatment settings (6).

In an orthotropic breast tumor model, fractioned does of PSs (MV6401) exhibited higher effectiveness in producing tumor suspension than the similar total dosage delivered as a single treatment, according to Dolmans et al. (34). The metronomic photodynamic treatment (mPDT) is a technique where the PSs and both light is administered at low rates over a prolong time for the management of preclinical brain tumor rat models. The light source was a laser diode or LED, and the 5-aminolevulinic acid (ALA) was administered with drinking water. Ex-vivo spectrofluorimetric results of 9L-gliosarcoma bearing mice were showed four-fold increased tumor accumulation and apoptosis (35).

8.3 PHOTOTHERMAL THERAPY (PTT) IN CANCER TREATMENTS

8.3.1 Mechanism

Photothermal therapy (PTT) is a therapeutic strategy that involves heating the tumors site with a NIR laser/light and causes cancer cells death. A biocompatible PT agent with a strong absorption capacity in the NIR regions, as well as an NIR light source, is important for PTT. As a result, NIR absorbed surface-modified metal, carbon, and semiconductor NPs could be appropriate PT agents (1). The NIR absorption coefficient and the intensity of the excitation light have a significant impact on the percentage increasing temperature throughout PTT. Visible light, radio waves, microwaves, and ultrasound waves are examples of other radiation sources that can cause hyperthermia. When compared to traditional therapy techniques, PTT provides a number of advantages, such as less severity and excellent specificity (2). As seen in Figure 8.3, increasing kinetic energy enables tumors to heat the surrounding microenvironment throughout local heating or hyperthermia. When a tissue's temperature is raised to 41 °C, a heat-shock response is triggered, which results in a sequence of fluctuation in gene expression patterns, such as the production of heat-shock proteins, in order to reduce the effects of the early thermal injury (36). The temperature rises to 42°C, causing irreparable tissue damage; heating tissues to 42–46 °C for 10 minutes causes cell necrosis. Microvascular thrombosis and ischemia cause cells to die quickly at 46–52 °C. Because of plasma membrane disintegration and protein denaturation at temperatures above 60 °C, that are commonly obtained using PTT, cell death is almost dynamic (37,38). However, such thermal treatments also damage healthy tissues. PTT technique, in general, focus on two mechanisms: The first involves subjecting the tumor site in high temperatures (more than 45 °C) in a few minutes, resulting in thermal ablation causing cellular death. This method frequently causes tumor vascular occlusion and hemorrhage, which is precluded by combined with other treatments. The second is moderate hyperthermia, which entails increasing and sustaining 42 to 43 °C temperature causing cell injury and increasing tumor vascular penetrability, that can be used to increase NP's bioavailability (7,39). Tumor tissues are more acidic and less oxygenated as compared to healthy tissues. These properties are intended to make tumors highly sensitive to temperature, enabling PTT to precisely eliminated cancer cells and protecting normal cells. Because cancer cells react to this range of temperature, this method can be used in combination with synergistic treatments (40).

8.3.2 Limitations of PTT

Although PTT have progressed rapidly in the field of cancer research, the majority of trials are still unable to be implemented into clinical practice. Figure 8.2 shows the contrast between PT agent deficiencies (e.g., thermal damage to normal tissue), heat-shock response, and insufficient therapy effectiveness.

8.3.2.1 Thermal Damage to Normal Tissue

The strong targeting potential of PSs to tumor site allows for the effective photothermal treatment. A high concentration of PSs at the tumor site and the absence in normal tissues produce the greatest clustering thermal impact, allowing for an adjusted temperature rise and then "heating" the tumor while causing no damage to normal tissues. Two strategies have been presented by researchers that ensure the tumor-killing effect while also ensuring safety of normal tissues. The first is to increase PSs targeting capabilities, and the second is to use low-temperature PTT. The photosensitizing compounds should be significantly concentrated in the tumor site to improve targeting capabilities. Only the tumor site where PSs are concentrated could generate a significant degree of thermal energy when excited by NIR light, that will not happen in normal tissues (4).

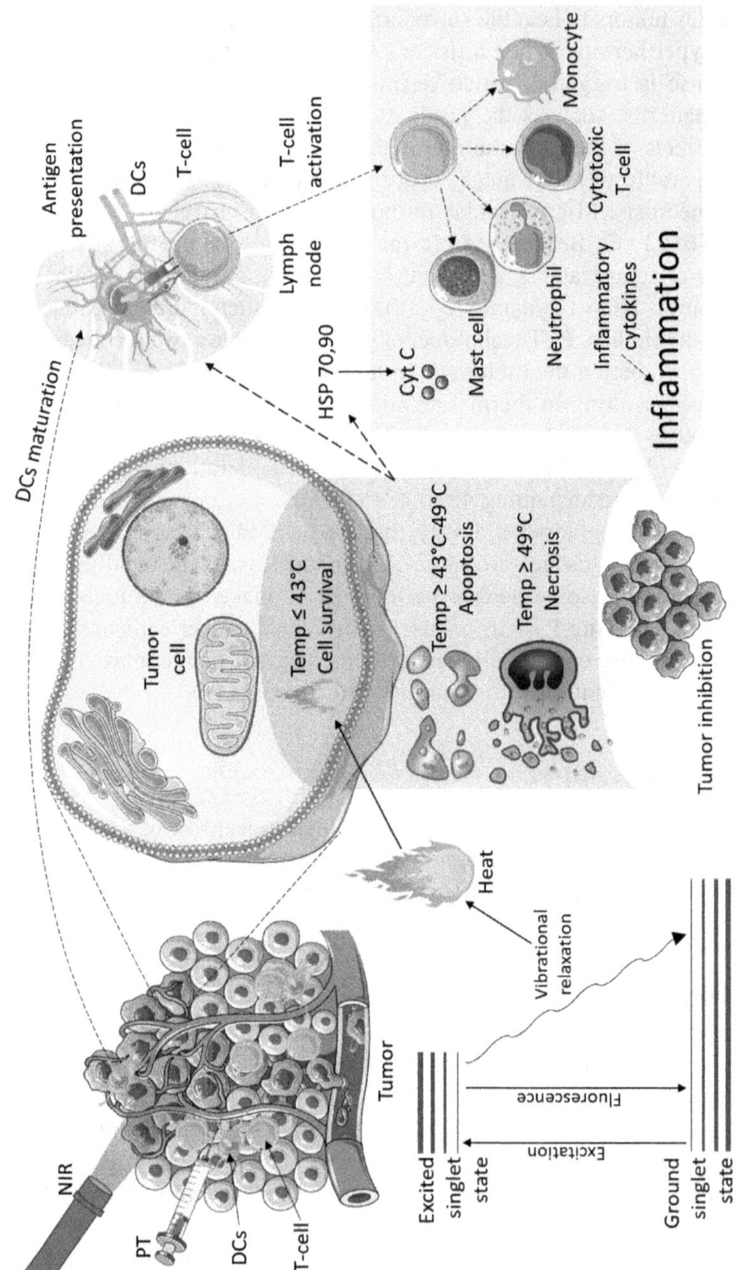

FIGURE 8.3 Hyperthermia and its associated immunological response to tumor inhibition.

8.3.2.2 Insufficient Photothermal Effect

The light's low penetration depth is the most critical problem of PTT, which causes deficiency in tumor treatment beyond the radiation range. In most cases, monotherapy is insufficient to completely eliminate the tumor, and PTT is no exception. Even if PTT has a significant therapeutic impact, its deficiencies may lead to incomplete tumor cells elimination, that can indirectly contribute to tumor recurrence and metastasis. When PTT is combined with other therapeutic techniques, the overall therapeutic efficacy improves. In several circumstances, combining several therapeutic techniques results in a synergistic treatment impact rather than just a basic supplement (4,41).

8.3.3 Solution to the Drawbacks of PTT

In absence of tumor-specific targeting molecule, most PT agents' tumor-targeting capabilities will rely on the enhanced permeability and retention (EPR) effect, that indicates macromolecules and NPs of smaller sizes (including vesicles and liposomes) are even more likely to infiltrate tumor tissues and stay there for a longer duration compared to normal tissues (7, 42). Tumor cells release vascular endothelial growth factor (VEGF) because they require more nutrients and oxygen to sustain their rapid growth. It would be very reliant on the nutrition and oxygen supply provided by tumor blood vessels, especially as the tumor grows 150–200 μm in diameter. The freshly formed tumor blood vessels have a shape and morphology that differs significantly from normal blood vessels at this stage. The smooth muscle layer of the microvasculature is absent, the vascular endothelial gap is broad, and the angiotensin receptor activity is lost. The returning of lymph is inhibited by tumor tissue due to a lack of lymphatic vessels (43, 44). These characteristics allow macromolecules to flow naturally through the blood vascular endothelium and accumulate inside tumor cells, because these substances are not taken away by lymph fluid, and remain in the tumor tissues for a longer duration. If the PT agents' properties, size, surface features, composition, porosity, and targeting ligands are maximized, the EPR effect would enhance the targeting function, resulting in greater antitumor activity (45).

Low-temperature PTT reduces tumor cells' heat resistance by suppressing the expression of heat shock proteins (HSPs), enhances photothermal treatment's efficacy, and reduces the treatment temperature. Liu's group researchers were demonstrating this concept and later much scientific research since has reinforced and optimized it to provide a stronger tumorigenic impact while minimizing thermal damage to normal tissues (46).

Qian et al. designed nanomaterials based on the breast tumor specific receptor for improve the targeting effect. They developed surface modified NPs with the Lyp-1 ligand to preferentially target p32 protein, which is overexpressed on breast cancer cells (surface and intracellular) and also tumor-associated lymphocytes. The combined PTT/PDT/chemotherapy of the in-vitro and in-vivo mouse model reveal that Lyp-1 modification increases targeted delivery and significantly suppresses the tumor growth of breast cancer metastasis. Thus, the targeting effect of NPs to tumors could be increased by controlling the size and properties and reinforcing the surface modification, thereby resolving the PTT induced adverse effects of thermal damage to normal tissues. There may be less drugs and active components in the system with more ligands loaded in NPs, reducing the anticancer effect. As a result, throughout the nanomaterials design phase, the balance between targeting abilities and antitumor efficiency must be considered (47).

Du and his colleagues created the "iCluster" concerted nanoplatform; it can compress their size from 100 to 5 nanometers inside the tumor. The feature improves particle permeability in the primary tumor stroma, and promotes particle aggregation in tumor tissue as well as lymphatic vessels. This is a prominent method for improving PTT efficiency by increasing the targeting ability of NPs by altering their size. Another approach to increase tumor targeting is to encapsulate NPs with tumor targeting ligands (48). Folic acid and hyaluronic acid are the most commonly employed targeted ligands because their receptors are overexpressed mostly in solid tumors due to their rapid proliferative nature. Some unique receptors can only be found on the surface of specific tumor cells (49, 50).

8.4 RECENT ADVANCEMENTS OF MAGNETIC QUANTUM DOTS IN CANCER PHOTOTHERAPIES

Nanocrystals of magnetic quantum dots (M-QDs) are magnetic semiconductor nanomaterials with inherent chemical and physical characteristics. These have distinct semiconductor energy levels that can be adopted by altering their size, shape, and charge potential. Quantum confinement is a feature of semiconductors in which the NPs diameter approaches the Bohr exciton radius. Because of the quantum confinement effect, these NPs have unique electronic and optical characteristics, such as size-tunable absorption bands and bright fluorescence emission (51). M-QDs are potential candidates for bioimaging and cancer phototherapy due to their large surface to volume ratio, exceptional photostability, and huge two-photon excitation. Since the previous three decades, researchers have been developing a variety of M-QDs for cancer phototherapies.

8.4.1 Metallic M-QDs

Because of the unique properties of metals such as significant optical field augmentation due to absorption of light and high scattering, noble metal nanostructures are attractive applicants in different fields of physics, chemistry, and biology. They have more surface modification potential and better magnetic and physical properties. The specific interaction of metallic NPs with light causes the optical and PTT enhancements (52). The valance electrons of metal NPs suffer a collective coherent oscillation with regard to the matrix when irradiated. This oscillation is resonating at a specific frequency of electromagnetic field of light, according to Mie Theory, and this effect is known as localized surface plasmon resonance (LSPR) (53). Metal NPs' LSPR wavelength is greatly influenced by their metal type, size, morphology, inter-particle distance, and surface dielectric constant. The LSPR will red-shift as the size of the nanospheres grows due to the electromagnetic delay, furthermore the plasmon band position fluctuates between spherical and non-spherical nanomaterials (shell, cube, rod, triangle, tetrapod, prism, etc.) (54). The researcher was interested in different types of metallic NPs and their composites, such as iron, gold, silver, manganese, molybdenum, zinc, and silica, as prospective candidates for cancer phototherapies.

8.4.1.1 Iron Oxide M-QDs

In the biological milieu, iron oxide NPs have a high photothermal conversion efficiency, strong chemical stability, and minimal toxicity. The USFDA has cleared the use of iron oxide NPs in *in-vivo*. However, due to the low tumor targeting efficiency of NPs, their application in many clinical procedures is limited. Magnetite (Fe_3O_4), maghemite (γ-Fe_2O_3), or a combination of the two are commonly used to make magnetic NPs (55). These nanosystems have developed as effective MRI contrast agents (CAs) and therapeutic applications because of their unique superparamagnetism properties. Magnetic NPs can be employed as targeting drug delivery in cancer therapy by using alternating magnetic fields to target specific tumor and eliminated by generating confined moderate hyperthermia (56, 57).

Orderly to produce multifunctional nanomaterials for drug delivery systems and phototherapy, scientists have combined phytochemical substances with magnetic NPs. Kharey et al. used a medicinal plant, Pimenta dioica, to produce 15 nm eugenate (4-allyl-2-methoxyphenolate)-capped iron oxide NPs (E-capped IONPs). In-vitro HeLa and HEK 293 cell lines study these NPs showed high biocompatibility and great hyperthermia generating efficacy when laser irradiated at NIR wavelength (58).

The PEG-coated USPIO-coupled sialyl Lewis X (USPIO-PEG-sLex) with good photothermal conversion characteristics was produced by Liu and colleagues (59). In a mouse model, the NPs were used for PTT-MRI in human nasopharyngeal cancer (NPC) xenografts. The cytotoxicity data demonstrated a decrease in cell viability of NPC 5-8F cells upon introduction to an 808 nm NIR at appropriate doses of USPIO-PEG-sLex NPs. Furthermore, after *in-vivo* post-injection, the NPs showed they were able to terminate xenograft tumor progression.

Wang and his team designed an Ac-Arg-Val-ArgArg-Cys (StBu)-Lys-CBT probe that was combined with monodispersed carboxyl-decorated SPIO-NPs to form SPIO@1NPs in order to improve the tumor delivery of iron oxide NPs. The cross-linking of SPIO@NPs aggregates is due to the overexpression of furin after accumulation of NPs inside tumor cells resulting formation of chain reaction. Increasing tumor site retention after self-aggregation of NPs, resulting in more effective PTT and superior T2 imaging of cancer cells at low dosages (60).

Cabana and colleagues examined the use of magnetic multicore nanoflowers (NFs) for PTT vs magnetic nanospheres induced magnetic hyperthermia (MHT). The effectuation of the NPs in PTT and MHT was evaluated in water and cancer cells. NFs were discovered to be more potent heaters for both modes. PTT performed well in the cell milieu at low dosages, however MHT remained lost for all NPs. Moreover, after NIR treatment (0.3 W/cm^2), magnetite nanoflowers showed the maximum cellular internalization and anticancer efficacy (61).

Ashkbar and colleagues developed curcumin loaded silica-coated Fe_3O_4 magnetic NPs for in-vitro and in-vivo studies to induce hyperthermia and singlet oxygen production. Curcumin (CUR) is a natural chemical that comes to the polyphenol family and is renowned for its photosensitizing and anticancer properties. PDT was measured with 450 nm diode lasers, and PTT was measured with an 808 nm NIR laser. The results of a post-treatment breast tumor mouse model exhibited that CUR + PDT reduced tumor size by roughly 58% compared to the untreated group, although a NPs + PDT + PTT treated group reduced tumor volume by more than 80 % compared to other treatment groups. These findings suggested that use of NPs + PDT + PTT therapy regimen, tumor development could be inhibited with[in] 10th days. The combined effect of hyperthermia and ROS production in the tumor site was responsible for such an outcome (62).

For spatiotemporal PTT-immunotherapy interaction in cancer, a delivery nanosystem consisting of R837-laden polyphenols coated Indocyanine green (ICG) loaded magnetic NPs (MIRDs) were developed. This approach reduced tumor growth, invasion, and recurrent, resulting in significant anticancer treatment results with minimal adverse effects (63).

Prussian blue-coated Fe_3O_4 hollow magnetic NPs (HMNP-PB-Pent-DOX) loaded with DOX were developed by Li et al. for simultaneous drug delivery and multimodal imaging in *in-vivo* and *in-vitro* PTT-chemotherapy. 1-pentadecanol, a phase-change material (PCM), was used as a "drug-janitor" in a temperature-induced controlled release mechanism. After NIR irradiation in a photoacoustic computed tomography scanner, higher drug accumulation and PTT activity were seen in HepG2 cells and tumor-bearing mice in in-vitro and in-vivo models (64).

8.4.1.2 Gold M-QDs

Since Gold (Au) exhibit excellent local heating following light irradiation, they have gained significant heed as PTs agents for cancer phototherapies. In the influence of electromagnetic radiation (EMR), the photoconversion mechanism in AuNPs is associated with fundamental vibration of unpaired electrons at the surface termed as Surface Plasmon Resonance (SPR). This unique thermodynamic feature provides substantial localized heating surrounding AuNPs due to electron excitation and relaxation, eliminating cancer tissue necrosis (65). AuNPs NPs have a significantly stronger SPR spectrum than other metals. The spectrum of the SPR wavelength is adjusted between the visible to the NIR area by altering sizes and shapes, and optical properties can be easy to control (66). The surface functionalization of the AuNPs with tumor specific ligand can improve their targeting, increasing therapy efficiency while reducing potential harmful effects in normal tissue. The cellular uptake and PTT conversion efficacy significantly influence by the size of AuNPs (67).

Smaller NPs convert heat more efficiently than larger ones according to Mie's hypothesis. The conversion efficiency of 20 nm gold nanospheres was reported to be 97–103%. The smaller diameter of NPs (<5 nm) is owing to their ability to be eliminated via urine, which improves their clearance from the body (68, 69). Saw et al. evaluated the PTT of breast cancer cells using four sizes (30, 60, 80, and 100 nm) of cystine/citric acid-coated confeito-like gold NPs (confeito-AuNPs). The researchers showed that the confeito-AuNPs with the lowest diameters (30 and 60 nm) had higher cellular uptake, better PTT efficacy and cytotoxicity in MDA-MB-231 cells than those with bigger sizes (70).

Sun et al. employed AuNPs coated with paclitaxel (PTX) grafted Pluronic-b-poly(L-lysine) co-polymeric micelles for the PT-chemotherapy of solid tumors. The PTX-Pluronic-PLL-AuNPs micelle showed efficient tumor targeting hyperthermic proficiencies upon 808 nm NIR irradiation and a synergistic effect of combined PTT-chemotherapy in *in-vivo* and *in-vitro* MDA-MB-231 tumor bearing mice model (71).

PEGylated hollow AuNPs (mPEG-AuNPs) were produced by Wang et al. for combined PTT and X-ray irradiation of breast carcinoma cells. *In-vitro* data indicated a synergistic effect in the MTT assay of 4T1 cells after exposure to 808 nm NIR and X-ray irradiation, with cell viability reducing to 65 %. The significant accumulation of mPEG-AuNPs into the tumor by the EPR effect and combined hyperthermia effects was seen in CT imaging of BALB/c mice in a xenografted breast tumor model (72).

The pH-triggered tumor targeted Glutathione (GSH) coated AuNPs was developed by Barram et al. for *in-vitro* rhabdomyosarcoma (RD) tumor model (73). Two low-power NIR irradiations (532 nm and 808 nm) were used to apply the PTT. The *in-vitro* results showed that both types of NIR lasers decreased cancer cell viability, which was associated to the prolonged exposure, leading to increased concentrations of GSH-AuNPs inside tumor cells.

Yuan et al. developed highly hydrophobic Au@T self-assembling nanodots using modified lipoic acid-PEG grafted with biotin, 4-mercaptobenzoic acid and p-hydroxythiophenol. The nanodots rapidly increase their size from 33 nm to 250 nm after being added in PBS pH 6 buffer. The PTT efficacy of nanodots were measured in HepG2 cells under different NIR irradiation. The *in-vitro* results were demonstrated that after 8 h of incubation, high accumulation and aggregation of Au@T-SiO_2 can prolong residence time of NP in cell. Next 3 min of NIR exposure the temperature was raised up to 50 °C at 1.0 W/cm^2, PTT effectiveness of aggregated Au@T-SiO_2 was 30.48% higher than the non-aggregated NPs (74).

PEG grafted Diazirine (DA) loaded covalently cross-linked photolabile AuNPs was prepared by Cheng et al. After laser irradiation at 405 nm, diazirine decorated AuNPs (dAuNPs) with a diameter of 20.5 nm were produced. The influence of hyperthermia on decreasing tumor size in different NPs treatment groups of Balb/c tumor bearing mice was investigated. *In-vivo* study was confirmed that dAuNPs and dAuNPs+405 nm NIR treatment group showed minimal tumor inhibition, while dAuNPs+808 nm NIR treatment grouped exhibited greater tumor inhibition. *In-vitro* cytotoxic assay in 4T1 cells shows aggregation of NPs upon 405 nm NIR and significant photothermal ablation effects was exhibited after 808 nm NIR irradiation (75).

8.4.1.3 Silver M-QDs

The SPR of AgNPs could be controlled toward the IR region, just as gold NPs, by changing their size and shape (76). The use of AgNPs as and in biomedicine is currently limited due to concerns about their inherent toxicity. AgNP interactions with the human body are yet unknown. Many researchers have discovered strategies for enhancing the biocompatibility of AgNPs by surface modification with biodegradable polymers and molecules or taking such hybridized NPs systems (77).

Wu et al. employed a hybridization chain reaction (HCR) to prepare fluorescent nanoprobe for cell surface glycans label-free imaging made up of DNA/silver nanoclusters (DNA/AgNCs). After 10 minutes of 808 nm NIR laser irradiation at 1 W/cm^{-2}, the NCs were able to convert light to heat, reaching 53.6 °C. The confocal microscopic results showed that the DNA/AgNCs could be used to tagged glycans on the surface of tumor cells. Furthermore, *in-vivo* investigations revealed that DNA/AgNCs may ablate and prevent tumor growth after exposed to NIR laser (78).

ICG loaded BSA coated PEGylated AgNPs was developed by Park et al. were investigated for PPT-anticancer activity in B16F10 cells. The cytotoxicity assay results showed a cell viability of 6% was observed after raised 50 °C temperature. The plasma half-life of PEG-BSA-AgNP/ ICG was prolonged, resulting in enhanced tumor accumulation. The post-administration of PEG-BSA-AgNPs-ICG and 0.95 W NIR exposure, after 4 h temperature of B16F10 mice was increased to 49.6 °C. *In-vitro* PTT activity of NPs treated group showed significant tumor inhibition (79).

8.4.1.4 Manganese M-QDs

Manganese oxide NPs (MONs) have gained popularity as PTs and MRI contrast agents in recent years owing to strong T1-weighted MRI signals and low toxicity, making them a viable replacement of standard PTT agents (80). The pH-sensitive, and biocompatible MnO-loaded carbonaceous nanospheres (MnO-CNSs) were produced by Xiang and colleagues for simultaneous PTT and MRI (81). In different biological conditions (pH 5.0, 6.5 and 7.4), the simulated pH-responsive release of Mn^{2+} was observed. They discovered that MnO-CNSs were stable in neutral solution (pH 7.4) but rapidly released Mn^{2+} ions in acidic pH 6.5 and 5. In-vivo investigations demonstrating that T1-MRI signal rates were increased in the acid environment. The PTT effect of MnO-CNSs was examined for 10 minutes using an 808 nm NIR laser (2 W/cm^{-2}). The MnO-CNSs were known to have a greater hyperthermia ablation efficiency in *in-vivo* tumor, enabling them as useful nanotheranostic tools.

A multifunctional nanoprobe (H-MnO_2/DOX/BPQDs) was created for synergistic PTT-chemotherapy and dual-modality cancer imaging using the unique properties of black phosphorus quantum dots (BPQDs). An anionic polymer poly (acrylic acid) (PAA) and a cationic polymer poly (allylamine hydrochloride) (PAH) are successively coated on hollow mesoporous MnO_2 (H-MnO_2) NPs. The final nanoprobe H-MnO_2/DOX/BPQDs was produced by DOX loading on H-MnO_2-PHA-PAA carbodiimide cross-linked BPQDs-PEG-NH_2. The H-MnO_2/DOX/BPQDs is destroyed lastly in the tumor microenvironment, releasing encapsulated active components DOX and BPQDs. Under dual NIR irradiation of 630 and 808 nm BPQDs confers the nanoprobe with PDT and PTT properties. H-MnO_2 diminished tumor hypoxia and induce PDT activity through the conversation of H_2O_2 to singlet oxygen, and offers the MRI contrast agents (82).

Mitochondrial targeted dye IR808 loaded PEGylated MnO multifunctional pH triggered NPs were prepared by Zhou et al. for synergistic PDT-MRI (83). The most of tumor cells overexpresses organic-anion transporting polypeptides (OATPs) receptors and several heptamethine cyanine dyes have good affinity toward it. From them IR808 has shown more tumor specific accumulation in the *in-vitro* MCF-7 cells significantly exhibited expression of OATP1B3 receptors. The IR808@MnO-PEG NPs was induced apoptotic mitochondrial dysfunction and ROS generation after NIR irradiation resulting secretion of cytochrome c. The MRI of MCF-7 bearing nude mice showed more accumulation of IR808@MnO-PEG NPs at the tumor site after 48 of IV administration due to EPR effects. The *in-vivo* tumor treated with NPs showed rapid increase in temperature about 56.7 °C, which causes complete tumor elimination and no recurrence after 18 days of treatment.

Photosensitive Mn^2 substituted BODIPY (4,4-Difluoro-1,3,5,7- tetramethyl-4-bora-3a,4adiaza-s-indacene) byproducts such as DBA nanoassembly coated with BSA through hydrophobic interaction was made by Lu and group for simultaneous PDT/PTT-MRI fluorescence imaging of tumor tissues (84). The resulting protein dye-based nanoassemblies further surface functionalized with PEGylated folate acid ligand in order to increase tumor targeting affinity and biocompatibility. The Mn-DBA@BSA-PF NPs showed high ROS turnover frequency (TOF) under NIR irradiation in acidic pH 5.0 with increasing photothermal conversion efficiency by 17.9 %. In-vitro MTT assay in HepG2 cells shows significant cytotoxicity in dose-dependent manner with or without NIR irradiation. *In-vivo* T_1-MR imaging of Mn-DBA@BSA-PF NPs in HepG2 bearing tumor mice model was demonstrated that higher accumulation and PDT/PTT effects NPs due to significant tumor growth inhibition in acidic tumor microenvironment by hyperthermia and ROS induced Mn-DBA release.

8.4.1.5 Molybdenum M-QDs

Biocompatibility and biodegradability of molybdenum oxide nanostructures have been demonstrated as a safe field of cancer phototherapy. MoO_3 NPs have good NIR absorption and can create singlet oxygen when exposed to NIR light, allowing them to be used for both PDT and PTT (85).

The pH-dependent and biocompatible multifunctional DOX loaded MnOX-F127 composite was made by surface modification of MnO_2 nanosheets with Pluronic F127 through single-pot hydrothermal method by Chen et al. for simultaneous PTT-chemotherapy (86). The stability of MoOx-F127 was shown to be good at pH 5.4 and reduces quickly at pH 7.4, revealing that entire NPs might penetrate into tumor via the EPR effect. Later 808 nm NIR irradiation for 5 minutes, the efficacy of MoOx-F127/DOX to eliminate MCF-7 tumor cells *in-vitro* was evaluated. *In-vivo* cytotoxicity data results showed that about 60 % of the cell death after treatment. Secondly, *in-vivo* tests revealed that tumor temperature was raised higher than 50 °C in mice injected with MoOx-F127/DOX, demonstrating that the NPs had a high hyperthermic efficiency.

Zhang et al. prepared targeting cyclic arginine-glycine-aspartate (cRGD) modified DOX loaded PEGylated molybdenum telluride nanosheets ($MoTe_2$-PEG-RGD-DOX) for synergistic PTT-chemotherapy. The obtained NCs showed high hyperthermia efficacy and cytotoxicity in in-vitro 4T1 cells after 808 nm NIR irradiation. In-vivo PTT imaging and antitumor activity in 4T1 tumor bearing mice showed that $MoTe_2$-PEG-RGD-DOX/NIR possesses strong PTT conversion and high tumor killing efficiency (87).

8.4.1.6 Zinc M-QDs

Zinc has a wide range of biomedical applications. Zinc oxide (ZnO) is a potential alternative to PTT because of its optical, electrical, anticancer properties, chemical stability, and low toxicity. One of ZnO-NPs' cytotoxic mechanisms is the intracellular

ROS generation (88). For combine PTT-chemotherapy with DOX, gene DNAzyme (DZ), Liu et al. developed a core-shell nanoplatform with polydopamine (PDA) shell covered a zinc oxide (ZnO) core (89). After NIR irradiation of 808 nm for 500 seconds, the NPs showed excellent stability and PTT efficacy. Confocal imaging also revealed that ZnO-PDA-DOX/DZ were significantly accumulated into cells, inducing gene-silencing activity through delivering DZ. In-vivo tumor bearing mice exhibited effective NPs accumulation at the tumor site. The tumor suppression following cancer cell death was observed when temperature of tumor tissue reached up to 47.3 °C. Finally, the authors performed Western blotting to measure the concentrations of survivin in tumor tissue. The survivin levels were found to be significantly low, indicating that DZ is triggered for in-vivo gene silencing.

Kim et al. developed berberine (BER) loaded ZnO-NPs for lung cancer PTT-chemotherapy. The in-vitro studies showed a significant cytotoxicity against A549 cells unless causing serious toxicity in rat blood fluid (90).

8.4.2 Metallic Semi-Conductor Quantum Dots (MSCQDs)

MSCQDs are synthetically prepared by elements from groups II-IV and III-V of periodic table such as Cd, Te, Se, and In. Typically, these nanoscale structures have a very less dimension about 2–10 nm acquire them more trustworthy and significant alternatives for several industrial applications (91). The surface to core atom ratio is high due to its small diameter. These properties of surface atoms influence the properties of the total particle when the ratio of surface atoms to core atoms has increases. The semiconductor lattice of QDs elapses on the surface, so the surface atoms behave very different chemically from the core atoms. As a result, MSCQDs become more useful in biomedical applications (92). CdSe, CdTe, CdSe/ZnS, and InP/ZnS are some of the MSCQDs employed in PDT of cancer models. The direct pyrolysis of organometallic composition of precursors (Cd, Se, Te, and in) generates MSCQDs (93, 94, 95).

Burda and colleagues showed the use of QDs in PDT for the first time by coupling a silicon phthalocyanine (Pc4) to MSCQDs via axial substituent of an alkyl group on Pc4's. MSCQDs have unique Förster resonance energy transfer (FRET) properties, which indirectly stimulate PS drug and enabling them for the generation of $^1O_2$ in toluene with a 5% quantum yield. The NIR fluorescence of $^1O_2$ itself at 1270 nm is used to monitor its production. Following that, high quantum yields of $^1O_2$ are produced using peptide coated MSCQDs-PS conjugates (96).

The two photon-excitation is another approach to design water soluble and biocompatible CdSe-porphyrin composite QDs. The water solubility of MSCQDs was increased by encapsulating them in micelles via EDC/NHS conjugation without changing the inorganic shell. The disodium salt of 9,10-anthracenedipropionic acid (ADPA) reacts rapidly with $^1O_2$ and forms the endoperoxide of ADPA, is used to measure $^1O_2$ generation. The quantum yield of $^1O_2$ is nearly increase about two-fold by two-photon excitation, when compared to free porphyrin (97).

Due to its excellent generation of $^1O_2$ under two-photon stimulation, another CdSe/ZnS MSCQD coupled with rose Bengal (RB) is employed for PDT. RB molecules are attached to the surface via the EDC/NHS conjugation after the CdSe/ZnS are phase-transferred into water using mercaptopropionic acid. The photo-oxidation of 1,3-diphenylisobenzofuran (DPBF) into its corresponding diketonic form is used to determine the $^1O_2$ production. The cell viability of HeLa cells is significantly reduced after treated with PS-conjugated QDs excited by two-photon light stimulation (98).

It is also observed that PbS MSCQDs coated with S-nitrosocysteine and adsorbed on TiO_2 nanotubes generate $^1O_2$. Using thiolatic acid and Pb $(NO_3)_2$, PbSe QDs are grafted on the surface of TiO2 nanotubes, and then surface functionalization was done by l-cysteine and loaded with NO. Following that, photoactivation of the hybrid system causes NO to be released and $^1O_2$ to be produced. Direct illumination of CdSe/ZnS alone without the introduction of any PS drug is also known to create ROI such as $^1O_2$, hydroxyl radical, and cause DNA damage, as seen in gel electrophoresis implicating base-excision repair enzymes (99).

8.4.3 Carbon-based M-QDs

Existence of the π-π stacking interactions in structure and ability to generate ROS as a result of functioning as a phototherapy agent in PDT and PTT, carbon-based allotropes such as nanotubes, fullerenes, and graphenes have been widely investigated for biomedical applications (100). Due to its unique physicochemical properties, graphene became one of the most studied materials with applications in different field as material science, energy, electronics, and biomedicine including the treatment of COVID-19 disease (101). In terms of bioapplications, graphene's versatility has enabled it to be explored as an antibacterial agent, sensors, drug and gene delivery, bioimaging, regenerative medicine, cancer treatment, PDT and PTT (102,103,104).

8.4.3.1 Graphene Quantum Dots (GQDs)

GQDs having zero dimensions are novel graphene analogues. A maximum of 10 stacked layers of graphene sheets makes up their thickness of less than 100 nm. Quantum confinement and surface edge effect properties are triggered by the smaller dimensions. GQDs offer a wide range of applications due to their low toxicity, biocompatibility, and photostability, including cell imaging, drug/gene delivery, and biosensors (105).

Ge et al. used GQDs and protoporphyrin-IX (PpIX) to investigate the PDT efficacy and cytotoxicity in HeLa cells exposed to 630 nm light for 10 min. A 60 % cell viability was found at the concentration 0.036 μM of GQDs, which further decrease with increase concentration. GQDs, on the other hand, had little effect on HeLa cell viability, showing that they are biocompatible and have low cytotoxicity. At the 1.8 μM of PpIX a low cell viability of 55% was observed in the darkness, and more than 35% of the cells survived even after irradiation. GQDs surpassed PpIX in terms of PDT efficiency and cytotoxicity. GQDs-PpIX shows elevated ROS levels in PDT *in-vivo* studies. After 9 and 17 days of post-treatment decrease of the tumor volume was confirmed in female BALB/c mice (106).

Li et al. prepared Folic acid (FA) EDC/NHS covalently conjugated GQDs loaded with IR780 dye for simultaneous NIR imaging and PTT after 808 nm NIR irradiation for 5 min. The cellular uptake of IR780/GQDs-FA was measured in HeLa cells using laser confocal scanning microscopy (LCSM). The results showed that higher accumulation of NPs inside the cells as compared to free IR780 and FA, while free FA competitively inhibited the IR780/GQDs-FA interaction with folate receptor which overexpressed in HeLa cells. The photothermal conversation efficacy of IR780/GQDs-FA was found to be 87.9% and exhibited higher raise in temperature as compared to free IR780 and GQDs. In-vitro cell cytotoxicity in HeLa cells was performed by CCK-8 assay, the results indicated that NPs-conjugate showed higher cytotoxicity upon NIR irradiation. The in-vivo Balb/c nude mice bearing HeLa cells showed greater tumor imaging and targeting efficacy. The tumor site can be scanned and identified precisely, only tumors will be irradiated during PTT, causing minimal damage to surrounding normal tissue, enhancing the therapeutic potential of IR780/GQDs-FA and reducing systemic side effects (107).

Surface functionalization with chimeric functional group and atoms has been suggested in order to broaden the range of GQDs photoluminescence from visible and IR regions. It is worth mentioning that GQDs efficiency in PDT could be linked to their structural characteristics. Chen et al. was measured photoactivity of single-atomic-layered GQDs and negligible $^1O_2$ production under halogen and NIR irradiation, which may be attributed to the morphology of the GQDs. The surface doping with sulfur and nitrogen on GQDs, improving phototherapy performance. Some experiments showed the influence of N-doped GQDs on ROS production (108).

Campbell and colleagues introduced a nanocomposite with three covalently bound molecules: hyaluronic acid, nitrogen-doped GQDs, and ferrocene. In-vitro cytotoxicity assay in HEK-293 cells showed insignificant responses at concentration 1 mg/ml, but in HeLa cells increased cytotoxicity by up to 20% after 72 h of incubation. Furthermore, analeptic ROS production was three-folds greater as compared to free ferrocene (109).

PTT studies have also estimated treatments under the second NIR window (1000–1700 nm) to improve infiltration and increase tumor impairment. Wang and group were prepared nitrogen and boron-doped GQDs for simultaneous *in-vivo* fluorescence imaging and PTT. In-vitro cytotoxicity was studied in three different cell lines SF763, 4T1 and B16F10, the results was showed that N-B-GQDs showed nontoxic behavior after 72 h of treatment, while after NIR-II irradiation N-B-GQDs exhibited highly photothermal conversation efficacy and significant decreased in cell viability as compared to non-irradiated cells. These indicated that combined N-B-GQDs/NIR-II showed enhanced PTT efficacy in-vitro. In-vivo and ex-vivo NIR-II fluorescence imaging and PTT was evaluated in C6 glioblastoma bearing nude mice xenograft model. The NIR fluorescence imaging showed higher accumulation of NPs inside the tumor, and significant decreased in tumor volume was found after 14 days of post-treatment (110).

Thakur et al. was used specifically withered leaves of Ficus racemose (Indian fig tree) as a waste carbon source to produce GQDs. The cell cycle analysis and biocompatibility of GQDs was measured in two cell lines: MDA-MB-231 and L292. The results demonstrated that GQDs discontinue the S- and G_2-Phase through terminating the DNA replications. Moreover, it was demonstrated that upon irradiation of 808 nm NIR irradiation (0.5 W/cm^2), the production of ROS and the concentration dependence of PTT response were studied in MDA-MB-231 cells (111).

Wang and coworkers developed DOX loaded with cRGD (Cyclic Arg-Gly-Asp peptide) conjugated GQDs to evaluate its PTT-chemotherapy against H460 and SK–mel–5 cells under 808 nm NIR irradiation. Moreover, its chemotherapy capacity was confirmed, signifying an IC_{50} decrease up to 39.63 μg/mL and 53.75 μg/mL, respectively. The composite was showed pH-responsive drug release and greater PTT efficacy (112).

8.4.3.2 Carbon Dots (C-DOTs)

C-DOTs are carbon-based quasi-0D materials with diameters less than 10 nm. According to the nature of sp^2 hybridization between core of carbon atom and inter surface functional groups, they have spherical or hemispherical morphology (113). Owing to their small size, fluorescence, low toxicity, high quantum yield, high biocompatibility, excellent and tunable photoluminescence, and low-cost synthesis, C-DOTs become newer emerging candidate in the carbon family, which received great interest, ensuring adequate applications in various fields, such as biomedicine, photocatalyst, optoelectronics, and anticounterfeiting (45, 100). C-DOTs have been found to be suitable for phototherapy applications as both nanocarriers and photo-absorption agents. In recent years, several hybrids produced through π-π stacking, covalent binding, or electrostatic interactions have been studied (114).

C-DOTs made from medicinal plants with a concentration of 0.5 mg/mL and exposed to 750 nm light for 10 minutes reached a temperature of up to 46 °C, demonstrating their effectiveness in PTT, according to Meena and colleagues. In addition, no apparent toxicity was found in NIH-3T3 normal cells. In-vitro photothermal conversation efficacy was measured in two different sources of cell lines: HeLa cancer cells and E. coli bacterial cells, the results show significant hyperthermia effect influence cytotoxicity in both cell lines. Hence prepared C-DOTs exhibited NIR irradiation mediated cancer as well as bacterial phototherapy (115).

Qin et al. employed o-phenylenediamine to synthesize C-DOTs, which showed a broad absorption peak at 380–500 nm and generated strong yellow fluorescence at 550 nm wavelength. HeLa cancer cells quickly absorbed the C-DOTs and reduced relative cell viability by up to 40%. Upon UV blue light excitation, a powerful ROS and bright yellow fluorescence were created proficiently, allowing for simultaneous PDT and fluorescent cancer cell imaging (116).

Li et al. used a simple one-pot hydrothermal method to produce porphyrin-containing C-DOTs from chitosan and mono-hydroxylphenyl triphenylporphyrin. *In-vitro* MTT assay in HepG2 cells showed significant cytotoxicity and greater PDT efficacy after 1 hr 625 nm LED irradiation. Cells undergoes lysis, severe apoptosis, and membrane rupture at a dosage of 0.5 mg/ml. *In-vivo* H22 tumor-bearing KM mice model demonstrated that tumor size of irradiated mice was decreases from 100 to 56 mm^3, but without irradiation, the tumor size increased to 800 mm^3 (117).

Guo and coworkers employed simple hydrothermal method to produce Cu, N-doped C-DOTs employing different temperatures condition to evaluate the synergistic PDT and PTT abilities. The in-vitro MTT studies in B16 cells showed concentration dependent cytotoxicity after NIR irradiation, in presence of antioxidant vitamin C reduces ROS generation and prevent cell death even upon NIR irradiation.

In-vivo biodistribution studies in Balb/c nude mice bearing B16 cells showed less tumor accumulation and high metabolic clearance from the body. In-vivo phototherapy was showed significant ROS generation and elevated tumor temperature up to 53 °C under 808 nm NIR irradiation, indicating anti-tumor activity (118).

C-DOTs were created by Hua et al. using a hydrothermal reaction of L-cysteine and m-phenylenediamine. C-DOTs and the PpIX were tested for cytotoxicity in HeLa cells. After 24 hr in the dark, free PpIX and C-DOTs-PpIX showed non-toxic against the cells, suggesting their good cytocompatibility. The in-vivo photoexcitation-based PDT effect was studied in U14 tumor bearing mice and C-DOTs-PpIX showed more tumor retention, nucleus targeting, and effective tumor inhibition compared to free PpIX (119).

Yang et al. produced Hemin-C-Dots composite, which raises the temperature of HepG2 cells to 26 °C under laser irradiation which significantly produces ROS and great PDT efficacy using the DCFH-DA probe. After 10 min of laser treatment more than 90% of cancer cells die. H-CQDs own hemin endowed synergistic dual modal PDT-PTT activity (120).

8.4.4 Hybrid M-QDs

The DOX-loaded Folic acid (FA) and riboflavin (Rf) conjugated Genipin (GP) cross-linked Fe, N-carboxymethyl-chitosan magneto-fluorescent targeted CQDs was developed by Zhang et al. The Gp-Rf-FA-FeN@CQDs-DOX nanocomposites shows synergistic PDT-PTT-chemotherapy activity after NIR irradiation. The in-vitro MTT assay in HeLa and HepG2 cells demonstrated concentration-dependent cytotoxicity. Due to over expression of folate receptor on the cellular membrane HeLa cells exhibited greater cytotoxicity as compared to HepG2 cells. In tumor-bearing mice, the *in-vivo* combination treatment revealed that the GP-Rf-FA-FeN@ CQDs-DOX with laser treatment generated apparent tumor inhibition but no clear injury to important organs, providing a promising platform for treating malignant tumors (121).

Liu et al. developed Gold and Silver doped hybrid CQDs successfully using an extreme facile mechanism which have been intended for PTT in both *in-vitro* and *in-vivo*. The Au/Ag-CQDs hybrid nanocomposites showed insignificant dose-dependent cytotoxicity in C6 glioma cells. The PTT activity in HeLa cells was reveals that Au/Ag-CQDs nanocomposite has induced photocatalytic process caused heat targeting and killing of tumor cells. Effective targeting at the site of tumor in *in-vivo* Balb/c nude mice model exhibited blackening of tumor due to raise in local hyperthermia, and reduction in tumor size (122).

Nafuijjaman et al. prepared GQDs immobilized on Mn_3O_4 coated with PDA NP for real-time MRI with simultaneous PDT activity. In-vitro cytotoxicity in MDCK cells and A549 cells demonstrated that GQD-PAD- Mn_3O_4 NPs showed partial of cell viability after NIR irradiation. The PDT-induced breakdown of cellular membranes and a higher cellular uptake of NPs may contribute to the enhanced death of cancer cells by laser irradiation. In-vivo PDT efficacy in A549 tumor bearing mice model reveals that the average size and weight of tumor has reduced after NIR irradiation. The GQD-PDA-Mn_3O_4 NPs exhibited prolonged circulation, high specificity for tumor cells visualization by MRI, due to excellent colloidal stability and accumulation by EPR effect (123).

Zhang et al. was developed multifunctional MOF made up of Paclitaxel (PTX) loaded porous coordination network (PCN) complexed Fe_3O nanocluster (mesoporous PCN-Fe (III)-PTX-NC) generate ROS and iron dependent Fenton like reaction enhancing PDT efficacy and anti-tumor activity in in-vivo and in-vitro colorectal cancer model. In-vitro drug release and real-time MRI studies in PANC-1 cells showed pH-dependent drug release at the tumor site. In-vitro CCK/8 cytotoxicity assay in PANC-1 cells shows that significant reduction in cell-viability after NIR irradiation. Peroxide degradation to single oxygen was estimated by intracellular MAK164 assay result was demonstrated that significant reduction in fluorescence intensity after NC incubation due to ROS generation in hypoxic tumor. In-vivo anti-tumor activity in Balb/c mice bearing allograft tumor model showed that simultaneous targeted drug release and PDT activity causes complete tumor regression after NIR radiation (124).

Yao et al. developed DOX loaded mesoporous Silica coated GQDs capped magnetic Fe_2O_3 NPs for simultaneous PTT-chemotherapy. Cytotoxicity was observed using CCK-8 assay in in-vitro 4T1 cells, cell viability was significantly reduced after combined NIR irradiation and magnetic field treatment. The DOX-MMSN/GQDs exhibited significant hyperthermia effects and higher drug release in acidic environment in in-vitro drug release studies. In comparison to chemotherapy and magnetic PTT alone, the combination chemo-magnetic PTT therapy had synergistic effects, leading to a greater cancer cell killing efficacy (125).

Zhang et al. prepared magneto-fluorescence multifunctional tumor targeted theranostic NCs comprising DOX loaded sgc8 aptamer conjugated SWCNT coated PEGylated Fe_3O_4@C-DOTs for synergistic PDT-PTT-chemotherapy. The in-vitro cytotoxicity in HeLa cells of SWCNTs-PEG- Fe_3O_4@CQDs/DOX-Apt transformed 808 nm NIR light into thermal energy, produced ROS, pH triggered DOX release and destroyed cancer cells. In-vivo HeLa tumor bearing mice was showed targeted tumor inhibition of NCs after post-IV injection. These nanocarriers could be useful in the treatment of cervical cancer and other disorders that require precision therapeutic targeting (126).

8.5 CLINICAL INTERFERENCE OF PHOTOTHERAPIES

Several clinical trials were approved and currently employed nanomedicines-based cancer therapy in recent years. Notwithstanding clinical functions are currently scares and face significant challenges, which must be overcome before post-approval surveillance (Figure 8.4). Laser systems are currently used in clinical PTT therapy since they may produce thermal ablation by effectively influencing chromophores located inside the endogenous tissue. This technique decreases the regulatory burden and PTT agent development costs. PTT agent augmented thermal ablation provides significant benefits, as well as greater tumor targeting affinity and easy fabrication by using lower-power lasers (37). Although, major trials employing light-absorbing gold-base nanomaterials as PTT agent, they have been limited to preclinical research. Table 8.1 brief a few early-phase pilot clinical trials of patented gold induced PTT-thermal ablation technologies.

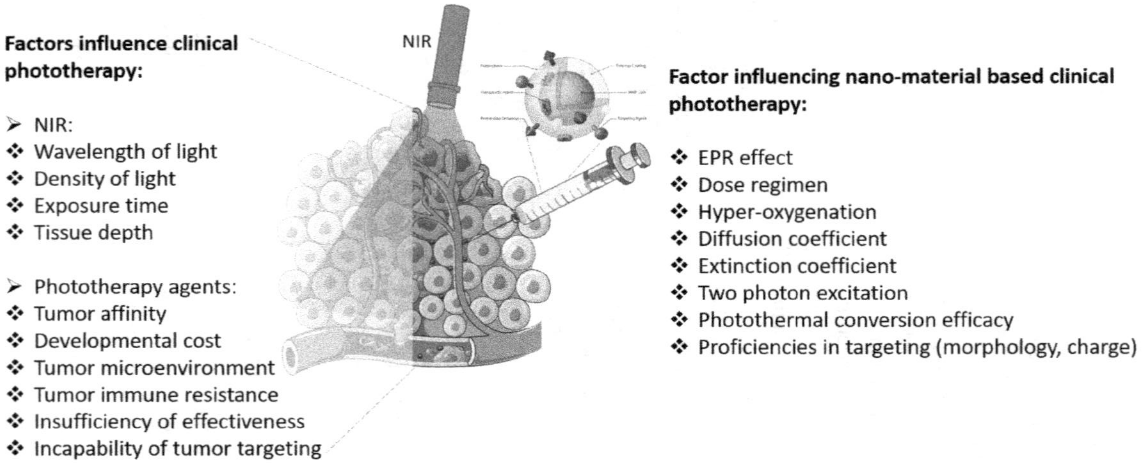

Factors influence clinical phototherapy:

➢ NIR:
- Wavelength of light
- Density of light
- Exposure time
- Tissue depth

➢ Phototherapy agents:
- Tumor affinity
- Developmental cost
- Tumor microenvironment
- Tumor immune resistance
- Insufficiency of effectiveness
- Incapability of tumor targeting

NIR

Factor influencing nano-material based clinical phototherapy:

- EPR effect
- Dose regimen
- Hyper-oxygenation
- Diffusion coefficient
- Extinction coefficient
- Two photon excitation
- Photothermal conversion efficacy
- Proficiencies in targeting (morphology, charge)

FIGURE 8.4 Challenges of clinical phototherapy applications.

TABLE 8.1
On ClinicalTrials.gov, There Have Been Four Clinical Studies Based on This PTT Platform So Far

Clinical Trial No.	Materials Employed	Therapeutics Modality	Cancer Type	Population Size (n)	Phase	Trial Status	Ref.
NCT00848042	AuroShell particle (non-	PTT	Head and Neck	11	NA	Completed	(127)
NCT02680535	conducting silica core		Prostate	45		Completed	(128)
NCT04240639	with gold coating)			60		Recruiting	(129)
NCT03945162	TLD1433 (Ru (II) polypyridyl complexes)	PDT and PCT	Non-Muscle Invasive Bladder Cancer	125	II	Recruiting	(130)

Nanospectra Biosciences designed Aurolase®, the most well-known PTT system included PEGylated 120-nm silica dielectric core coated with gold 150-nm gold nanoshell for NIR triggered thermal ablation (131). AuroShellTM produces hyperthermia induce cell death after NIR irradiation with minimal injury to normal peripheral tissue. The clinical trials evidence that NPs were accumulated inside the tumor tissues through EPR effect after IV administration, and induce PTT effect causes cell death and tumor suppression (132,133).

REFERENCES

1. Sebastian R, Brown T. Phototherapy in cancer prevention and treatment. J Cancer Prev Curr Res. 2017;7(1):22–4.
2. Sidharth E, Hamada M, Murase N, Biju V. Nanomaterials formulations for photothermal and photodynamic therapy of cancer. J Photochem Photobiol C Photochem Rev. 2013;15:53–72.
3. Shi H, Sadler PJ. How promising is phototherapy for cancer. Br J Cancer. 2020;123(6):871–3.
4. Deng X, Shao Z, Zhao Y. Solutions to the drawbacks of photothermal and photodynamic cancer therapy. Vol. 8, Advanced Science. 2021. 2002504.
5. Wen X, Li Y, Hamblin MR. Photodynamic therapy in dermatology beyond non-melanoma cancer: An update. Photodiagnosis Photodyn Ther. 2017 Sep 1;19:140–52.
6. Gunaydin G, Gedik M, Ayan S. PDT current limitations and novel approaches. Front Chem. 2021;9:691697.
7. Pivetta TP, Botteon CEA, Ribeiro PA, Marcato PD, Raposo M. Nanoparticle systems for cancer phototherapy: An overview. Nanomaterials. 2021;11:3132.
8. Oniszczuk A, Wojtunik-Kulesza KA, Oniszczuk T, Kasprzak K. The potential of photodynamic therapy (PDT)—Experimental investigations and clinical use. Biomed Pharmacother. 2016 Oct 1;83:912–29.
9. Dolmans DEJGJ, Fukumura D, Jain RK. Photodynamic therapy for cancer. Nat Rev Cancer 2003 35 [Internet]. 2003 May [cited 2022 May 29];3(5):380–7. Available from: www.nature.com/articles/nrc1071
10. Kelly JF, Snell ME, Berenbauai MC. Photodynamic destruction of human bladder carcinoma. Br J Cancer [Internet]. 1975 [cited 2022 May 29];31(2):237–44. Available from: https://pubmed.ncbi.nlm.nih.gov/1164470/
11. Stüttgen G. The risk of photochemotherapy [Internet]. Vol. 21, International Journal of Dermatology. Int J Dermatol; 1982 [cited 2022 May 29]. 198–202. Available from: https://pubmed.ncbi.nlm.nih.gov/7047417/
12. Zhang J, Jiang C, Figueiró Longo JP, Azevedo RB, Zhang H, Muehlmann LA. An updated overview on the development of new photosensitizers for anticancer photodynamic therapy. Acta Pharm Sin B. 2018 Mar 1;8(2):137–46.
13. Xiong H, Zhou K, Yan Y, Miller JB, Siegwart DJ. Tumor-activated water-soluble photosensitizers for near-infrared photodynamic cancer therapy. ACS Appl Mater Interfaces [Internet]. 2018 May 16 [cited 2022 May 31];10(19):16335–43. Available from: https://pubs.acs.org/doi/abs/10.1021/acsami.8b04710
14. Yang X, Shi X, D'arcy R, Tirelli N, Zhai G. Amphiphilic polysaccharides as building blocks for self-assembled nanosystems: molecular design and application in cancer and inflammatory diseases. J Control Release. 2018 Feb 28;272:114–44.
15. Patel A, Sant S. Hypoxic tumor microenvironment: Opportunities to develop targeted therapies. Biotechnol Adv. 2016 Sep 1;34(5):803–12.

16. Xu Y, Liu R, Yang H, Qu S, Qian L, Dai Z. Enhancing photodynamic therapy efficacy against cancer metastasis by ultrasound-mediated oxygen microbubble destruction to boost tumor-targeted delivery of oxygen and renal-clearable photosensitizer micelles. ACS Appl Mater Interfaces [Internet]. 2022 May 26 [cited 2022 May 31]; Available from: https://pubs.acs.org/doi/abs/10.1021/acsami.2c06655
17. Jin C, Yuan M, Bu H, Jin C. Antiangiogenic strategies in epithelial ovarian cancer: Mechanism, resistance, and combination therapy. Tomao F, editor. J Oncol [Internet]. 2022 Apr 12 [cited 2022 May 31];2022:1–15. Available from: www.hindawi.com/journals/jo/2022/4880355/
18. Zhang S, Li Y, Li T, Zhang Y, Li H, Cheng Z, et al. Activable targeted protein degradation platform based on light-triggered singlet oxygen. J Med Chem [Internet]. 2022 Feb 24 [cited 2022 May 31];65(4):3632–43. Available from: https://pubs.acs.org/doi/abs/10.1021/acs.jmedchem.1c02037
19. van Straten D, Mashayekhi V, de Bruijn HS, Oliveira S, Robinson DJ. Oncologic photodynamic therapy: Basic principles, current clinical status and future directions [Internet]. Vol. 9, Cancers. Multidisciplinary Digital Publishing Institute; 2017 [cited 2022 May 31]. p. 19. Available from: www.mdpi.com/2072-6694/9/2/19/htm
20. Kim MM, Darafsheh A. Light sources and dosimetry techniques for photodynamic therapy. Photochem Photobiol [Internet]. 2020 Mar 1 [cited 2022 May 31];96(2):280–94. Available from: https://onlinelibrary.wiley.com/doi/full/10.1111/php.13219
21. Brancaleon L, Moseley H. Laser and non-laser light sources for photodynamic therapy. Lasers Med Sci [Internet]. 2002 Feb 12 [cited 2022 May 31];17(3):173–86. Available from: https://link.springer.com/article/10.1007/s101030200027
22. Sharman WM, Allen CM, Van Lier JE. Role of activated oxygen species in photodynamic therapy. Methods Enzymol. 2000 Jan 1;319:376–400.
23. Kessel D, Reiners JJ. Promotion of proapoptotic signals by lysosomal photodamage. Photochem Photobiol [Internet]. 2015 Jul 1 [cited 2022 May 31];91(4):931–6. Available from: https://onlinelibrary.wiley.com/doi/full/10.1111/php.12456
24. Karaman O, Almammadov T, Emre Gedik M, Gunaydin G, Kolemen S, Gunbas G. Mitochondria-targeting selenophene-modified bodipy-based photosensitizers for the treatment of hypoxic cancer cells. ChemMedChem [Internet]. 2019 Nov 20 [cited 2022 May 31];14(22):1879–86. Available from: https://onlinelibrary.wiley.com/doi/full/10.1002/cmdc.201900380
25. Chen W, Liu J, Wang Y, Jiang C, Yu B, Sun Z, et al. A C5N2 nanoparticle based direct nucleus delivery platform for synergistic cancer therapy. Angew Chemie – Int Ed [Internet]. 2019 May 6 [cited 2022 May 31];58(19):6290–4. Available from: https://onlinelibrary.wiley.com/doi/full/10.1002/anie.201900884
26. Ozdemir T, Lu YC, Kolemen S, Tanriverdi-Ecik E, Akkaya EU. Generation of singlet oxygen by persistent luminescent nanoparticle–photosensitizer conjugates: A proof of principle for photodynamic therapy without light. ChemPhotoChem [Internet]. 2017 May 1 [cited 2022 May 31];1(5):183–7. Available from: https://onlinelibrary.wiley.com/doi/full/10.1002/cptc.201600049
27. Bolze F, Jenni S, Sour A, Heitz V. Molecular photosensitisers for two-photon photodynamic therapy. Chem Commun [Internet]. 2017 Nov 30 [cited 2022 May 31];53(96):12857–77. Available from: https://pubs.rsc.org/en/content/articlehtml/2017/cc/c7cc06133a
28. Turan IS, Cakmak FP, Yildirim DC, Cetin-Atalay R, Akkaya EU. Near-IR absorbing BODIPY derivatives as glutathione-activated photosensitizers for selective photodynamic action. Chem – A Eur J [Internet]. 2014 Dec 1 [cited 2022

May 31];20(49):16088–92. Available from: https://onlinelibrary.wiley.com/doi/full/10.1002/chem.201405450

29. Ni K, Lan G, Veroneau SS, Duan X, Song Y, Lin W. Nanoscale metal-organic frameworks for mitochondria-targeted radiotherapy-radiodynamic therapy. Nat Commun [Internet]. 2018 Oct 17 [cited 2022 May 31];9(1):1–13. Available from: www.nature.com/articles/s41467-018-06655-7
30. Blum NT, Zhang Y, Qu J, Lin J, Huang P. Recent advances in self-exciting photodynamic therapy. Vol. 8, Frontiers in Bioengineering and Biotechnology. Frontiers Media S.A.; 2020. 1136.
31. Abrahamse H, Hamblin MR. New photosensitizers for photodynamic therapy [Internet]. Vol. 473, Biochemical Journal. Portland Press; 2016 [cited 2022 May 31]. p. 347–64. Available from: https://portlandpress.com/biochemj/article-abstract/473/4/347/49333/New-photosensitizers-for-photodynamic-therapy
32. Chen Q, Huang Z, Chen H, Shapiro H, Beckers J, Hetzel FW. Improvement of tumor response by manipulation of tumor oxygenation during photodynamic therapy. Photochem Photobiol [Internet]. 2007 Aug 1 [cited 2022 May 31];76(2):197–203. Available from: https://onlinelibrary.wiley.com/doi/full/10.1562/0031-8655%282002%290760197IOTRBM2.0.CO2
33. Henderson BW, Busch TM, Snyder JW. Fluence rate as a modulator of PDT mechanisms. Lasers Surg Med [Internet]. 2006 Jun 1 [cited 2022 May 31];38(5):489–93. Available from: https://onlinelibrary.wiley.com/doi/full/10.1002/lsm.20327
34. Dolmans DEJGJ, Kadambi A, Hill JS, Flores KR, Gerber JN, Walker JP, et al. Targeting tumor vasculature and cancer cells in orthotopic breast tumor by fractionated photosensitizer dosing photodynamic therapy. Cancer Res [Internet]. 2002 [cited 2022 May 31];62(15):4289–94. Available from: https://pubmed.ncbi.nlm.nih.gov/12154031/
35. Bisland, SK, Lilge L, Lin A, Rusnov R, Wilson BC. Metronomic photodynamic therapy as a new paradigm for photodynamic therapy: Rationale and preclinical evaluation of technical feasibility for treating malignant brain tumors. Photochem Photobiol [Internet]. 2004 [cited 2022 May 31];80(1):22. Available from: https://pubmed.ncbi.nlm.nih.gov/15339204/
36. Knavel E, Brace C. Tumor ablation: common modalities and general practices. Tech Vasc Interv rediology [Internet]. 2013 [cited 2022 Jun 1];16(4):192–200. Available from: www.sciencedirect.com/science/article/pii/S1089251613000632
37. Han HS, Choi KY. Advances in nanomaterial-mediated photothermal cancer therapies: Toward clinical applications. Biomedicines. 2021;9(3):1–15.
38. Richter K, Haslbeck M, Buchner J. The heat shock response: Life on the verge of death. Mol Cell [Internet]. 2010 [cited 2022 Jun 1];40(2):253–66. Available from: www.sciencedirect.com/science/article/pii/S1097276510007823
39. Fernandes N, Rodrigues CF, Moreira AF, Correia IJ. Overview of the application of inorganic nanomaterials in cancer photothermal therapy. Biomater Sci [Internet]. 2020 Jun 2 [cited 2022 Jun 1];8(11):2990–3020. Available from: https://pubs.rsc.org/en/content/articlehtml/2020/bm/d0bm00222d
40. Li X, Lovell JF, Yoon J, Chen X. Clinical development and potential of photothermal and photodynamic therapies for cancer. Nat Rev Clin Oncol. 2020;17(11):657–74.
41. Liu Y, Bhattarai P, Dai Z, Chen X. Photothermal therapy and photoacoustic imaging via nanotheranostics in fighting cancer. Chem Soc Rev [Internet]. 2019 Apr 1 [cited 2022 Jun 1];48(7):2053–108. Available from: https://pubs.rsc.org/en/content/articlehtml/2019/cs/c8cs00618k

42. Yao C, Zhang L, Wang J, He Y, Xin J, Wang S, et al. Gold nanoparticle mediated phototherapy for cancer. Vol. 2016, Journal of Nanomaterials. Hindawi Limited; 2016.
43. Matsumoto Y, Nichols JW, Toh K, Nomoto T, Cabral H, Miura Y, et al. Vascular bursts enhance permeability of tumour blood vessels and improve nanoparticle delivery. Nat Nanotechnol [Internet]. 2016 Feb 15 [cited 2022 Jun 1];11(6):533–8. Available from: www.nature.com/articles/nnano.2015.342
44. Simon T, Gagliano T, Giamas G. Direct effects of anti-angiogenic therapies on tumor cells: VEGF signaling. Trends Mol Med. 2017 Mar 1;23(3):282–92.
45. Gao D, Guo X, Zhang X, Chen S, Wang Y, Chen T, et al. Multifunctional phototheranostic nanomedicine for cancer imaging and treatment. Mater Today Bio [Internet]. 2020;5:100035. Available from: https://doi.org/10.1016/j.mtbio.2019.100035
46. Yang Y, Zhu W, Dong Z, Chao Y, Xu L, Chen M, et al. 1D coordination polymer nanofibers for low-temperature photothermal therapy. Adv Mater [Internet]. 2017 Oct 1 [cited 2022 Jun 1];29(40):1703588. Available from: https://onlinelibrary.wiley.com/doi/full/10.1002/adma.201703588
47. Li WT, Peng JR, Tan LW, Wu J, Shi K, Qu Y, et al. Mild photothermal therapy/photodynamic therapy/chemotherapy of breast cancer by Lyp-1 modified Docetaxel/IR820 Co-loaded micelles. Biomaterials. 2016 Nov 1;106:119–33.
48. Liu J, Li HJ, Luo YL, Xu CF, Du XJ, Du JZ, et al. Enhanced primary tumor penetration facilitates nanoparticle draining into lymph nodes after systemic injection for tumor metastasis inhibition. ACS Nano [Internet]. 2019 Aug 27 [cited 2022 Jun 1];13(8):8648–58. Available from: https://pubs.acs.org/doi/abs/10.1021/acsnano.9b03472
49. Agabeigi R, Rasta SH, Rahmati-Yamchi M, Salehi R, Alizadeh E. Novel chemo-photothermal therapy in breast cancer using methotrexate-loaded folic acid conjugated Au@SiO2 nanoparticles. Nanoscale Res Lett [Internet]. 2020 Mar 19 [cited 2022 Jun 1];15(1):1–14. Available from: https://nanoscalereslett.springeropen.com/articles/10.1186/s11671-020-3295-1
50. Jacinto TA, Rodrigues CF, Moreira AF, Miguel SP, Costa EC, Ferreira P, et al. Hyaluronic acid and vitamin E polyethylene glycol succinate functionalized gold-core silica shell nanorods for cancer targeted photothermal therapy. Colloids Surfaces B Biointerfaces. 2020 Apr 1;188:110778.
51. Takagahara T, Takeda K. Theory of the quantum confinement effect on excitons in quantum dots of indirect-gap materials. Phys Rev B [Internet]. 1992 Dec 15 [cited 2022 Jun 1];46(23):15578–81. Available from: https://journals.aps.org/prb/abstract/10.1103/PhysRevB.46.15578
52. Jain PK, Huang X, El-Sayed IH, El-Sayed MA. Noble metals on the nanoscale: Optical and photothermal properties and some applications in imaging, sensing, biology, and medicine. Acc Chem Res. 2008 Dec 16;41(12):1578–86.
53. Hutter E, Fendler JH. Exploitation of localized surface plasmon resonance. Adv Mater [Internet]. 2004 Oct 4 [cited 2022 May 29];16(19):1685–706. Available from: https://onlinelibrary.wiley.com/doi/full/10.1002/adma.200400271
54. Petryayeva E, Krull UJ. Localized surface plasmon resonance: Nanostructures, bioassays and biosensing - A review. Anal Chim Acta. 2011 Nov 7;706(1):8–24.
55. Estelrich J, Antònia Busquets M. Iron oxide nanoparticles in photothermal therapy. Mol 2018, Vol 23, Page 1567 [Internet]. 2018 Jun 28 [cited 2022 May 30];23(7):1567. Available from: www.mdpi.com/1420-3049/23/7/1567/htm
56. Alphandéry E. Bio-synthesized iron oxide nanoparticles for cancer treatment. Int J Pharm. 2020 Aug 30;586:119472.

57. Munasinghe E, Aththapaththu M, Jayarathne L. Magnetic and Quantum Dot Nanoparticles for Drug Delivery and Diagnostic Systems. In: Colloid Science in Pharmaceutical Nanotechnology. London: IntechOpen. 2020. pp. 1–15.
58. Kharey P, Dutta SB, Manikandan M, Palani IA, Majumder SK, Gupta S. Green synthesis of near-infrared absorbing eugenate capped iron oxide nanoparticles for photothermal application. Nanotechnology [Internet]. 2020 [cited 2022 May 30];31(9). Available from: https://pubmed.ncbi.nlm.nih.gov/31715590/
59. Liu Q, Liu L, Mo C, Zhou X, Chen D, He Y, et al. Polyethylene glycol-coated ultrasmall superparamagnetic iron oxide nanoparticles-coupled sialyl Lewis X nanotheranostic platform for nasopharyngeal carcinoma imaging and photothermal therapy. J Nanobiotechnology [Internet]. 2021 Dec 1 [cited 2022 May 30];19(1):1–14. Available from: https://jnanobiotechnology.biomedcentral.com/articles/10.1186/s12951-021-00918-0
60. Wang Y, Li X, Chen P, Dong Y, Liang G, Yu Y. Enzyme-instructed self-aggregation of Fe3O4 nanoparticles for enhanced MRI T 2 imaging and photothermal therapy of tumors. Nanoscale [Internet]. 2020 Jan 23 [cited 2022 May 30];12(3):1886–93. Available from: https://pubs.rsc.org/en/content/articlehtml/2020/nr/c9nr09235h
61. Cabana S, Curcio A, Michel A, Wilhelm C, Abou-Hassan A. Iron oxide mediated photothermal therapy in the second biological window: A comparative study between magnetite/maghemite nanospheres and nanoflowers. Nanomaterials [Internet]. 2020 Aug 7 [cited 2022 May 30];10(8):1–17. Available from: www.mdpi.com/2079-4991/10/8/1548/htm
62. Ailioaie LM, Litscher G. Curcumin and photobiomodulation in chronic viral hepatitis and hepatocellular carcinoma. Int J Mol Sci [Internet]. 2020 Sep 28 [cited 2022 May 30];21(19):1–25. Available from: www.mdpi.com/1422-0067/21/19/7150/htm
63. Zhang F, Lu G, Wen X, Li F, Ji X, Li Q, et al. Magnetic nanoparticles coated with polyphenols for spatio-temporally controlled cancer photothermal/immunotherapy. J Control Release. 2020 Oct 10;326:131–9.
64. Li J, Zhang F, Hu Z, Song W, Li G, Liang G, et al. Drug "pent-up" in hollow magnetic prussian blue nanoparticles for NIR-induced chemo-photothermal tumor therapy with trimodal imaging. Adv Healthc Mater. 2017;6(14):1–12.
65. Ferro-Flores G, Ocampo-García B, Santos-Cuevas C, María Ramírez F, Azorín-Vega E, Meléndez-Alafort L. Theranostic radiopharmaceuticals based on gold nanoparticles labeled with 177lu and conjugated to peptides. Curr Radiopharm. 2015 Aug 27;8(2):150–9.
66. Guo J, Rahme K, He Y, Li LL, Holmes JD, O'Driscoll CM. Gold nanoparticles enlighten the future of cancer theranostics. Int J Nanomedicine [Internet]. 2017 Aug 22 [cited 2022 May 30];12:6131–52. Available from: www.dovepress.com/gold-nanoparticles-enlighten-the-future-of-cancer-theranostics-peer-reviewed-fulltext-article-IJN
67. Santos-Martinez MJ, Rahme K, Corbalan JJ, Faulkner C, Holmes JD, Tajber L, et al. Pegylation increases platelet biocompatibility of gold nanoparticles. J Biomed Nanotechnol. 2014;10(6):1004–15.
68. Carrillo-Cazares A, Jiménez-Mancilla NP, Luna-Gutiérrez MA, Isaac-Olivé K, Camacho-López MA. Study of the optical properties of functionalized gold nanoparticles in different tissues and their correlation with the temperature increase. J Nanomater. 2017;2017.
69. Giustini AJ, Petryk AA, Cassim SM, Tate JA, Baker I, Hoopes PJ. Magnetic nanoparticle hyperthermia in cancer treatment. Nano Life [Internet]. 2010 Mar [cited 2022 May 30];1:17–32. Available from: /pmc/articles/PMC3859910/

70. Saw WS, Ujihara M, Chong WY, Voon SH, Imae T, Kiew LV, et al. Size-dependent effect of cystine/citric acid-capped confeito-like gold nanoparticles on cellular uptake and photothermal cancer therapy. Colloids Surfaces B Biointerfaces. 2018 Jan 1;161:365–74.
71. Sun Y, Wang Q, Chen J, Liu L, Ding L, Shen M, et al. Temperature-sensitive gold nanoparticle-coated Pluronic-PLL nanoparticles for drug delivery and chemo-photothermal therapy. Theranostics [Internet]. 2017 [cited 2022 May 30];7(18):4424–44. Available from: https://pubmed.ncbi.nlm.nih.gov/29158837/
72. Wang R, Deng J, He D, Yang E, Yang W, Shi D, et al. PEGylated hollow gold nanoparticles for combined X-ray radiation and photothermal therapy in vitro and enhanced CT imaging in vivo. Nanomedicine Nanotechnology, Biol Med. 2019 Feb 1;16:195–205.
73. AL-Barram LFA. Laser enhancement of cancer cell destruction by photothermal therapy conjugated glutathione (GSH)-coated small-sized gold nanoparticles. Lasers Med Sci [Internet]. 2021 May 12 [cited 2022 May 30];36(2):325–37. Available from: https://link.springer.com/article/10.1007/s10103-020-03033-y
74. Cheng M, Zhang Y, Zhang X, Wang W, Yuan Z. One-pot synthesis of acid-induced: In situ aggregating theranostic gold nanoparticles with enhanced retention in tumor cells. Biomater Sci [Internet]. 2019 Apr 23 [cited 2022 May 30];7(5):2009–22. Available from: https://pubs.rsc.org/en/content/articlehtml/2019/bm/c9bm00014c
75. Cheng X, Sun R, Yin L, Chai Z, Shi H, Gao M. Light-triggered assembly of gold nanoparticles for photothermal therapy and photoacoustic imaging of tumors in vivo. Adv Mater [Internet]. 2017 Feb 1 [cited 2022 May 30];29(6):1604894. Available from: https://onlinelibrary.wiley.com/doi/full/10.1002/adma.201604894
76. Thompson EA, Graham E, Macneill CM, Young M, Donati G, Wailes EM, et al. Differential response of MCF7, MDA-MB-231, and MCF 10A cells to hyperthermia, silver nanoparticles and silver nanoparticle-induced photothermal therapy. Int J Hyperth [Internet]. 2014 [cited 2022 May 30];30(5):312–23. Available from: www.tandfonline.com/doi/abs/10.3109/02656736.2014.936051
77. Dos Santos CA, Seckler MM, Ingle AP, Gupta I, Galdiero S, Galdiero M, et al. Silver nanoparticles: Therapeutical uses, toxicity, and safety issues. J Pharm Sci [Internet]. 2014 Jul 1 [cited 2022 May 30];103(7):1931–44. Available from: http://jpharmsci.org/article/S0022354915305128/fulltext
78. Wu J, Li N, Yao Y, Tang D, Yang D, Ong'Achwa Machuki J, et al. DNA-stabilized silver nanoclusters for label-free fluorescence imaging of cell surface glycans and fluorescence guided photothermal therapy. Anal Chem [Internet]. 2018 Dec 18 [cited 2022 May 30];90(24):14368–75. Available from: https://pubs.acs.org/doi/abs/10.1021/acs.analchem.8b03837
79. Park T, Lee S, Amatya R, Cheong H, Moon C, Kwak HD, et al. ICG-loaded pegylated BSA-silver nanoparticles for effective photothermal cancer therapy. Int J Nanomedicine [Internet]. 2020 Jul 31 [cited 2022 May 30];15:5459–71. Available from: www.dovepress.com/icg-loaded-pegylated-bsa-silver-nanoparticles-for-effective-phototherm-peer-reviewed-fulltext-article-IJN
80. Ding B, Zheng P, Ma P, Lin J. Manganese oxide nanomaterials: Synthesis, properties, and theranostic applications [Internet]. Vol. 32, Advanced Materials. John Wiley & Sons, Ltd; 2020 [cited 2022 May 30]. 1905823. Available from: https://onlinelibrary.wiley.com/doi/full/10.1002/adma.201905823
81. Xiang Y, Li N, Guo L, Wang H, Sun H, Li R, et al. Biocompatible and pH-sensitive MnO-loaded carbonaceous nanospheres (MnO@CNSs): A theranostic agent for magnetic resonance imaging-guided photothermal therapy. Carbon N Y. 2018 Sep 1;136:113–24.

82. Wu Y, Chen Z, Yao Z, Zhao K, Shao F, Su J, et al. Black phosphorus quantum dots encapsulated biodegradable hollow mesoporous MnO2: Dual-modality cancer imaging and synergistic chemo-phototherapy. Adv Funct Mater. 2021;31:2104643.
83. Zhou L, Wu Y, Meng X, Li S, Zhang J, Gong P, et al. Dye-anchored MnO nanoparticles targeting tumor and inducing enhanced phototherapy effect via mitochondria-mediated pathway. Small [Internet]. 2018 Sep 1 [cited 2022 May 30];14(36):1801008. Available from: https://onlinelibrary.wiley.com/doi/full/10.1002/smll.201801008
84. Lu WL, Lan YQ, Xiao KJ, Xu QM, Qu LL, Chen QY, et al. BODIPY-Mn nanoassemblies for accurate MRI and phototherapy of hypoxic cancer. J Mater Chem B. 2017;5(6):1275–83.
85. Liu M, Zhu H, Wang Y, Sevencan C, Li BL. Functionalized MoS2-based nanomaterials for cancer phototherapy and other biomedical applications [Internet]. Vol. 3, ACS Materials Letters. American Chemical Society; 2021 [cited 2022 May 31]. p. 462–96. Available from: https://pubs.acs.org/doi/abs/10.1021/acsmaterialslett.1c00073
86. Chen Y, Khan AR, Yu D, Zhai Y, Ji J, Shi Y, et al. Pluronic F127-functionalized molybdenum oxide nanosheets with pH-dependent degradability for chemo-photothermal cancer therapy. J Colloid Interface Sci. 2019 Oct 1;553:567–80.
87. Ma N, Zhang MK, Wang XS, Zhang L, Feng J, Zhang XZ. NIR light-triggered degradable MoTe2 nanosheets for combined photothermal and chemotherapy of cancer. Adv Funct Mater. 2018;28(31):1–11.
88. Vasuki K, Manimekalai R. NIR light active ternary modified ZnO nanocomposites for combined cancer therapy. Heliyon [Internet]. 2019 Nov 1 [cited 2022 May 31];5(11):e02729. Available from: www.cell.com/article/S2405844019363893/fulltext
89. Liu M, Peng Y, Nie Y, Liu P, Hu S, Ding J, et al. Co-delivery of doxorubicin and DNAzyme using ZnO@polydopamine core-shell nanocomposites for chemo/gene/photothermal therapy. Acta Biomater. 2020 Jul 1;110:242–53.
90. Kim S, Lee SY, Cho HJ. Berberine and zinc oxide-based nanoparticles for the chemo-photothermal therapy of lung adenocarcinoma. Biochem Biophys Res Commun. 2018 Jun 27;501(3):765–70.
91. Su H, Wang Z, Liu G. Near-infrared fluorescence imaging probes for cancer diagnosis and treatment. In: Cancer Theranostics. Amsterdam: Elsevier; 2014. pp. 55–67.
92. Allen PM, Bawendi MG. Ternary I-III-VI quantum dots luminescent in the red to near-infrared. J Am Chem Soc [Internet]. 2008 Jul 23 [cited 2022 May 31];130(29):9240–1. Available from: https://pubs.acs.org/doi/abs/10.1021/ja8036349
93. Chu M, Pan X, Zhang D, Wu Q, Peng J, Hai W. The therapeutic efficacy of CdTe and CdSe quantum dots for photothermal cancer therapy. Biomaterials. 2012 Oct 1;33(29):7071–83.
94. Li JJ, Wang YA, Guo W, Keay JC, Mishima TD, Johnson MB, et al. Large-scale synthesis of nearly monodisperse CdSe/CdS core/shell nanocrystals using air-stable reagents via successive ion layer adsorption and reaction. J Am Chem Soc [Internet]. 2003 Oct 15 [cited 2022 May 31];125(41):12567–75. Available from: https://pubmed.ncbi.nlm.nih.gov/14531702/
95. Kim SW, Zimmer JP, Ohnishi S, Tracy JB, Frangioni J V., Bawendi MG. Engineering InAsxP1-x/InP/ZnSe III-V alloyed core/shell quantum dots for the near-infrared. J Am Chem Soc [Internet]. 2005 Aug 3 [cited 2022 May 31];127(30):10526–32. Available from: https://pubs.acs.org/doi/abs/10.1021/ja0434331
96. Samia ACS, Chen X, Burda C. Semiconductor quantum dots for photodynamic therapy. J Am Chem Soc [Internet]. 2003 Dec 24 [cited 2022 May 31];125(51):15736–7. Available from: https://pubs.acs.org/doi/abs/10.1021/ja0386905

97. Moeno S, Antunes E, Nyokong T. The determination of the photosensitizing properties of mercapto substituted phthalocyanine derivatives in the presence of quantum dots capped with mercaptopropionic acid. J Photochem Photobiol A Chem [Internet]. 2011 [cited 2022 May 31];218(1):101–10. Available from: www.sciencedirect.com/science/article/pii/S1010603010004995
98. Charron G, Stuchinskaya T, Edwards DR, Russell DA, Nann T. Insights into the mechanism of quantum dot-sensitized singlet oxygen production for photodynamic therapy. J Phys Chem C [Internet]. 2012 Apr 26 [cited 2022 May 31];116(16):9334–42. Available from: https://pubs.acs.org/doi/abs/10.1021/jp301103f
99. Anas AA, Akita H, Harashima H, Itoh T, Ishikawa M, Biju V. Photosensitized breakage and damage of DNA by CdSe-ZnS quantum dots. J Phys Chem B [Internet]. 2008 Aug 14 [cited 2022 May 31];112(32):10005–11. Available from: https://pubmed.ncbi.nlm.nih.gov/18582008/
100. Chavda V, Patel V. The multifarious medical applications of carbon curvatures: A cohort review. Curr Bioact Compd. 2020;16:1–5.
101. Seifi T, Reza Kamali A. Antiviral performance of graphene-based materials with emphasis on COVID-19: A review. Med Drug Discov. 2021 Sep 1;11:100099.
102. Patel V, Chavda V, Shah J. Nanotherapeutics in neuropathologies: Obstacles, challenges and recent advancements in cns targeted drug delivery systems. Curr Neuropharmacol [Internet]. 2020 [cited 2021 Mar 11];18. Available from: https://europepmc.org/article/med/32851949
103. Zhao H, Ding R, Zhao X, Li Y, Qu L, Pei H, et al. Graphene-based nanomaterials for drug and/or gene delivery, bioimaging, and tissue engineering. Drug Discov Today [Internet]. 2017;22(9):1302–17. Available from: http://dx.doi.org/10.1016/j.drudis.2017.04.002
104. Viseu T, Lopes CM, Fernandes E, Real Oliveira MECD, Lúcio M. A systematic review and critical analysis of the role of graphene-based nanomaterials in cancer theranostics. Pharmaceutics. 2018;10(4):2–45.
105. Patel V, Shah J, Gupta A. Design and in-silico study of bioimaging fluorescence graphene quantum dot-bovine serum albumin complex synthesized by diimide-activated amidation. Comput Biol Chem [Internet]. 2021 [cited 2021 Jul 15];93:107543. Available from: www.sciencedirect.com/science/article/pii/S1476927121001109
106. Ge J, Lan M, Zhou B, Liu W, Guo L, Wang H, et al. A graphene quantum dot photodynamic therapy agent with high singlet oxygen generation. Nat Commun [Internet]. 2014 Aug 8 [cited 2022 May 31];5(1):1–8. Available from: www.nature.com/articles/ncomms5596
107. Li S, Zhou S, Li Y, Li X, Zhu J, Fan L, et al. Exceptionally high payload of the IR780 iodide on folic acid-functionalized graphene quantum dots for targeted photothermal therapy. ACS Appl Mater Interfaces [Internet]. 2017 Jul 12 [cited 2022 May 31];9(27):22332–41. Available from: https://pubs.acs.org/doi/abs/10.1021/acsami.7b07267
108. Fan H yang, Yu X hua, Wang K, Yin Y jia, Tang Y jie, Tang Y ling, et al. Graphene quantum dots (GQDs)-based nanomaterials for improving photodynamic therapy in cancer treatment. Vol. 182, European Journal of Medicinal Chemistry. Elsevier Masson; 2019. 111620.
109. Campbell E, Hasan MT, Gonzalez-Rodriguez R, Truly T, Lee BH, Green KN, et al. Graphene quantum dot formulation for cancer imaging and redox-based drug delivery. Nanomedicine Nanotechnology, Biol Med. 2021 Oct 1;37:102408.

110. Wang H, Mu Q, Wang K, Revia RA, Yen C, Gu X, et al. Nitrogen and boron dual-doped graphene quantum dots for near-infrared second window imaging and photothermal therapy. Appl Mater Today. 2019 Mar 1;14:108–17.
111. Thakur M, Kumawat MK, Srivastava R. Multifunctional graphene quantum dots for combined photothermal and photodynamic therapy coupled with cancer cell tracking applications. RSC Adv [Internet]. 2017 Jan 17 [cited 2022 May 31];7(9):5251–61. Available from: https://pubs.rsc.org/en/content/articlehtml/2017/ra/c6ra25976f
112. Wang C, Chen Y, Xu Z, Chen B, Zhang Y, Yi X, et al. Fabrication and characterization of novel cRGD modified graphene quantum dots for chemo-photothermal combination therapy. Sensors Actuators, B Chem. 2020 Apr 15;309:127732.
113. Liu J, Li R, Yang B. Carbon Dots: A new type of carbon-based nanomaterial with wide applications. ACS Cent Sci [Internet]. 2020 Dec 23 [cited 2022 May 31];6(12):2179–95. Available from: https://pubs.acs.org/doi/full/10.1021/acscentsci.0c01306
114. Lagos KJ, Buzzá HH, Bagnato VS, Romero MP. Carbon-based materials in photodynamic and photothermal therapies applied to tumor destruction. Int J Mol Sci. 2022;23(1):22.
115. Meena R, Singh R, Marappan G, Kushwaha G, Gupta N, Meena R, et al. Fluorescent carbon dots driven from ayurvedic medicinal plants for cancer cell imaging and phototherapy. Heliyon [Internet]. 2019 Sep 1 [cited 2022 May 31];5(9):e2483. Available from: www.cell.com/article/S2405844019361432/fulltext
116. Qin X, Liu J, Zhang Q, Chen W, Zhong X, He J. Synthesis of yellow-fluorescent carbon nano-dots by microplasma for imaging and photocatalytic inactivation of cancer cells. Nanoscale Res Lett [Internet]. 2021 Jan 21 [cited 2022 May 31];16(1):1–9. Available from: https://nanoscalereslett.springeropen.com/articles/10.1186/s11671-021-03478-2
117. Li Y, Zheng X, Zhang X, Liu S, Pei Q, Zheng M, et al. Porphyrin-based carbon dots for photodynamic therapy of hepatoma. Adv Healthc Mater [Internet]. 2017 Jan 1 [cited 2022 May 31];6(1):1600924. Available from: https://onlinelibrary.wiley.com/doi/full/10.1002/adhm.201600924
118. Guo XL, Ding ZY, Deng SM, Wen CC, Shen XC, Jiang BP, et al. A novel strategy of transition-metal doping to engineer absorption of carbon dots for near-infrared photothermal/photodynamic therapies. Carbon N Y. 2018 Aug 1;134:519–30.
119. Hua XW, Bao YW, Wu FG. Fluorescent carbon quantum dots with intrinsic nucleolus-targeting capability for nucleolus imaging and enhanced cytosolic and nuclear drug delivery. ACS Appl Mater Interfaces [Internet]. 2018 Apr 4 [cited 2022 May 31];10(13):10664–77. Available from: https://pubs.acs.org/doi/abs/10.1021/acsami.7b19549
120. Yang W, Wei B, Yang Z, Sheng L. Facile synthesis of novel carbon-dots/hemin nanoplatforms for synergistic photo-thermal and photo-dynamic therapies. J Inorg Biochem. 2019 Apr 1;193:166–72.
121. Zhang M, Wang W, Zhou N, Yuan P, Su Y, Shao M, et al. Near-infrared light triggered photo-therapy, in combination with chemotherapy using magnetofluorescent carbon quantum dots for effective cancer treating. Carbon N Y. 2017 Jul 1;118:752–64.
122. Liu F, Wang X di, Du S yu. Production of gold/silver doped carbon nanocomposites for effective photothermal therapy of colon cancer. Sci Rep. 2020;10(1):1–9.
123. Mohammed N, Mohammed N, Kang S, Reeck G, Khan H, Lee Y. Ternary graphene quantum dot-polydopamine-Mn3o4 nanoparticles for optical imaging guided photodynamic therapy and T1-weighted magnetic resonance imaging. J Mater Chem B. 2015;3(28):5815–23.

124. Zhang T, Jiang Z, Chen L, Pan C, Sun S, Liu C, et al. PCN-Fe(III)-PTX nanoparticles for MRI guided high efficiency chemo-photodynamic therapy in pancreatic cancer through alleviating tumor hypoxia. Nano Res. 2020;13(1):273–81.
125. Yao X, Niu X, Ma K, Huang P, Grothe J, Kaskel S, et al. Graphene quantum dots-capped magnetic mesoporous silica nanoparticles as a multifunctional platform for controlled drug delivery, magnetic hyperthermia, and photothermal therapy. Small. 2017;13(2):1–11.
126. Zhang M, Wang W, Cui Y, Chu X, Sun B, Zhou N, et al. Magnetofluorescent Fe3O4/carbon quantum dots coated single-walled carbon nanotubes as dual-modal targeted imaging and chemo/photodynamic/photothermal triple-modal therapeutic agents. Chem Eng J. 2018 Apr 15;338:526–38.
127. Pilot study of aurolase(tm) therapy in refractory and/or recurrent tumors of the head and neck. ClinicalTrials.gov: National Library of Medicine (US). 2017 [cited 2022 May 31]. Available from: https://clinicaltrials.gov/ct2/show/NCT00848042
128. MRI/US fusion imaging and biopsy in combination with nanoparticle directed focal therapy for ablation of prostate tissue [Internet]. ClinicalTrials.gov: National Library of Medicine (US). 2021 [cited 2022 May 31]. Available from: https://clinicaltrials.gov/ct2/show/NCT02680535
129. An extension Study MRI/US fusion imaging and biopsy in combination with nanoparticle directed focal therapy for ablation of prostate tissue [Internet]. ClinicalTrials.gov: National Library of Medicine (US). 2021 [cited 2022 May 31]. Available from: https://clinicaltrials.gov/ct2/show/NCT04240639
130. Intravesical photodynamic therapy (PDT) in BCG refractory/intolerant non-muscle invasive bladder cancer (NMIBC) patients [Internet]. ClinicalTrials.gov: National Library of Medicine (US). 2022 [cited 2022 May 31]. Available from: https://clinicaltrials.gov/ct2/show/NCT03945162
131. Stern JM, Kibanov Solomonov V V., Sazykina E, Schwartz JA, Gad SC, Goodrich GP. Initial evaluation of the safety of nanoshell-directed photothermal therapy in the treatment of prostate disease. Int J Toxicol [Internet]. 2016 Jan 1 [cited 2022 May 31];35(1):38–46. Available from: https://journals.sagepub.com/doi/10.1177/1091581815600170
132. Gad SC, Sharp KL, Montgomery C, Payne JD, Goodrich GP. Evaluation of the toxicity of intravenous delivery of auroshell particles (Gold-Silica Nanoshells). Int J Toxicol [Internet]. 2012 Nov 4 [cited 2022 May 31];31(6):584–94. Available from: https://journals.sagepub.com/doi/10.1177/1091581812465969
133. Rastinehad AR, Anastos H, Wajswol E, Winoker JS, Sfakianos JP, Doppalapudi SK, et al. Gold nanoshell-localized photothermal ablation of prostate tumors in a clinical pilot device study. Proc Natl Acad Sci [Internet]. 2019 Sep 10 [cited 2022 May 31];116(37):18590–6. Available from: www.pnas.org/cgi/doi/10.1073/pnas.1906929116

9 Magnetic Quantum Dots for In-Vitro Imaging

Preeti Kush[1], *Ranjit Singh*[1], *and Parveen Kumar*[2]
[1]Adarsh Vijendra Institute of Pharmaceutical Sciences, Shobhit University Gangoh, Saharanpur, Uttar Pradesh, India
[2]Exigo Recycling Pvt Ltd, Noida, Uttar Pradesh, India

CONTENTS

9.1 INTRODUCTION

In-vitro imaging is performed with cells/tissue culture, microorganisms, and biomolecules isolated from their biological context. This technique is simple, convenient, and can be used for detailed analysis even without using whole animals. Traditionally, optical absorption microscopy, electron microscopy, fluorescence microscopy, and magnetic resonance imaging has been extensively used for the in-vitro imaging of cells. Among these, the fluorescence technique using the fluorophores is the most effective but possesses certain limitations like variation in signal intensity during the multicolor experiments, small excitation ranges, wide emission spectra, photobleaching, and small lifetime due to the optical properties of fluorophores. Moreover, these fluorophores are not able to encounter the requirement for more complex and sensitive in-vitro imaging requirements within the visible range [1].

The semiconductor nanocrystals (zero-dimensional nanostructures) also known as quantum dots (QDs) or artificial atoms have been widely used for in-vitro and in-vivo imaging (single or multicolor experiments) owing to their unique physical, chemical, electrical and optical properties [2, 3, 4, 5]. QDs-based probes are favorable for fluorescent imaging due to size-regulated absorbance and emission, large Stokes shift, increased lifetime, and the high-surface area-to-volume ratio [6]. Moreover, QDs exhibit a wide excitation and small emission range, photostability along with intense fluorescence with different colors [7]. Additionally, these can also be used as fluorescent dyes for cell labelling and motility assays. The typical QDs are composed of a semiconductor core of combinations of elements from groups II-VI (CdTe, CdSe,

DOI: 10.1201/9781003319870-9

CdS), groups IV-VI (PbS, PbSe), or groups III-V (GaAs, InP, GaN, InAs) of periodic table [8], that is surrounded by a semiconductor shell such as ZnS, ZnSe to minimized the poor quantum yields [4, 9]. Various type of QDs such as environmentally friendly QDs (Ag_2S, CuInSe, Si), core/shell QDs (CdTe/ZnS, CdSe/ZnS, CdTe/ZnSe), and core/shell/shell QDs (CdSe/CdTe/ZnSe, CdTe/CdS/ZnS,) have been developed [1]. Recently, near-infrared (NIR) luminescent QDs with emission wavelengths of 700-900 nm have been developed to apply in biological imaging. The light within this wavelength has its maximum penetration depth in tissues along with minimal interference of tissue autofluorescence. Despite multiple advantages, QDs face certain limitations like random on/off behavior, incompatibility with the polar solvents due to their inorganic nature [10, 11]. Moreover, these are more prone to reticuloendothelial system uptake (RES) or liver and spleen entrapment leading to rapid clearance resulting in poor imaging quality and increased background noise. Additionally, QDs are also susceptible to liberating toxic metal ions via oxidation and photolysis leading to cause adverse effects [12]. Recently multiplex competence QDs like magnetically modified QDs/magnetic QDs (MQDs), radiolabelled QDs, QD polymer conjugates, multicolor QD-antibody conjugates, QD aptamer conjugates, etc. have been developed via the functionalization of QDs enabling increased in-vitro and in-vivo imaging quality and sensitivity [2, 13, 14]. Moreover, QDs, doped and modified with biocompatible molecules are more particular, biocompatible, and target specific in contrast to bare QDs [1].

Among all, MQDs are a vital part of the QDs family that can deal with electromagnetic interference (EMI) owing to their powerful exchange coupling effect and small size. MQDs are dual-functional nanoparticles (NPs) composed of inorganic heterodimers of QDs and magnetic nanomaterials/magnetic nanoparticles (MNPs) enabling magnetic, optical, and magneto-optical properties (Figure 9.1) [15]. These can be synthesized cost-effectively by using a one-step strategy without damaging the fluorescence property of QDs [16]. These bifunctional NPs are widely used in various applications such as photothermal therapy (PTT), photodynamic therapy (PDT), molecular detection and separation, theranostics, and in-vitro/in-vivo imaging owing to their fluorescence/magnetic resonance imaging (MRI) dual-mode imaging. [17, 18]. Moreover, these are also used as co-catalyst enabling the visible light absorbability enhancement of composite photocatalysts [19]. This chapter will briefly review the synthesis methods and biological application of MQDs. Further, this chapter will focus on the specific application of MQDs for in-vitro imaging, challenges, and future perspectives.

9.2 SYNTHESIS OF MAGNETIC QUANTUM DOTS

Various techniques have been adopted to synthesize MQDs such as (i) core/shell nanostructure (Figure 9.2a) where magnetic core material (e.g., Co, FeP, Fe_3O_4, FePt, Fe_2O_3) is synthesized first and subsequently coated with the QDs e.g., FePt/CdSe [20], Co/CdSe [21], or heterodimer structures/ dumbbell-like NPs (Figure 9.2b) of magnetic nanomaterials and QDs via heterocrystalline growth, e.g., FePt-CdSe [22], Fe_2O_3-CdSe [23], FeP-CdTe; (ii) doping of QDs with paramagnetic ions (e.g., Mn^{2+}, Gd^{3+}, Co^{2+}, Ni^{2+}) [24] either in shell (e.g., CdSe/$Zn_{1-x}Mn_{xS}$ [25], Figure 9.2c) or core (e.g., CdS:Mn/ZnS QDs [26], Figure 9.2d); (iii) encapsulation of composite nanostructure

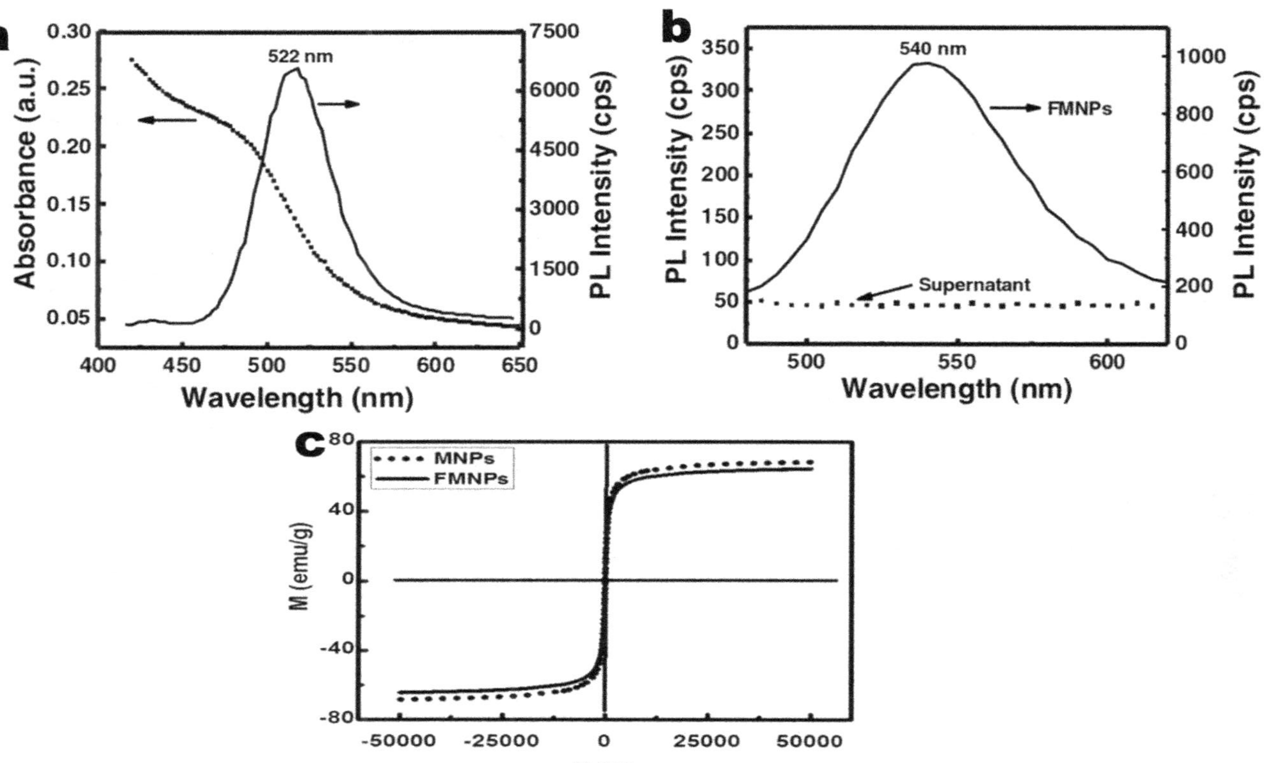

FIGURE 9.1 Optical and magnetic properties on magnetic quantum dots. a) absorbance and photoluminescence (PL) of CdTe quantum dot; b) PL spectra of magnetic quantum dot (Fe_3O_4NPs/ CdTe); c) hysteresis loops of the magnetic quantum dot. Reproduced from [15].

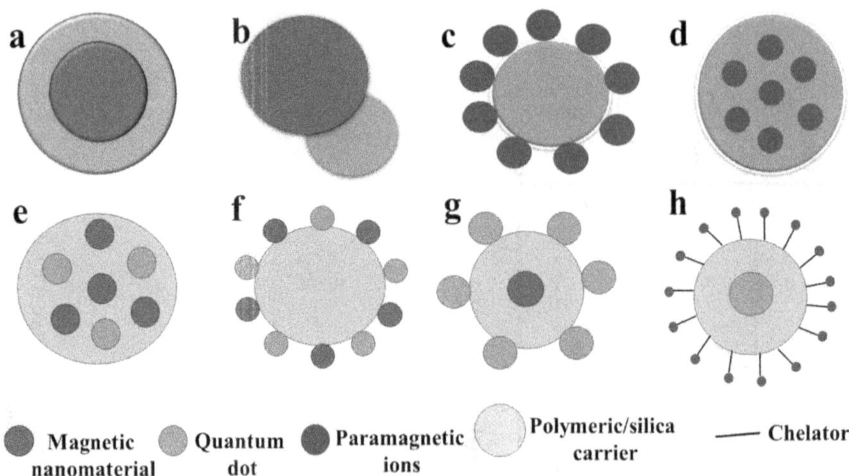

FIGURE 9.2 Synthesis of magnetic quantum dots using various approaches. a) core-shell nanostructure; b) heterodimer or dumbbell-shaped nanostructure; c) doping of paramagnetic ions on the shell of QD; d) doping of paramagnetic ions into the core of QD; e) encapsulation of magnetic nanomaterial and QDs inside the carrier material; f) integration of magnetic nanomaterial and QD outside the carrier material; g) combination of e and f approaches; h) coating of QD with Gd chelates.

of MNPs and QDs either inside the carrier material (polymer e.g., PbS QDs and Fe_3O_4 NPs encapsulated in poly(lactic-co-glycolic-acid) (PLGA) [27] or silica matrix e.g., Fe_3O_4/SiO_2-CdSe/ZnS MQDs [28]) (Figure 9.2e), attached outside the carrier material (Figure 9.2f), or a combination of both (Figure 9.2g); (iv) QDs with a paramagnetic coating of gadolinium (Gd)-chelates (Figure 9.2h) [18, 29, 30].

The biomedical application of MQDs is expected to heighten in coming years owing to their multifunctionality in various areas such as fast point of care-based diagnostic systems, in-vitro and in-vivo imaging, theranostic nanorobots, magnetic hyperthermia, biosensing, magnetic separation, drug delivery, and site-specific targeting.

9.3 MAGNETIC QUANTUM DOTS FOR IN-VITRO IMAGING

MQDs are an efficient, selective, and specific tool for in-vitro imaging via the integration of fluorescence properties of semiconductor QDs and the paramagnetic properties of MNPs. MNPs are potential contrast agents (CAs) and increase MRI resolution by shortening the T1 and T2 relaxation time of adjoining protons [31]. Therefore, MRI/fluorescence detectable MQDs have been developed to broaden the in-vitro imaging applications. Various types of MQDs such as core/shell, heterodimers, QDs doped with MNPs, the composite nanostructure of MNPs and QDs encapsulated in carrier matrix, Gd-chelates MQDs have been reported for in-vitro imaging applications. These strategies facilitate the aqueous solubility enhancement, a versatile surface for further functionalization of different biomolecules enabling the target drug delivery and selective imaging. Moreover, MQDs can also be used for enhanced cellular

TABLE 9.1
Application of Core/Shell and Heterodimers for In-Vitro Imaging

Type of Nanostructure	Quantum Dot	Magnetic Material	Cell Line	Ref
Core/shell	CdTe	Fe_3O_4 NPs	LS174T	[15]
	CdSe-ZnS	Si-coated Fe_3O_4 NPs	Panc-1	[28]
	CdTe	Si-coated Fe_3O_4 NPs	HeLa	[32]
	GOQD	Amine functionalized Fe_3O_4 NPs	Hep G2 liver cancer	[33]
	CdSe/CdS/ZnS, and CdZnSe/CdZnS/ZnS	Dextran coated Fe_3O_4 NPs	SK-BR3	[34]
Heterodimer	CdTe	CNT filled with Fe_3O_4 NPs	HeLa	[35]
	QD 605, QD 800	MNP	HeLa	[36]
	rGO	Fe_3O_4	Fibroblast	[37]

Abbreviations: CdS: cadmium sulfide; CdSe: cadmium selenide; CdTe: cadmium telluride; CNT: carbon nanotube; Fe_3O_4: iron oxide; GOQDs: graphene oxide quantum dots; MNPs: magnetic nanoparticles; NPs: nanoparticles; rGO: reduced graphene oxide; Si: silica; ZnS: zinc sulfide.

uptake under the external magnetic field (EMF) and tracked optically with fluorescence microscopy [18, 29, 30]. This section will cover the in-vitro imaging application of MQDs based on their type.

Core/shell nanostructure presented enhanced relaxivity and increased quantum yield for MRI and fluorescent imaging (Table 9.1). Ahmed et al. developed QDs incorporated MNPs for in-vitro imaging of colon cancer cells. The developed MQDs were fluorescent (shell) and magnetic responsive (core) enabling optical and cellular imaging along with magnetic separation. Further, the developed NPs were coated with the Fab region of hCC49 antibodies for LS174T cells. The fluorescence results revealed that the developed probe was selectively bound to the target cancer cells due to the attachment of a specific antibody (Figure 9.3) [15].

Xu et al. developed biocompatible luminescent superparamagnetic nanocomposites for the imaging of pancreatic cancer cells (Panc-1). The composite was synthesized by the surface functionalization of silica-coated Fe_3O_4 NPs with water-soluble CdSe-ZnS QDs [28]. In another study, a water-soluble Fe_3O_4/CdTe magnetic fluorescent multifunctional nanocomposite was developed for labelling and imaging cancer cells. The developed nanocomposite exhibited fluorescent and magnetic properties favorable for magnetic separation and fluorescent imaging. Further, the developed magnetic nanocomposite was conjugated with an anti-CEACAM8 antibody for immunolabelling and imaging of HeLa cells. However, the microscopic results revealed the cellular toxicity of the nanocomposites [32]. Shi et al. developed multifunctional graphene oxide QDs (GOQDs) coated Fe_3O_4 NPs for the efficient and selective capture of glypican-3 (GPC3)-expressed Hep G2 liver cancer cells. The nanoprobe was developed by the surface coating of MNPs with GOQDs, and further functionalized with anti-GPC3-antibody. The results

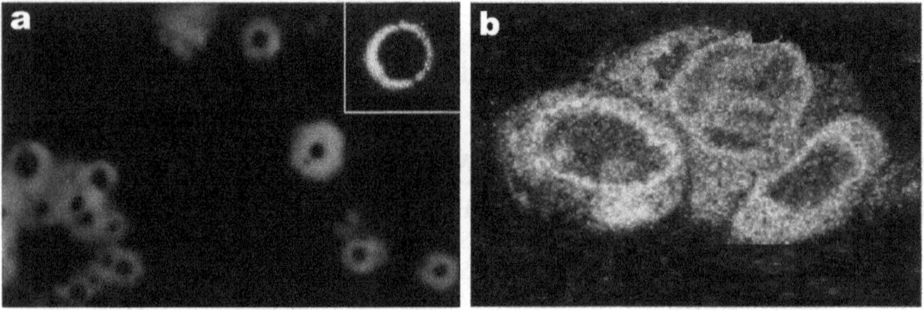

FIGURE 9.3 Multifunctional nanoparticles for in-vitro imaging of colon cancer cells. a) total internal reflection fluorescence image of Fe_3O_4 nanoparticles without any fluorescence, whereas CdTe quantum dots QDs covering the surface of nanoparticles display green fluorescence; b) fluorescence image of developed nanoparticles for LS174T cells. Adapted from [15].

revealed that the developed nanoprobe was very selective for the cancer cells due to the augmented two-photon absorption cross-section, and magnetism [33]. Recently, Tran et al. proposed a smartphone-based imaging platform integrated with magnetic-fluorescent NPs assemblies (MNP@QD: QD485; QD575; QD605; QD635) leading to rapid, simple, and selective cell isolation, fluorescent labelling, and counting. The MNP@QD assemblies were prepared by the surface functionalization of dextran-coated Fe_3O_4 NPs with luminescent QDs and further stabilized by dextran. Furthermore, the MNP@QD assemblies were conjugated with tetrameric antibody complexes (TACs) (Figure 9.4a). The cell isolation and imaging results revealed that the developed platform was able to selectively isolate, immunolabel, and count breast cancer (BC) cells that are positive for human epidermal growth factor receptor 2 (Figure 9.4b) [34].

Chen et al. developed a novel platform for in-vitro cancer-targeted imaging and drug delivery under the EMF. The platform was developed by the conjugation of silica-coated QDs with Fe_3O_4 NPs-filled carbon nanotubes (CNTs). The CNTs filled with MNPs facilitated the magnetically guided delivery leading to synergize targeting efficiency and magnified drug loading. The QDs attached to the exterior of the CNTs surface were responsible for magnified fluorescence due to bypassing the fluorescence quenching. Further, the developed probe was modified with transferring, enabling the in-vitro cancer-targeted imaging and drug delivery under EMF [35]. Lee et al. proposed an MNP-QD heterodimer probe for dual-mode imaging. The heterodimer was developed by the conjugation of platinum functionalized QD (QD-Pt) with guanine functionalized MNP (6G-MNP) via the metal-DNA conjugation method. The in-vitro bioimaging results revealed that the developed probe exhibited increased HeLa cell uptake and can be used for cellular labelling along with in-vitro and in-vivo imaging [36]. In another study, photoluminescent and paramagnetic reduced graphene oxide (rGO)–Fe_3O_4 QDs were developed for drug delivery, photothermal therapy, and dual imaging. The dual imaging capability of developed QDs was evaluated by T2 relaxation and in-vitro imaging of dermal fibroblast cells. The results revealed that the developed QDs were photoluminescent and magnetic leading to the dual imaging without any external fluorescent dye [37].

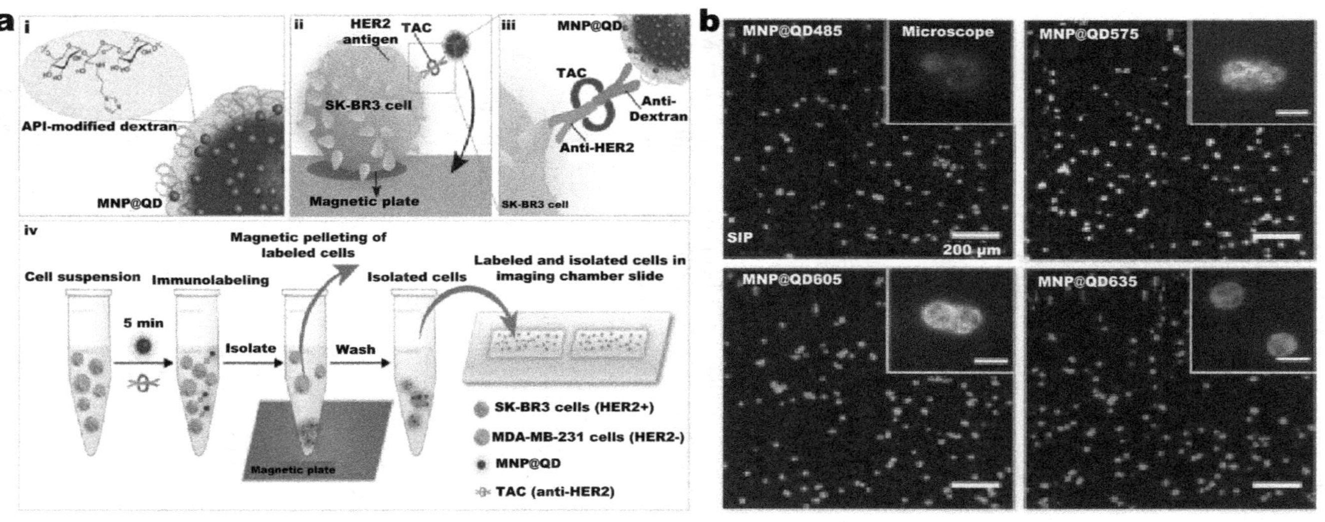

FIGURE 9.4 a) Magnetic nanoparticles and quantum dots assemblies (MNP@QDs) integrated with a smartphone-based imaging platform (SIP) for selective cell isolation and counting; (i) synthesis of MNP@QDs assemblies; (ii) tetrameric antibody complexes (TAC) mediated coupling of an MNP@QD to HER2 antigen on the SK-BR3 cell surface and isolated by magnetic pull-down (iii) magnified view of the TAC-mediated coupling; (iv) cell counting assay; b) SIP (main images, scale bar = 200 μm) and microscope images (insets, scale bar = 20 μm) of fixed SK-BR3 cells isolated with MNP@QDs of various colors (QD485 (blue), QD575 (yellow), QD605 (orange), and QD635 (red)). SIP images were achieved in RGB color format, and microscope images were pseudo-colored from the measured monochrome intensity values. Reprinted with permission from [34]. Copyright (2019) American Chemical Society.

TABLE 9.2
Quantum Dots Doped with Paramagnetic Material for In-Vitro Imaging

Quantum Dot	Paramagnetic Material	Cell Line	Ref
CdSe/ZnS	Mn	Macrophage	[25]
CdTeSe/CdS	Mn	Panc-1 and MiaPaCa	[39]
Si	Mn	P388D1macrophages	[40]
CdTeS	Fe	J774 macrophages	[41]
CuInS2	Mn	BXPC-3 cells	[42]
GQDs	Fe	3T3	[43]
ZnO	Gd	HeLa	[45]
CdTe	Gd	HepG2	[16]
CdZnSeS	Gd	N2a	[46]
GQDs	Gd	HepG2	[47]
CdSe/ZnS	Ni	MCF-7 and HeLa	[48]

Abbreviations: CdS: cadmium sulfide; CdSe: cadmium selenide; CdTe: cadmium telluride; CdTeS: cadmium sulfide telluride; CdTeSe: cadmium selenide telluride; CuInS2: copper indium sulfide; Fe: iron; Gd: gadolinium; GQDs: graphene quantum dots; Mn: manganese; Ni: nickel; Si: silica; ZnO: zinc oxide; ZnS: zinc sulfide.

QDs doped with paramagnetic materials have been used as efficient MRI/fluorescent imaging probes due to their structural stability, and small size. Particularly, Mn, Fe, and Gd are widely used as doping material owing to their higher paramagnetism (Table 9.2) [38]. Wang et al. synthesized Mn-doped CdSe/ZnS QDs by capping CdSe cores with Mn-doped ZnS shells [25]. Yong et al. developed NIR-emitting Mn-doped CdTeSe/CdS QDs for in-vitro and in-vivo imaging. For in-vitro imaging the functionalized QDs were conjugated with monoclonal antibody (mAb) specific for pancreatic cancer cells. The results revealed that the developed probe was able to early detection of pancreatic cancer due to the dual NIR fluorescence and MR imaging [39]. Tu et al. synthesized Mn-doped silica QDs for MRI and two-photon imaging of macrophages. In-vitro results revealed that the developed QDs were non-toxic, and accumulated into macrophages [40]. In another study, water dispersible, NIR-emitting Fe-doped CdTeS QDs were developed for bioimaging application. The developed MQDs exhibited bimodal optical/MR imaging for in-vitro/in-vivo application [41]. Lin et al. synthesized Mn-doped multifunctional core-shell QDs (CuInS2–Zn1-xMnxS) (QDs) for in-vitro dual-mode imaging of cancer cells. The developed MQDs exhibited magnified quantum yield, augmented relaxivity, and photoluminescence due to the capping of luminescent QDs cores with Mn-doped ZnS surface. In-vitro imaging and MRI results revealed that the developed MQDs were useful for optical/MR imaging of pancreatic cancer cell lines (BXPC-3). Moreover, it was observed that the increased quantum yield and Mn concentration were capable to generate contrast for both modalities even at low concentrations of QDs [42].

In contrast to fluorescence/MR imaging, Li et al. developed reported on paramagnetic graphene QDs (PGQDs) as a dual-mode CA for fluorescence/optical coherence tomography imaging (OCT). The PGQDs were synthesized doping of Fe on the surface of graphene QDs (GQDs). The in-vitro imaging results revealed that the developed MQDs were internalized into 3T3 cells enabling blue fluorescence, whereas the OCT results revealed that the developed probe exhibited a significant magnetomotive signal (320 Hz) resulting in cellular visualization. Hence, the developed probe was a promising CA for dual-mode imaging due to the fluorescence and paramagnetic property of PGQDs [43].

Gd is a commonly used CA for achieving T1-weighted MR enhancement of QDs owing to its paramagnetic property. Moreover, various Gd-based nanomedicines have been also developed due to the unique electrical, catalytic, and optical properties of Gd. Various Gd-based nanomaterials such as sodium gadolinium fluoride ($NaGdF_4$) NPs, Gd-chelates, Gd^{3+}metal-organic frameworks (Gd-MOF), gadolinium oxide (Gd_2O_3) NPs, and other Gd-based nanocomposites have been developed and evaluated as a CAs for imaging and imaging-guided therapy [44]. QDs directly doped by Gd could minimize the size of QDs enabling rapid excretion of administered agents and augmented biosafety. Moreover, direct doping with Gd-based NPs is promising CAs owing to their superior electromagnetic properties, magnified chemical reactivity, facile surface modification, easy intracellular uptake, and adequate stability in vivo. Liu et al. synthesized Gd doped ZnO QDs for MRI/fluorescent imaging by a simple, environment-friendly, and versatile methodology. The developed probe was highly fluorescent and can label HeLa cells successfully in a short time without any toxicity due to Gd doping [45]. In another study, Gd doped CdTe QDs were synthesized via a facile, and one-pot method. Further, the developed probe was conjugated with folic acid for specific cellular imaging of HepG2 cells without any toxicity [16]. Du et al. developed paramagnetic QDs (pQDs) functionalized with biotinylated dextran amine (BDA) as a multifunctional probe for nervous system imaging. The probe was developed by the surface coating of QDs with lipid-Gd and pegylated phospholipids, and further functionalized with BDA (pQDs-BDA). It was observed that the synthesized QDs were more biocompatible and can be easily taken up by the cells. Additionally, the developed probe exhibited magnified MR properties in contrast to the commercial CAs. The developed probe exhibited enhanced nervous system imaging due to the three different functionalities such as paramagnetism (Gd), fluorescence (CdZnSeS), and axonal tracing (BDA) in one nanosystem [46].

Recently, a uniform-sized GQDs anchored with single-atom Gd (SAGd-GQDs) was prepared for MRI signal amplification (Figure 9.5a). The developed SAGd-GQDs exhibited an enhanced longitudinal relaxation rate (86.08 m/M.s) at a low concentration of Gd, which is 25 times higher than the commercial CA (Figure 9.5b). The higher relaxation rate was due to the SAGd anchoring on GQDs surface leading to facilitate maximum EMF resulting in the exchange of a second sphere of the water molecule in the MRI. In-vitro cytotoxicity results revealed that the developed QDs were biocompatible. Further, in-vitro MRI results revealed that the SAGd-GQD exhibited significantly higher brightness in contrast to the commercial CA (Magnevist) at the same concentration of Gd^{3+} (Figure 9.5c). Furthermore, uniform and bright green fluorescence were observed in the cytoplasm of HepG2 cells suggesting that SAGd-GQD and GQD samples can be easily taken by cells. In conclusion, in-vitro/in-vivo results revealed that the developed SAGd-GQDs were a novel and advanced CA for fluorescence/MR imaging leading to improved diagnosis and therapy [47]. Very

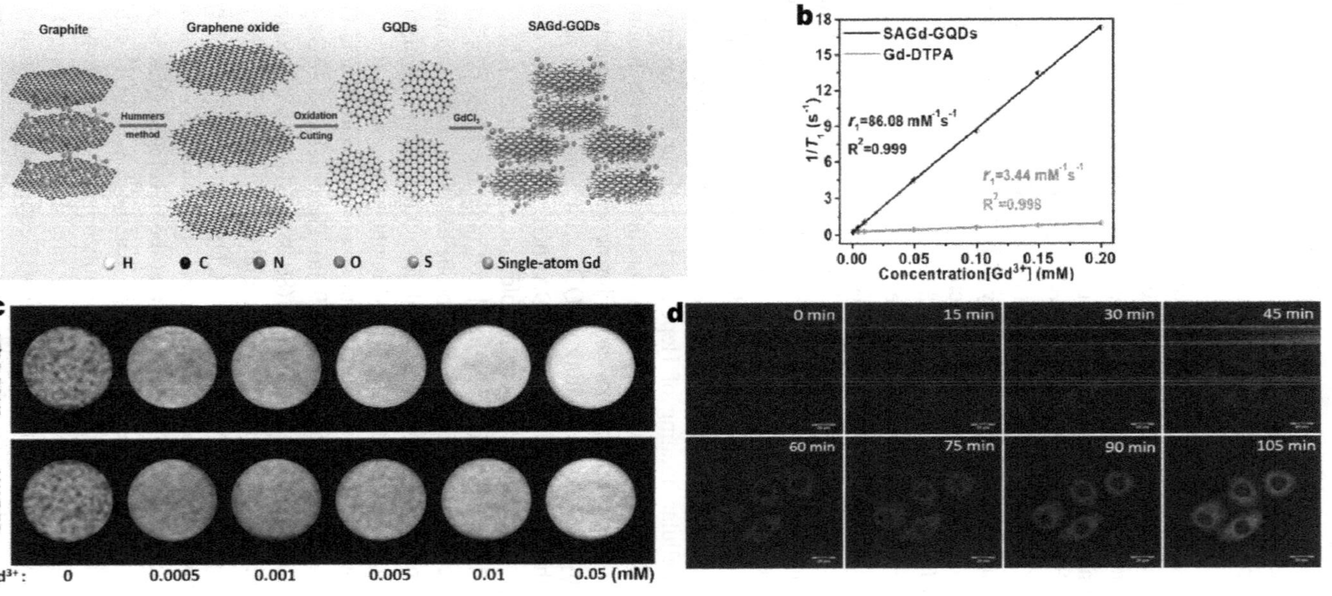

FIGURE 9.5 A uniform-sized graphene quantum dots (GQDs) anchored with single-atom Gd (SAGd-GQDs) as a signal amplifier for magnetic resolution imaging (MRI). a) synthesis of SAGd-GQDs; b) 1/T1 vs. concentration results of the Gd^{3+}-diethylenetriamine-pentaacetic acid (Gd-DTPA) and SAGd-GQD samples; c) in-vitro MRI results of the SAGd-GQDs and Gd-DTPA samples; d) time-lapse fluorescence images of cellular uptake of the SAGd-GQD sample at 10 min interval (scale bar: 20 μm). Reprinted with permission from [47]. Copyright (2021) American Chemical Society.

recently, a biocompatible Ni-doped CdSe/ZnS QDS was reported for cellular sorting and imaging. In-vitro results revealed that the developed MQDs exhibited significant cellular uptake enabling biological sorting and imaging [48].

Doping of QDs with paramagnetic material featured dual imaging but the concentration of doping material is relatively low for QDs fluorescence amplification leading to poor MRI signals. The MRI/optical dual-mode imaging can be enhanced by the encapsulation of composite nanostructure of MNPs and QDs either inside the carrier material, attached outside the carrier material, or a combination of both (). In a study, a biocompatible magnetofluorescent probe was developed by encapsulation of the silicon QdS (SiQDs) and Fe_3O_4 NPs into phospholipid-polyethyleneglycol micelles. The developed multiprobe featured a single platform with multimodal properties due to the luminescent properties of SiQDs, and superparamagnetic properties of Fe_3O_4 NPs. Moreover, it was able to overcome the challenges associated with heavy metal toxicity and can be used for drug delivery and macrophage cell imaging. The cellular imaging results revealed that the developed probe exhibited enhanced cellular uptake under the EMF and tracked optically with fluorescence microscopy [8]. In another study, the cellular imaging of BC was enhanced by the development of a multi-layered core-shell MQD-probe. The development of the probe was based on the incorporation of MNPs (Fe_3O_4), visible fluorescent QDs (CdSe/ZnS), NIR-QDs (CdSeTe/CdS) into the silica matrix. Further, the developed MQD-probe was functionalized with a specific antibody (anti-HER2) for cellular imaging of the KPL-4 BC cell line. The developed multifunctional probe exhibited enhanced photostability and brightness of QDs enabling fluorescence and MR imaging [49].

Zhou et al. proposed a fluorescent enhancement strategy via the incorporation of Mn-doped ZnSe QDs into the pores of the mesoporous silica NPs. The developed nanocomposite featured magnified fluorescent/MRI dual-imaging due to the increased concentration of Mn^{2+}. Moreover, the fluorescence of Mn-doped QDs was also expected to be enhanced without appreciable aggregation-caused quenching due to their large Stokes shift, and zero self-absorption. The imaging performance was analyzed via both in-vitro and in-vivo fluorescence and MRI imaging. The results revealed that the proposed strategy exhibited simultaneous signal improvement of the two imaging modes of Mn-doped ZnSe QDs [50]. Xu et al. developed an integrated liposome composed of superparamagnetic iron oxide nanoparticles (SPIONs), chemotherapeutics, and QDs for dual imaging-guided glioma-targeted delivery under EMF [51]. Recently, Zhang et al. developed an immuno-fluorescent magnetic nanobead system (iFMNS) functionalized with an anti-EpCAM antibody for the concomitant capturing, isolation, and imaging of circulatory tumor cells (CTCs). The iFMNS was developed by the encapsulation of CdSe/ZnS QDs and MNPs into poly(styrene-co-maleic anhydride) (PSMA) solution (Figure 9.6a), and further functionalized with an anti-EpCAM antibody (Figure 9.6b). The developed system was able to isolate and quantify CTCs from the whole blood sample based on fluorescence, and magnetism generated by QDs, and MNPs, respectively (Figure 9.6c). Further, the captured cells were identified by the developed system owing to the magnetic/fluorescence nature (Figure 9.6d) [52].

Gd chelates are primary CAs used to magnify the MRI contrast due to the enhancement of water proton relaxation rate, instigated by the neighboring Gd ion.

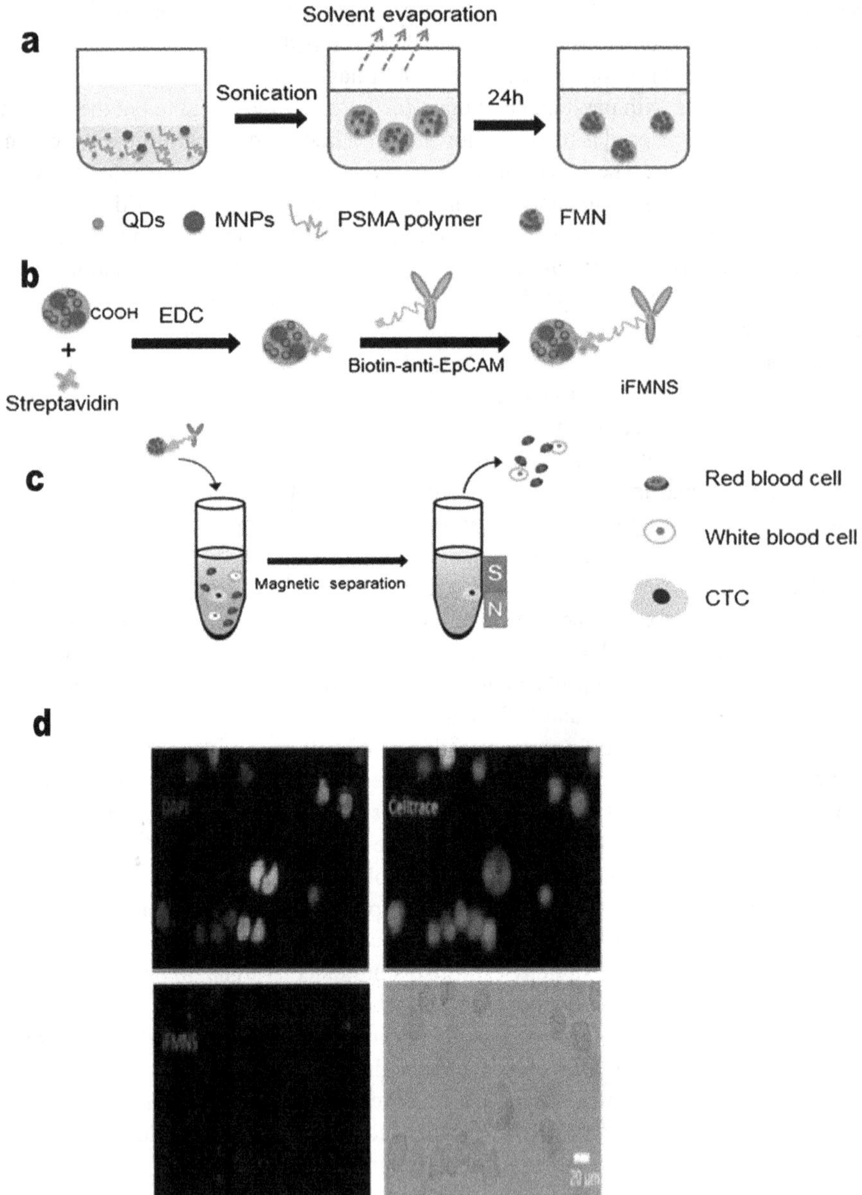

FIGURE 9.6 Capture and identification of circulatory tumor cells using immuno-fluorescent magnetic nanobead system (iFMNS). a) encapsulation of CdSe/ZnS QDs and MNPs into poly(styrene-co-maleic anhydride) (PSMA) matrix; b) functionalization of iFMNS with an anti-EpCAM antibody; c) recognition and isolation of CTCs by iFMNS; d) fluorescence imaging of Celltrace pre-labeled gastric cancer cells (SGC-790) captured by iFMNS (red) and 4′,6-diamidino-2-phenylindole (DAPI, blue) stained-cells. Reprinted from [52].

TABLE 9.3
Encapsulation of Quantum Dots/Doped Quantum Dots and Magnetic Nanomaterial in Biocompatible Carrier Material for In-Vitro Imaging

Quantum Dots	Paramagnetic Material	Carrier	Cell Line	Ref
Si	Fe_3O_4 NPs	PEG- DSPE	Macrophage cells	[8]
CdSe/ZnS, CdSeTe/CdS	Fe_3O_4 NPs	Si	KPL-4	[49]
ZnSe	Mn^{2+}	Si	4T1 cells	[50]
CdSe/ZnS	Fe_3O_4 NPs	Liposomes composed of PEG2000-DSPE and egg yolk lecithin (PC-98T)	C6 glioma cells	[51]
CdSe/ZnS	MNPs	PSMA	SGC-7901	[52]

Abbreviations: CdS: cadmium sulfide; CdSe: cadmium selenide; CdTeS: cadmium sulfide telluride; CdTeSe: cadmium selenide telluride; Fe_3O_4: iron oxide; Mn: manganese; MNPs: magnetic nanoparticles; NPs: nanoparticles; PEG: polyethylene glycol; PEG-DSPE: (1,2- distearoyl-sn-glycero-3-phosphoethanolamine-N-[methoxy- (poly(ethylene glycol))-2000]) PSMA: poly(styrene-co-maleic anhydride); Si: silica; ZnS: zinc sulfide; ZnSe: zinc selenide.

TABLE 9.4
Quantum Dots with a Paramagnetic Coating of Gadolinium-Chelates for In-Vitro Imaging

Quantum Dots	Gd-Chelates	Cell Lines	Ref
CdSe/ZnS	Gd-DTPA-BSA	HUVEC	[54]
InP/ZnS	Gd-TCEP	CHO	[55]
CuInS2/ZnS	Gd-DTPA-BSA	SU2 stem cells	[56]
CuInS2/ZnS	Gd-DTDTPA	HeLa	[57]
CdTe	Gd-DOTA-NHS	HeLa	[58]
GQDs	Gd-DTPA	HeLa and HepG2	[59]
GQDs	Gd-DOTA-PEG	A-549	[60]

Abbreviations: BSA: bis(stearylamide); CdSe: cadmium selenide; CdTe: cadmium telluride; CHO: Chinese hamster ovarian cells; CuInS2: copper indium sulfide DOTA: (1,4,7,10-tetraazaciclododecano-1,4,7,10-tetraacetic acid); DTDTPA: dithiolated derivative of 2- [bis[2-[carboxymethyl-[2-oxo-2-(2-sulfanylethyl-amino)ethyl]- amino]ethyl]amino]acetic acid; DTPA: diethylenetriaminepentaacetic acid; Gd: gadolinium; GQDs: graphene quantum dots; HUVEC: human umbilical vein endothelial cells; InP: indium phosphide; NHS:N-hydroxysulfosuccinimide sodium salt; PEG: poly ethylene glycol; TCEP: tris(carboxyethyl)phosphine; ZnS: zinc sulfide.

Clinically Gd^{3+}-diethylenetriamine-pentaacetic acid (Gd-DTPA) and Gd^{3+}-1,4,7,10-tetraazacyclododecane-1,4,7,10-tetraacetic acid (Gd-DOTA) have been widely used as a CA [53]. However, the relaxivity of commercial CAs is very low to detect the events at the molecular level. The relaxivity of the CAs can be enhanced by the integration of paramagnetic Gd chelates with luminescent QDs (Table 9.4). MRI detectable, biocompatible, water-soluble MQDs was developed based on the surface modification of CdSe/ZnS QDs with pegylated phospholipid, and a paramagnetic lipid. Further, the composite was conjugated to cyclic arginine-glycine-aspartic acid (RGD) to make specific for activated and angiogenic vascular endothelium. The developed MQDs exhibited superior imaging due to the magnified relaxivity, specificity, and bimodal character [54]. Stasiuk et al. reported a dual probe based on InP/ZnS QDs coated with covalent grafting of Gd-chelates followed by simultaneous attachment of maurocalcine (cell-penetrating peptide) resulting in a cell-permeable multimodal CA. The developed probe exhibited cellular localization, sharp MRI signal, and magnified tissue retention in contrast to commercial CA due to augmented fluorescence and high relaxivity [55]. In another study, Cd-free CuInS2/ZnS core/shell QDs coupled with Gd^3 chelates were developed for fluorescence/MR imaging. The developed probe was further conjugated with anti-CD133 mAb for specific tumor targeting. The developed probe exhibited magnified fluorescence together with increased relaxivity (r_1 = 15.2 m/M.s) in contrast to the commercial Magnevist (r_1 = 3.12 m/M.s). Further, the in-vitro imaging results revealed that the developed probe was able to target and label the specific cell due to the attached mAb [56]. Another, Cd-free CuInS2/ZnS core/shell QDs coupled with Gd^3 chelates were developed for fluorescence/MR imaging of HeLa cells. The longitudinal relaxivity of the developed probe was 2.5 times higher than the clinically approved Gd-DTPA. Further, the in-vitro and in-vivo results revealed that the developed probe exhibited enhanced NIR fluorescence and T1-weighted MR imaging along with the magnified biocompatibility and fluorescence stability [57]. Pereira et al. developed a bimodal nanoprobe comprising hydrophilic CdTe QDs conjugated with Gd-chelates for in-vitro imaging. The developed nanoprobe was colloidal stable and preserved an intense fluorescence with successful labelling of HeLa cells without any cytotoxicity (Figure 9.7) [58].

PGQDs have been used for optical and MRI diagnosis owing to their compatibility, stable photoluminescence, and unique paramagnetic and fluorescence properties. Huang et al. reported a multifunctional nanoprobe comprising doxorubicin (Dox), folate, and PGQDs as chemotherapeutics, a ligand for specific targeting, and a carrier, respectively. The PGQDs were prepared by the covalent coupling of DTPA, Gd, and folic acid on the surface of GQDs. The developed probe was biocompatible and non-toxic to HeLa and HepG2 cells along with magnified MRI resolution. Moreover, the results revealed that the developed nanoprobe could be used as a drug delivery carrier for the detection and image-guided therapy of different cancers [59]. In another study, PGQDs were engineered with magnified relaxivity for tumor imaging. The PGQDs were synthesized by chelation of Gd^{3+} on the GQDs surface, and further modified with hyaluronic acid (HA) to selectively target the cancer cells. The developed composites exhibited magnified imaging contrast together with a carrier for drug delivery. In-vitro imaging results revealed that the developed nanocomposite was selective for A 549 [60].

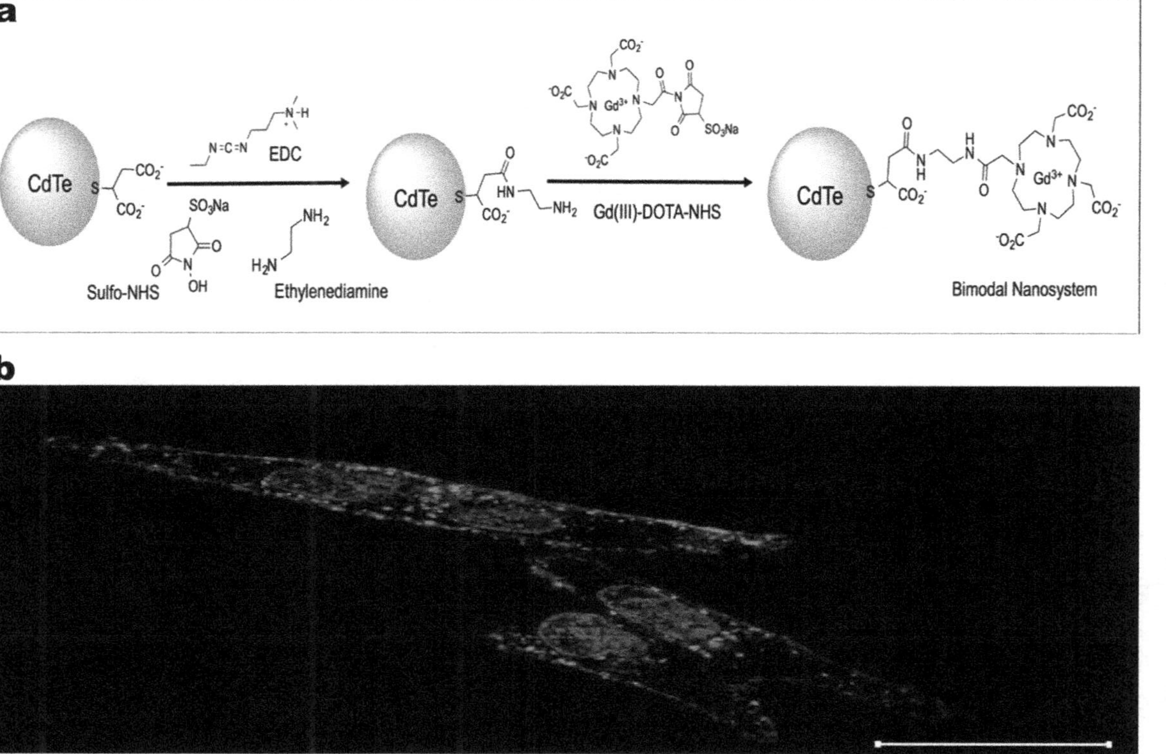

FIGURE 9.7 Bimodal probe based on quantum dot coated with gadolinium chelates for in-vitro imaging. a) conjugation of cadmium telluride quantum dot with gadolinium- DOTA (1,4,7,10-tetraazaciclododecano-1,4,7,10-tetraacetic acid-NHS (N-hydroxysulfosuccinimide sodium salt); b) cellular imaging of HeLa cells. Reprinted from [58] Scientific Reports 2019.

9.4 CHALLENGES AND FUTURE PERSPECTIVE

Although significant advancement has been made regarding biocompatible MQDs synthesis for improved biomedical applications, yet their clinical applications are limited due to some challenges such as quality, reproducibility, and comparability. Moreover, MQDs are mostly synthesized at a laboratory scale but at large scale production, they are not stable. Therefore, some critical points need to be readdressed further to overcome the challenges. Primarily, the synthesis, post-synthetic modification, and characterization techniques should be standardized to maintain the quality and quantity of the MQDs due to the various sources of QDs-based experimental protocols. Further, the quality of MQDs can be improved by the training of scientific and medical professionals with QDs-based advanced tools [30, 61].

Cellular toxicity and biosafety are the other challenges that limit the clinical application of MQDs especially for in-vivo imaging. The toxicity is due to the photon-induced free-radical generation, and colloidal effect of toxic metal ions. Therefore, more studies should be focused on the effect of MQDs on human health and the environment. Currently, the National Nanotechnology Initiative and the National Institutes of Health are trying to resolve these toxicity-related issues. Additionally, various efforts are also carried out to develop non-toxic QDs with magnified detection efficiency via modifying their sizes, constituents, and surface functionalization [61].

9.5 CONCLUSION

MQDs are synthesized by integration of inorganic heterodimers of QDs and magnetic nanomaterials/magnetic nanoparticles (MNPs) enabling fluorescence/magnetic resonance imaging (MRI) dual-mode imaging. These can be synthesized cost-effectively by using a one-step strategy without damaging the fluorescence property of QDs. In this chapter, we have covered the different strategies such as core/shell nanostructure where the magnetic core is synthesized first and subsequently coated with the QDs, or heterodimer structures/ dumbbell-like NPs, doping of QDs with paramagnetic ions either in the core, encapsulation of composite nanostructure of MNPs and QDs either inside the silica or polymeric carrier material or outside the carrier material, and QDs with a paramagnetic coating of Gd-chelates have been adopted for the development of MQDs regarding their in-vitro imaging application. Despite the recent advancement still, various challenges like poor quantum yield and toxicity associated with the use of MQDs need to be resolved. Moreover, the development of alternative probes is also restricted due to insufficient literature related to toxicity mechanisms, and poorly standardized protocols for toxicity assessment. Thus, it is mandatory to develop and establish the quality criteria or standardization for the synthesis, labelling, imaging capture, and toxicity analysis.

REFERENCES

[1] Moulick A, Blazkova I, Milosavljevic V, Fohlerova Z, Hubalek J, Kopel P, Vaculovicova M, Adam V, Kizek R. Application of CdTe/ZnSe quantum dots in in vitro imaging of chicken tissue and embryo. Photochemistry and photobiology. 2015; *91:* 417–423.

[2] Jin S, Hu Y, Gu Z, Liu L, Wu H-C. Application of quantum dots in biological imaging. Journal of Nanomaterials. 2011; *2011:* 834139.https://doi.org/10.1155/2011/834139
[3] Khaledian S, Nalaini F, Mehrbakhsh M, Abdoli M, Salehi Zahabi S. Applications of novel quantum dots derived from layered materials in cancer cell imaging. FlatChem. 2021; *27:* 100246.https://www.sciencedirect.com/science/article/pii/S2452262721000258
[4] Abbasi E, Kafshdooz T, Bakhtiary M, Nikzamir N, Nikzamir N, Nikzamir M, Mohammadian M, Akbarzadeh A. Biomedical and biological applications of quantum dots. Artificial Cells, Nanomedicine, and Biotechnology. 2016; *44:* 885–891.
[5] Kumar P, Sharma SC, Deep A. Bioconjugation of anti estrogen alpha antibody with CdSSe/ZnS quantum dots for molecular sensing of a breast cancer antigen. Sensors and Actuators B: Chemical. 2014; *202:* 404–409. www.sciencedirect.com/science/article/pii/S0925400514006492
[6] Gil HM, Price TW, Chelani K, Bouillard JG, Calaminus SDJ, Stasiuk GJ. NIR-quantum dots in biomedical imaging and their future. iScience. 2021; *24:* 102189.
[7] Kumar P, Deep A, Sharma SC, Bharadwaj LM. Bioconjugation of InGaP quantum dots for molecular sensing. Analytical Biochemistry. 2012; *421:* 285–290. www.sciencedirect.com/science/article/pii/S0003269711007032
[8] Erogbogbo F, Yong K-T, Hu R, Law W-C, Ding H, Chang C-W, Prasad PN, Swihart MT. Biocompatible magnetofluorescent probes: luminescent silicon quantum dots coupled with superparamagnetic iron(III) oxide. ACS Nano. 2010; *4:* 5131–5138.https://doi.org/10.1021/nn101016f
[9] Michalet X, Pinaud FF, Bentolila LA, Tsay JM, Doose S, Li JJ, Sundaresan G, Wu AM, Gambhir SS, Weiss S. Quantum dots for live cells, in vivo imaging, and diagnostics. Science (New York, N.Y.). 2005; *307:* 538–544.
[10] Schipper ML, Cheng Z, Lee SW, Bentolila LA, Iyer G, Rao J, Chen X, Wu AM, Weiss S, Gambhir SS. microPET-based biodistribution of quantum dots in living mice. Journal of Nuclear Medicine: Official Publication, Society of Nuclear Medicine. 2007; *48:* 1511–1518.
[11] Diagaradjane P, Deorukhkar A, Gelovani JG, Maru DM, Krishnan S. Gadolinium chloride augments tumor-specific imaging of targeted quantum dots in vivo. ACS nano. 2010; *4:* 4131–4141.
[12] Hardman R. A toxicologic review of quantum dots: toxicity depends on physicochemical and environmental factors. Environmental Health Perspectives. 2006; *114:* 165–172.
[13] Kumar P, Kumar P, Bharadwaj LM, Paul AK, Sharma SC, Kush P, Deep A. Aqueous synthesis of l-cysteine stabilized water-dispersible Cds:Mn quantum dots for biosensing applications. BioNanoScience. 2013; *3*: 95–101. https://doi.org/10.1007/s12668-013-0078-5
[14] Gao X, Yang L, Petros JA, Marshall FF, Simons JW, Nie S. In vivo molecular and cellular imaging with quantum dots. Current Opinion in Biotechnology. 2005; *16:* 63–72.
[15] Ahmed S, Dong J, Yui M, Kato T, Lee J, Park EY. Quantum dots incorporated magnetic nanoparticles for imaging colon carcinoma cells. J Nanobiotechnology. 2013; *11:* 28.
[16] Zhang F, Sun TT, Zhang Y, Li Q, Chai C, Lu L, Shen W, Yang J, He XW, Zhang YK, Li WY. Facile synthesis of functional gadolinium-doped CdTe quantum dots for tumor-targeted fluorescence and magnetic resonance dual-modality imaging. Journal of Materials Chemistry. B. 2014; *2:* 7201–7209.

[17] Gu H, Zheng R, Zhang X, Xu B. Facile one-pot synthesis of bifunctional heterodimers of nanoparticles: A conjugate of quantum dot and magnetic nanoparticles. Journal of the American Chemical Society. 2004; *126:* 5664–5665. https://doi.org/10.1021/ja0496423

[18] Mahajan KD, Fan Q, Dorcéna J, Ruan G, Winter JO. Magnetic quantum dots in biotechnology—synthesis and applications. Biotechnology Journal. 2013; *8:* 1424–1434.

[19] Qin Y, Li H, Lu J, Ma C, Liu X, Meng M, Yan Y. Fabrication of magnetic quantum dots modified Z-scheme Bi2O4/g-C3N4 photocatalysts with superior hydroxyl radical productivity for the degradation of rhodamine B. Applied Surface Science. 2019; *493*: 458–469. www.sciencedirect.com/science/article/pii/S0169433219320252

[20] Gao J, Zhang B, Gao Y, Pan Y, Zhang X, Xu B. Fluorescent magnetic nanocrystals by sequential addition of reagents in a one-pot reaction: a simple preparation for multifunctional nanostructures. J Am Chem Soc. 2007; *129:* 11928–11935.

[21] Kim H, Achermann M, Balet LP, Hollingsworth JA, Klimov VI. Synthesis and characterization of Co/CdSe core/shell nanocomposites: bifunctional magnetic-optical nanocrystals. J Am Chem Soc. 2005; *127:* 544–546.

[22] Gu H, Zheng R, Zhang X, Xu B. Facile one-pot synthesis of bifunctional heterodimers of nanoparticles: a conjugate of quantum dot and magnetic nanoparticles. J Am Chem Soc. 2004; *126:* 5664–5665.

[23] Selvan ST, Patra PK, Ang CY, Ying JY. Synthesis of silica-coated semiconductor and magnetic quantum dots and their use in the imaging of live cells. Angewandte Chemie (International ed. In English). 2007; *46:* 2448–2452.

[24] Malik P, Thareja R, Singh J, Kakkar R. II-VI core/shell quantum dots and doping with transition metal ions as a means of tuning the magnetoelectronic properties of CdS/ZnS core/shell QDs: A DFT study. Journal of Molecular Graphics & Modelling. 2022; *111:* 108099.

[25] Wang S, Jarrett BR, Kauzlarich SM, Louie AY. Core/shell quantum dots with high relaxivity and photoluminescence for multimodality imaging. J Am Chem Soc. 2007; *129:* 3848–3856.

[26] Santra S, Yang H, Holloway PH, Stanley JT, Mericle RA. Synthesis of water-dispersible fluorescent, radio-opaque, and paramagnetic CdS:Mn/ZnS quantum dots: a multifunctional probe for bioimaging. J Am Chem Soc. 2005; *127:* 1656–1657.

[27] *Ortgies DH, de la Cueva L, del Rosal B, Sanz-Rodríguez F, Fernández N, Iglesias-de la Cruz MC, Salas G, Cabrera D, Teran FJ, Jaque D, Martín Rodríguez E.* In vivo deep tissue fluorescence and magnetic imaging employing hybrid nanostructures. ACS Applied Materials & Interfaces. 2016; *8:* 1406–1414. https://doi.org/10.1021/acsami.5b10617

[28] Xu Y, Karmakar A, Wang D, Mahmood MW, Watanabe F, Zhang Y, Fejleh A, Fejleh P, Li Z, Kannarpady G, Ali S, Biris AR, Biris AS. Multifunctional Fe3O4 cored magnetic-quantum dot fluorescent nanocomposites for rf nanohyperthermia of cancer cells. The Journal of Physical Chemistry C. 2010; *114*: 5020–5026. https://doi.org/10.1021/jp9103036

[29] Koole R, Mulder WJ, van Schooneveld MM, Strijkers GJ, Meijerink A, Nicolay K. Magnetic quantum dots for multimodal imaging. Wiley Interdisciplinary Reviews. Nanomedicine and Nanobiotechnology. 2009; *1:* 475–491.

[30] Tufani A, Qureshi A, Niazi JH. Iron oxide nanoparticles based magnetic luminescent quantum dots (MQDs) synthesis and biomedical/biological applications: A review.

Materials Science & Engineering. C, Materials For Biological Applications. 2021; *118:* 111545.

[31] Kush P, Kumar P, Singh R, Kaushik A. Aspects of high-performance and bio-acceptable magnetic nanoparticles for biomedical application. Asian Journal of Pharmaceutical Sciences. 2021; *16:* 704–737.

[32] Sun P, Zhang H, Liu C, Fang J, Wang M, Chen J, Zhang J, Mao C, Xu S. Preparation and characterization of Fe3O4/CdTe magnetic/fluorescent nanocomposites and their applications in immuno-labeling and fluorescent imaging of cancer cells. Langmuir: The Acs Journal of Surfaces and Colloids. 2010; *26:* 1278–1284.

[33] Shi Y, Pramanik A, Tchounwou C, Pedraza F, Crouch RA, Chavva SR, Vangara A, Sinha SS, Jones S, Sardar D, Hawker C, Ray PC. Multifunctional biocompatible graphene oxide quantum dots decorated magnetic nanoplatform for efficient capture and two-photon imaging of rare tumor cells. ACS Applied Materials & Interfaces. 2015; *7:* 10935–10943.

[34] Tran MV, Susumu K, Medintz IL, Algar WR. Supraparticle assemblies of magnetic nanoparticles and quantum dots for selective cell isolation and counting on a smartphone-based imaging platform. Analytical Chemistry. 2019; *91:* 11963–11971.

[35] Chen ML, He YJ, Chen XW, Wang JH. Quantum dots conjugated with Fe3O4-filled carbon nanotubes for cancer-targeted imaging and magnetically guided drug delivery. Langmuir: The Acs Journal of Surfaces and Colloids. 2012; *28:* 16469–16476.

[36] Lee J, Hwang G, Hong YS, Sim T. One step synthesis of quantum dot-magnetic nanoparticle heterodimers for dual modal imaging applications. The Analyst. 2015; *140:* 2864–2868.

[37] Justin R, Tao K, Román S, Chen D, Xu Y, Geng X, Ross IM, Grant RT, Pearson A, Zhou G, MacNeil S, Sun K, Chen B. Photoluminescent and superparamagnetic reduced graphene oxide–iron oxide quantum dots for dual-modality imaging, drug delivery and photothermal therapy. Carbon. 2016; *97:* 54–70. www.sciencedirect.com/science/article/pii/S000862231530018X

[38] Irvine SE, Staudt T, Rittweger E, Engelhardt J, Hell SW. Direct light-driven modulation of luminescence from Mn-doped ZnSe quantum dots. Angewandte Chemie (International ed. In English). 2008; *47:* 2685–2688.

[39] Yong KT. Mn-doped near-infrared quantum dots as multimodal targeted probes for pancreatic cancer imaging. Nanotechnology. 2009; *20:* 015102.

[40] Tu C, Ma X, Pantazis P, Kauzlarich SM, Louie AY. Paramagnetic, silicon quantum dots for magnetic resonance and two-photon imaging of macrophages. J Am Chem Soc. 2010; *132:* 2016–2023.

[41] Saha AK, Sharma P, Sohn HB, Ghosh S, Das RK, Hebard AF, Zeng H, Baligand C, Walter GA, Moudgil BM. Fe doped CdTeS magnetic quantum dots for bioimaging. Journal of Materials Chemistry. B. 2013; *1:* 6312–6320.

[42] Lin B, Yao X, Zhu Y, Shen J, Yang X, Jiang H, Zhang X. Multifunctional manganese-doped core–shell quantum dots for magnetic resonance and fluorescence imaging of cancer cells. New Journal of Chemistry. 2013; *37*: 3076–3083.http://dx.doi.org/10.1039/C3NJ00407D

[43] Li W, Song W, Chen B, Matcher SJ. Superparamagnetic graphene quantum dot as a dual-modality contrast agent for confocal fluorescence microscopy and magnetomotive optical coherence tomography. Journal of Biophotonics. 2019; *12:* e201800219.

[44] Zeng Y, Li H, Li Z, Luo Q, Zhu H, Gu Z, Zhang H, Gong Q, Luo K. Engineered gadolinium-based nanomaterials as cancer imaging agents. Applied Materials Today. 2020; *20:* 100686. www.sciencedirect.com/science/article/pii/S2352940720301335

[45] Liu Y, Ai K, Yuan Q, Lu L. Fluorescence-enhanced gadolinium-doped zinc oxide quantum dots for magnetic resonance and fluorescence imaging. Biomaterials. 2011; *32:* 1185-1192.

[46] Du Y, Rajamanickam K, Stumpf TR, Qin Y, McCulloch H, Yang X, Zhang J, Tsai E, Cao X. Paramagnetic quantum dots as multimodal probes for potential applications in nervous system imaging. Journal of Inorganic and Organometallic Polymers and Materials. 2018; *28*: 711–720. https://doi.org/10.1007/s10904-017-0766-7

[47] Ding H, Wang D, Sadat A, Li Z, Hu X, Xu M, de Morais PC, Ge B, Sun S, Ge J, Chen Y, Qian Y, Shen C, Shi X, Huang X, Zhang RQ, Bi H. Single-atom gadolinium anchored on graphene quantum dots as a magnetic resonance signal amplifier. ACS Applied Bio Materials. 2021; *4:* 2798–2809.

[48] Vyshnava SS, Pandluru G, Kumar KD, Panjala SP, Paramasivam K, Banapuram S, Anupalli RR, Dowlatabad MR. Biocompatible Ni-doped CdSe/ZnS semiconductor nanocrystals for cellular imaging and sorting. Luminescence: The Journal of Biological and Chemical Luminescence. 2022; *37:* 490–499.

[49] Ma Q, Nakane Y, Mori Y, Hasegawa M, Yoshioka Y, Watanabe TM, Gonda K, Ohuchi N, Jin T. Multilayered, core/shell nanoprobes based on magnetic ferric oxide particles and quantum dots for multimodality imaging of breast cancer tumors. Biomaterials. 2012; *33:* 8486–8494.

[50] Zhou R, Sun S, Li C, Wu L, Hou X, Wu P. Enriching Mn-doped ZnSe quantum dots onto mesoporous silica nanoparticles for enhanced fluorescence/magnetic resonance imaging dual-modal bio-imaging. ACS Applied Materials & Interfaces. 2018; *10:* 34060–34067.

[51] Xu HL, Yang JJ, ZhuGe DL, Lin MT, Zhu QY, Jin BH, Tong MQ, Shen BX, Xiao J, Zhao YZ. Glioma-targeted delivery of a theranostic liposome integrated with quantum dots, superparamagnetic iron oxide, and cilengitide for dual-imaging guiding cancer surgery. Advanced Healthcare Materials. 2018; *7:* e1701130.

[52] Zhang P, Draz MS, Xiong A, Yan W, Han H, Chen W. Immunoengineered magnetic-quantum dot nanobead system for the isolation and detection of circulating tumor cells. Journal of Nanobiotechnology. 2021; *19*: 116. https://doi.org/10.1186/s12951-021-00860-1

[53] Rogosnitzky M, Branch S. Gadolinium-based contrast agent toxicity: a review of known and proposed mechanisms. Biometals: An International Journal on the Role of Metal Ions in Biology, Biochemistry, and Medicine. 2016; *29:* 365–376.

[54] Mulder WJ, Koole R, Brandwijk RJ, Storm G, Chin PT, Strijkers GJ, de Mello Donegá C, Nicolay K, Griffioen AW. Quantum dots with a paramagnetic coating as a bimodal molecular imaging probe. Nano Letters. 2006; *6*: 1–6.

[55] Stasiuk GJ, Tamang S, Imbert D, Poillot C, Giardiello M, Tisseyre C, Barbier EL, Fries PH, de Waard M, Reiss P, Mazzanti M. Cell-permeable Ln(III) chelate-functionalized InP quantum dots as multimodal imaging agents. ACS Nano. 2011; *5:* 8193–8201.

[56] Zhang J, Hao G, Yao C, Hu S, Hu C, Zhang B. Paramagnetic albumin decorated CuInS(2)/ZnS QDs for CD133(+) glioma bimodal MR/fluorescence targeted imaging. Journal of Materials Chemistry. B. 2016; *4:* 4110–4118.

[57] Yang Y, Lin L, Jing L, Yue X, Dai Z. CuInS2/ZnS quantum dots conjugating Gd(III) chelates for near-infrared fluorescence and magnetic resonance bimodal imaging. ACS Applied Materials & Interfaces. 2017; *9:* 23450–23457. https://doi.org/10.1021/acsami.7b05867

[58] Pereira MIA, Pereira G, Monteiro CAP, Geraldes CFGC, Cabral Filho PE, Cesar CL, de Thomaz AA, Santos BS, Pereira GAL, Fontes A. Hydrophilic quantum dots functionalized with Gd(III)-DO3A monoamide chelates as bright and effective T1-weighted bimodal nanoprobes. Scientific Reports. 2019; *9:* 2341. https://doi.org/10.1038/s41598-019-38772-8

[59] Huang CL, Huang CC, Mai FD, Yen CL, Tzing SH, Hsieh HT, Ling YC, Chang JY. Application of paramagnetic graphene quantum dots as a platform for simultaneous dual-modality bioimaging and tumor-targeted drug delivery. Journal of Materials Chemistry. B. 2015; *3:* 651–664.

[60] Yang Y, Chen S, Li H, Yuan Y, Zhang Z, Xie J, Hwang DW, Zhang A, Liu M, Zhou X. Engineered paramagnetic graphene quantum dots with enhanced relaxivity for tumor imaging. Nano Letters. 2019; *19:* 441–448.

[61] Fang M, Chen M, Liu L, Li Y. Applications of quantum dots in cancer detection and diagnosis: A Review. Journal of Biomedical Nanotechnology. 2017; *13:* 1–16.

10 Magnetic Quantum Dots for In-Vivo Imaging

Venu Yakati[a], *Swathi Vangala*[b], *VJ Reddy*[c], *Bina Gidwani*[d], *Amber Vyas*[e], *Vishal Jain*[e], *Veenu Joshi*[g], *Kapil Agrawal*[f], *and Gopikrishna Moku*[c]

[a]Department of Chemical and Biological Engineering, University of Alabama, Tuscaloosa, Alabama, USA
[b]Telangana Social Welfare Residential Degree College for Women, Bhupalapally, Telangana, India
[c]Department of Physical Sciences, Kakatiya Institute of Technology and Science, Warangal, Telangana, India
[d]Columbia Institute of Pharmacy, Raipur, India
[e]University Institute of Pharmacy, Pt. Ravishankar Shukla University, Raipur, Chhattisgarh, India
[f]R. C. Patel College of Pharmacy, Shirpur, Maharashtra, India
[g] Center for Basic Science, Pt. Ravishankar Shukla University, Raipur, India

CONTENTS

DOI: 10.1201/9781003319870-10

10.1 INTRODUCTION

As one of the world's top causes of mortality in the twenty-first century, cancer seriously threatens public health [1]. It is yet unknown how cancer develops, invades the body, and spreads. To uncover the underlying molecular causes of cancer, real-time monitoring is essential. As a result, the development of a unique approach to cancer detection is urgent [2]. Organic fluorophores have been used for the detection of cancer. Unfortunately, they are subject to certain limitations, such as narrow excitation spectra, broad emission spectra, and insufficient sensitivity. Quantum dots (QDs) are nanometer-size luminescent semiconductor nanocrystals with sizes 2–10 nm that has attracted much interest because of their unique optical properties, such as broad excitation spectra, narrow emission bandwidth, and enhanced photobleaching [3]. Generally, QDs composed of II-VI or III–V group atoms and are observed as outstanding inorganic fluorophores that show luminescence and shine and contribute substantial benefits when excited by a particular frequency light source [4]. Due to their distinctive optical and physicochemical properties, QDs have applications in cellular labelling, deep-tissue imaging and diagnosis, assay labelling, biosensors, and energy materials [5].

As QDs have developed, there has been emerging interest in joining fluorescent QDs with different materials to makc multimodal nanocomposites [6]. Specifically, the fusion of fluorescent QDs with magnetic materials has expected applications in biological imaging, diagnostic, and theranostics. Even though some magnetic materials are conceivable (e.g., iron phosphide, gadolinium), the most explored materials are the iron oxides, precisely superparamagnetic iron oxide nanoparticles or SPIONs. SPIONs are magnetic nanoparticles made out of iron oxide, either maghemite (Fe_2O_3) or magnetite (Fe_3O_4), whose size is lesser than the magnetic crystal domain [7]. This implies that the nanoparticles show paramagnetic nature in solution, significantly diminishing their true capacity for aggregation, yet inducible magnetic nature within the sight of a magnetic field. SPIONs additionally display small or no hysteresis failures during magnetic cycling. SPIONs have been utilized in different biomedical applications, including cell separations (for instance magnetic-activated cell separation, MACS), single-molecule control, and MRI (magnetic resonance imaging). The last application was approved by the FDA (Feridex®) [8].

In clinical diagnosis, molecular imaging is a rapidly growing area that allows the visualization, measurement, and characterization of biological processes at the in-vitro and in-vivo levels [9]. With the advancement of nanoparticle-based molecular imaging probes, significant enrichment of the specificity and sensitivity of diagnostic imaging has been done by allowing the non-invasive and quantitative finding of accurate biomolecules in living subjects [10]. Various available imaging techniques are in the clinical diagnosis including magnetic resonance imaging (MRI), computed tomography (CT), optical imaging, ultrasound, positron emission tomography, and single photon emission computed tomography (SPECT), and these are having their benefits and limits. This simplifies the need of promoting multimodal imaging probes that can give a more consistent and precise location of the infectious area [11].

In this regard, the combination of fluorescence and magnetic hybrid nanoparticles (magnetic quantum dots) can result in MRI and optical imaging probes [12]. In this book chapter, we have discussed the properties of quantum dots, the advantages of QDs in bioimaging, some of the common synthetic methods of magnetic quantum dots, nanoprobes for multimodal imaging using fluorescent and magnetic quantum dots, and MQDs biological applications in *in-vivo* imaging.

10.2 PHOTO PHYSICAL PROPERTIES OF QUANTUM DOTS

Due to their unique properties, which can be customized to fit the demands of a variety of experiments, quantum dots have found application in the biological sciences. The important properties include sharp emission profiles, controllable emission wavelengths, robust signal strength, and the use of a single excitation source [13]. The core size, core composition, shell composition, and surface coating of the quantum dots are only a few of the variables that might affect their optical properties. The emission profile of a quantum dot can be customized to have a certain maximum anywhere along the electromagnetic spectrum, commencing in the ultraviolet (UV) area and extending to the near-infrared (near-IR) region, by changing either the size of the quantum dot core or its composition [14]. The stability of the core is impacted when the shell composition and/or surface coating are changed, and this increases photoluminescence while having a minimal impact on the emission range. When the excitation energy is higher than the band gap energy, semiconductor nanoparticles emit narrow, symmetric peaks that are different from one another. Due to this property, a single source can stimulate quantum dots of different sizes and compositions. However, depending on the quantum efficiency of the quantum dots at a given wavelength, the relative intensities of various quantum dot emission profiles will change with excitation wavelength independently of one another. Furthermore, more quantum dot emissions can be resolved within the visible spectrum than is achievable with conventional fluorophores because the peaks of quantum dot emission are far narrower than those of organic dyes [15].

10.3 ADVANTAGES OF QDs IN BIOIMAGING

One factor contributing to the high interest in QDs for bioimaging is that fluorescent dyes and fluorescent proteins (FPs) have several drawbacks, including limited excitation spectra, modest effective Stokes shifts, broad fluorescence bands, and susceptibility to photobleaching [16]. Additionally, unlike biological autofluorescence, whose photoluminescence (PL) lifetimes are in the same range (a few nanoseconds), they cannot be detected using time-gated detection, which improves signal-to-noise ratio (SNR) and allows for better differentiation between different target molecules, as was demonstrated for QDs [17]. Due to optical crosstalk, dye emission spectra exhibit a distinct red shoulder that restricts spectral multiplexing [18]. Furthermore, it is difficult to effectively excite many dyes at once due to the narrow absorption spectra. In contrast, multiple-QD excitation from a single excitation source (wavelength) and simultaneous multiple-QD PL detection gives the potential for multiplexing with essentially zero spectral crosstalk. In a direct comparison of photostability, QDs may be demonstrated to be 20 times brighter and 100 times more stable than Rhodamine 6G [19]. Additional comparisons with various dyes revealed that QDs were superior in terms of their photostability, even when compared to AlexaFluor 488, one of the most stable organic dyes [20]. Another benefit of using QDs over conventional fluorophores is their nearly two orders of magnitude greater two-photon cross-section, which is particularly intriguing for in-vivo applications using near-infrared (NIR) irradiation [21].

10.4 SYNTHETIC METHODS OF MAGNETIC QUANTUM DOTS

At present, the synthetic methods for magnetic quantum dots have been addressed in various pieces of literature [22]. Furthermore, its properties such as dispersibility, stability, biocompatibility, and desirable surface fictionalization are also related to its practical applications. While the configuration and morphologies of these MQDs are executed separately, the synthetic approaches are divided into the following various categories. These are hetero-crystalline growth, doping, cross-linking, and encapsulation method [23].

10.4.1 SYNTHESIS OF MQDs BY HETERO-CRYSTALLINE GROWTH

The MQDs can be synthesized in this method by combining the magnetic nanoparticles and quantum dots of either core-shell or hetero dimmers (two asymmetric nanoparticles). The deposition of semiconductor materials on the pre-assembled magnetic nanocrystals by disintegrating the antecedents at high temperatures produces the development of MQDs with diverse functional spheres. These are the previously reported core-shell structures, Fe_3O_4@PANI/CQDs, Fe_3O_4/CdSe/ZnS, and Co@CdSe,and the heterodimers areFe_3O_4-CdS (Se), Fe_2O_3-CdSe, FePt@CdS, and FePt-Pb (S, Se) [24]. Even though hetero-crystalline growth is the first reported technique for the synthesis of MQDs (or magnetic nanoparticles & quantum dots), this method has some failures such as inadequate optimization properties, loss of functional groups, and low quantum yield, and unwanted magnetic responsiveness. An important reason for the hetero-crystalline method is the crystal lattice mismatch between magnetic nanoparticles and quantum dots, the MQDs show low quantum yield and photoluminescence. Additional examination is required that should be focused on protecting the native photoluminescence and quantum yield of quantum dots within MQDs while keeping its heterodimer structure intact, which will allow for achieving efficient multimodal functionality for biological applications [25]. Here are some examples of MQDs synthesized by the hetero-crystalline growth method.

Lee et al. reported the synthesis of colloidal nanocomposites by combining the core-shell nanostructures (PbS&PbSe) with magnetic FePt particles [26]. Initially, the synthesized magnetic nanoparticles (FePt) were dissolved in toluene and injected into a lead oleate complex under a nitrogen atmosphere. For making the PbS shell around the FePt nanoparticles, the bis(trimethylsilyl)sulfide (TMS_2S) solution was injected rapidly. The injection of bis(trimethylsilyl)sulfide solution into the oleate complex leads to the formation of two nanostructures at a time with varying temperatures that are FePt-PbS core-shell nanostructures at 100 °C and cubic core-shell nanostructures at 150°C. Meanwhile, the FePt-PbSe nanostructures were prepared using trioctylphosphine instead of bis(trimethylsilyl)sulfide. To synthesize the FePt-PbS dumbbell-like nanostructures, the magnetic nanoparticles (FePt) were extensively washed from allylamine ligands and repeated with less labile oleate ligands. These MQDs showed semiconductor-type transport properties with magnetoresistance and magnetic tunnel separations. Nonetheless, lead-based quantum dots are more appropriate for optoelectronic applications because of IR emission attributes but are not suitable for biological applications due to the toxic nature of lead to the living cells [27].

Bandari et al. developed multimodal nanocomposites using SPION and ZnS quantum dots (Fe_3O_4 –ZnS). In these, the surface of Fe_3O_4nanoparticles was occupied by the cysteine-capped quantum dots. To form an octahedral ZnQ2 complex on the surface of ZnS quantum dots, the nanoparticles were treated with 8-hydroxyquinoline (HQ). The newly formed octahedral complex (ZnQ_2) synthesized by complexation of Fe_3O_4 –ZnS, showed high quantum yield, photoluminescence, photobleaching, superparamagnetic, high luminescence stability in human blood serum, non-toxic to the living cell, and highly useful for cellular imaging [28]. Lin et al. synthesized the silicon-coated MQDs nanoparticles and successfully labelled the cell membrane of liver cancer (HepG2) and breast cancer (4T1) cells [29].

10.4.2 Synthesis of MQDs by Doping

Doping is another technique to synthesize the MQDs, in which the quantum dots are doped with dopant ions (paramagnetic transition metal ions). The main challenge in the doping process is introducing some atoms within the crystal lattice of two or three hundred particles of metal nanoparticles without clustering dopant ions on the surface or inside the quantum dots [30]. Previously reported manganese-doped nanoparticles are the best examples of this method, in which magnetic resonance imaging, magnetic spintronics, and photoluminescence imaging have been studied by using these nanomaterials [31]. Mn^{2+}can be doped in both II-VI and III–V nanocrystals, either in the core or the shell. The optical and magnetic properties of the material are determined by the position of Mn^{2+}ion. Lanthanides (Tb^{3+}, Er^{3+}, and Yb3^{+}) are different dopant materials that are doped in the host lattice. Doped materials offer benefits over the materials prepared by the hetero-crystalline method or high-temperature decomposition. Nonetheless, doped materials are restricted by the scope of accessible dopants and the trouble in synthesizing doped structures [32].

10.4.3 Synthesis of MQDs by Cross-Linking

Researchers have studied techniques that allow for the distinct creation of nanoparticles, which are then mixed using a different technique, to improve the qualities of constituent materials (e.g., cross-linking). Cross-linking is another method in which the florescent-magnetic quantum dots were cross-linked by SPIONs (magnetic materials and quantum dots) and fluorescent materials (fluorescent proteins or organic dyes) [33, 34]. The functional groups such as carboxyl, thiol, or siloxane groups are the coupling ligands that bind both the materials; however, proteins have also been used to mediate these interactions. In some of the cross-linking approaches, only electrostatic interactions were mediated instead of the chemical linker. For instance, positively charged Fe_3O_4 nanoparticles were electrostatically bound with the negatively charged carboxy-CdTe quantum dots [34]. This method has the advantage that the presence of the carboxyl group on the surface of quantum dots forms the chelation with Fe ions to reinforce the binding. Many limitations of later synthesis or doping techniques are circumvented by cross-linking, however, the structures that are created can be bulky and the stability of the binding ligands and cross-linkers might vary depending on the surrounding environment [35].

10.4.4 Synthesis of MQDs by the Encapsulation Method

In the encapsulation method, the pre-formed magnetic and fluorescent material is incorporated into a matrix. Due to the separate synthesis of fluorescent and magnetic nanoparticles, the encapsulation methods avoid the intrinsic obstacles of controlled crystallization and doping. This method avoids coupling reagents and not required any interaction between ligand and linker to form stable nanoparticles. The encapsulation method performs by two processes which are the top-down process and the bottom-up process [36]. An example of the top-down method is the single emulsion process was used to co-encapsulate the quantum dots and SPIONs with polymers (e.g., polylactic-co-glycolic acid) for the application in drug delivery). The major disadvantage of the top-down process is the particle size is relatively larger than expected and these particles do not penetrate the deeper tissues in biological imaging.

For instance, the preparation of smaller particles (<200 nm) with disaggregated PLGA is a very difficult process. In that case, energy addition is required to the top-down process to get the desired quantity of particles. The bottom-up is another method in the encapsulation process; this method uses thermodynamic driving forces to get the desired particles. Any self-assembled particle with a hydrophobic space greater than the sum of the encapsulant volumes can theoretically serve as the matrix. For instance, the previously reported micellar-based co-encapsulation method combined the SPIONs, quantum dots, and quantum dot rods to form fluorescent-magnetic nanoparticles [25, 37].

10.5 NANOPROBES FOR MULTIMODAL IMAGING USING FLUORESCENT AND MAGNETIC QUANTUM DOTS

QDs are a novel class of molecular imaging tools that have had a substantial impact on biological and medical research, paving the way for the creation of new applications. When combined with NIR emission, paramagnetic, or superparamagnetic nanomaterials, QDs appear to be a vital component for the continued fabrication of multifunctional nanostructures and nanodevices. The most recent uses of fluorescent and magnetic nanocomposites, which are used to create unique multiplexed nanoprobe models, are shown in the following section. A variety of materials, including silica-based, dye-functionalized MNPs, and QDs-MNP composites, are combined to form fluorescent-magnetic nanocomposites. In various papers, various methods for creating composites of fluorescent semiconductor QDs and MNPs have been described. These methods include mixing the fluorescent and magnetic materials to create a single heteromeric particle with optical and magnetic properties, encasing the single particles in a polymer or silica gel, using magnetically doped QDs, and using ionic aggregates. In the field of nanobiotechnology, these multifunctional fluorescent magnetic nanocomposites can be used for a variety of biological and biomedical processes, including drug delivery, cell tracking, and sorting, separation, and imaging and therapy [].

Fluorescence imaging and magnetic resonance imaging (MRI) are two very potent and highly complementary imaging modalities, and Koole et al. described four different techniques of obtaining a single nanoparticle that combines the fluorescence and magnetic properties and represents a new sensitive bimodal contrast instrument [39]. For two reasons—the first being that multimodal magnetic-fluorescent tests would be advantageous for in-vitro and in-vivo bioimaging applications (MRI and fluorescence microscopy) Corr et al. emphasized the fact that the combination of a magnetic and fluorescent entity offers novel two-in-one multifunctional nanomaterials with a wide range of practicable applications. To both diagnose and simultaneously treat various diseases, fluorescent magnetic nanocomposites are used.

New approaches to therapeutic procedures and imaging modalities (such as the technique of the correlation between MRI and ultrasensitive optical imaging) are demonstrated in some of the earlier reports to be developed and become standard clinical methods for the visual detection of microscopic tumors during an operation and the total elimination of the diseased cells and tissues. The authors of these studies demonstrate that while medical imaging techniques can identify diseases, they cannot provide a visual blueprint for a particular surgery. Some MQD probes can be used to solve this problem. MQDs are a type of magnetic contrast agent used in MRI. To improve image contrast in various MRI applications, paramagnetic and superparamagnetic agents Gd (III) and various types of iron oxide (Fe_2O_3) in molecular and nanoparticle form are linked to QDs [39,40].

A concept of multifunctional fluorescent magnetic nanocomposites, consisting of silica-coated Fe_3O_4 and TGA-capped CdTe QDs, was communicated after extensive research into paramagnetic QDs (pQDs). HeLa cells have been labelled and imaged using this kind of nanocomposite during magnetic separation. With the addition of a PEGylated phospholipid and a Gd lipid, biocompatible and MRI-active CdSe/ZnS core-shell QDs were transformed into enhanced paramagnetic QDs. In-vitro studies using human umbilical vein, endothelial cells demonstrated that the pQDs were additionally coupled by maleimide to cyclic RGD peptides for targeting angiogenic vascular endothelium (HUVECs) [39, 41].

In order to create QDs-enclosed MNPs for use in cancer cell imaging, Ahmed et al. invented a revolutionary layer-by-layer (LbL) self-assembly process. In a study, Park's team [45] reported long-circulating, micellar hybrid nanoparticles (MHNs), which combine MNs, QDs, and the chemotherapeutic drug doxorubicin (DOX) in a single poly(ethylene glycol) (PEG)-phospholipid micelle and provide the first models of simultaneous targeted drug delivery and dual-mode near-infrared fluorescence imaging and MRI of diseased tissue in vitro and in vivo.

10.6 APPLICATIONS OF MQDs IN IN-VIVO IMAGING

Nanotechnology is one of the most fast-growing biomedical types of research for in-vivo imaging, drug delivery, and theranostics due to the nanosized particles. Quantum dots are nanosized semiconductor materials with exceptional optical properties that

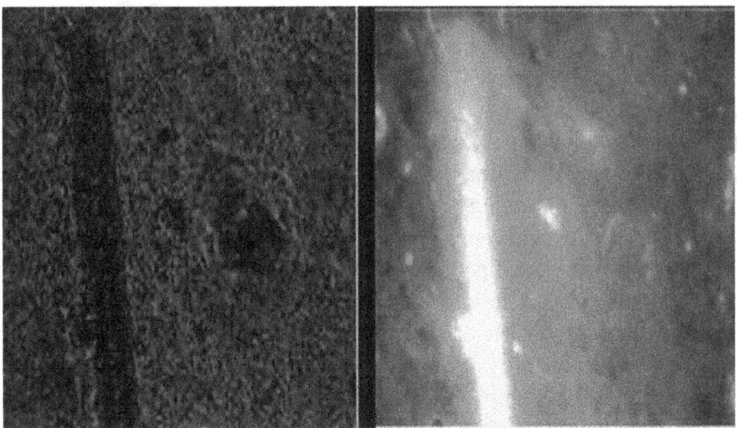

FIGURE 10.1 Diagram showing the histological images of rat brain cross sections when the TAT-conjugated quantum dots. *Copyright © 2005, American Chemical Society.*

greatly enhance fluorescence-based bioimaging. In the 19th century, Alivisatos, and Nie demonstrated the quantum dots for in-vivo imaging. These materials showed great achievement as imaging agents in the biomedical field. However, to get more effective imaging, researchers have incorporated the quantum dots with magnetic materials, including SPIONs to form dual functional nanocomposites for MRI and fluorescence microscopy. We have discussed here some of the applications of fluorescence magnetic quantum dots (or nanocomposites) in cancer imaging and other biological in-vivo imaging [42].

Santra et al. doped the quantum dots with paramagnetic manganese ions to form multifunctional magnetic quantum dots and these ultra-small (3.1 nm) semiconductor particles possess fluorescent, paramagnetic, and radio-opacity properties. These quantum dots were bioconjugated with TAT, a tumor-penetrating peptide, and administered the suspension administered through the right common carotid artery into a rat model. The histological analysis of brain slices confirmed the labelling of the cerebral and carotid arteries in the brain (Figure 10.1). These multimodal properties of magnetic quantum dots are useful for crossing the blood-brain barrier (BBB) and help in diagnosis and drug delivery to the brain [36].

Shi et al. developed a new class of nanohybrids that can be used in biomedical in-vivo imaging via hyperthermia. Initially, they prepared the magnetically modified polystyrene nanospheres (Fe_3O_4) and then coupled them with quantum dots (CdSeTe) with carbodiimide to form magnetic quantum dots (QD-MNSs). The suspension of QD-MNSs when administered intravenously into nude mice, significant fluorescence was reported in the non-sinusoidal region of the spleen with limited distribution to red pulp reticular meshwork. This study concluded that the fluorescence signals can be detectable by QD-MNSs when administered to live mice [37].

Early cancer diagnosis and treatment are critical challenges to nanotechnology for the development of multifunctional nanocomposites. In the extension of the above

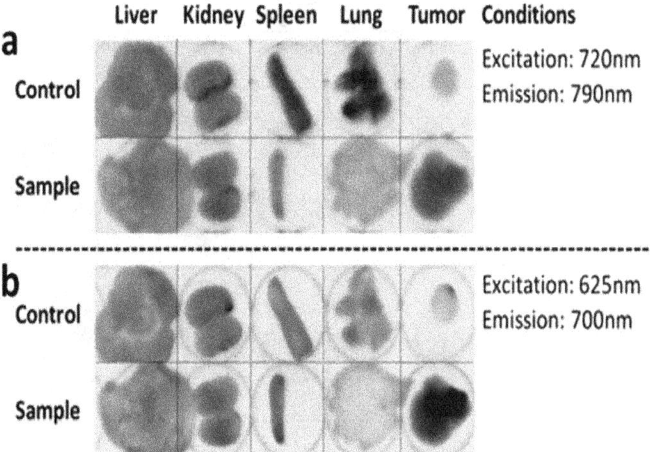

FIGURE 10.2 Illustration showing fluorescent images of *ex vivo* organs of PTX-loaded multifunctional quantum dots. *Copyright © 2010, American Chemical Society.*

study (by Shi et el.), Cho et al. developed the multifunctional quantum dots for diagnosis of the cancer tissue and simultaneously delivered the Paclitaxel (PTX)into tumor-bearing nude mice. Initially, the quantum dots were conjugated with iron oxide nanoparticles as discussed in the above application. For the preparation of PTX-loaded nanocomposites, the PTX was loaded onto the surface of multifunctional nanocarriers by using PLGA biodegradable polymer [38]. The biodistribution study showed (Figure 10.2) significant differences in fluorescent signals in the ex-vivo tumor regions with the targeted nanocarrier system and the non-targeted nanocarrier system.

Tan et al. co-encapsulated the iron oxide and fluorescent quantum dots in of poly(lactide)-tocopheryl polyethyleneglycol succinate (PLA-TPGS) nanoparticles. When these nanoparticles were administered into a xenograft mouse model, MRI imaging showed the maximum fluorescent intensity in the tumor region. The ex vivo images of the various organs of the mice injected also revealed similar results (Figure 10.3). These results were appropriately compared with the control experiments [43].

Ma et al. synthesized multilayered core-shell QDs based on magnetic nanoparticles and quantum dots for multimodal tumor imaging. The multilayered quantum dots (MQDs-probe) probe contains iron oxide magnetic nanoparticles, visible-fluorescent quantum dots, and near-IR fluorescent quantum dots in multiple silica layers. The MQDs-probe can be used as a contrast reagent for MRI and fluorescent probes. Anti-HER2 antibody was also conjugated to the surface of MQDs-probe for imaging the breast cancer tumor. The HER2-MQQ-probes are found mainly in the peripheral regions around the blood vessels of the tumor (Figure 10.4), suggesting the permeability of the probes across the blood vessels. These nanoprobes would be used for biological and biomedical applications such as cellular imaging and molecular diagnostics of various diseases in vitro and in vivo [44].

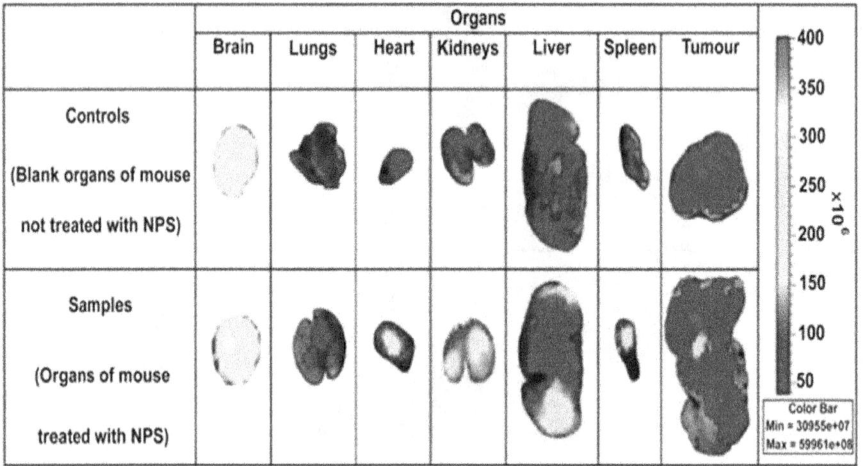

FIGURE 10.3 Diagrammatic representation of the fluorescent Images of the various organs. *Copyright © 2011 Elsevier Ltd.*

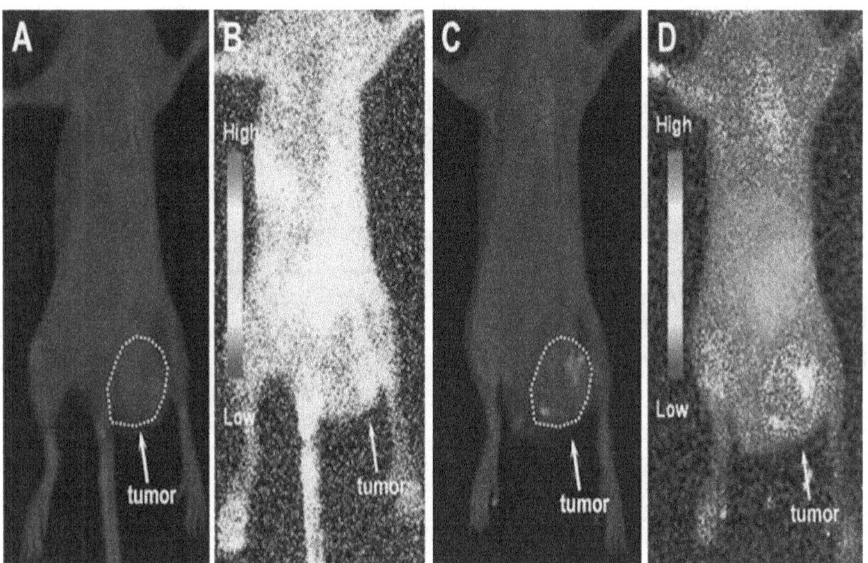

FIGURE 10.4 Illustration showing *in-vivo,* NIR fluorescence imaging of a HER2-MQQ-probe injected mice bearing human breast tumors. *Copyright © 2012 Elsevier Ltd.*

10.7 CONCLUSION

MQDs synthesized by a chemical combination of QDs and MNPs provide unique multimodal functionality because their excellent combination of optical and magnetic properties is the most desirable feature applicable in many biological and biomedical applications, such as biolabelling, MR imaging, hyperthermia treatments, chemotherapy, disease diagnostics, drug delivery, sorting and separation processes. In this book chapter, we have discussed the different preparation methods of magnetic quantum dots in which the quantum dots and magnetic nanoparticles are coupled or synthesized using various methods such as hetero-crystalline growth or high-temperature decomposition, doping process, cross-linking method, and co-encapsulation method. And also, we have explored some of the applications of MQDs in in-vivo imaging.

Since the toxicity of quantum dots is a serious problem, there is a lot of interest in creating non-toxic fluorescence probes and quantum dots. Despite these limitations, fluorescent-magnetic nanocomposites, particularly magnetic quantum dots, offer many exciting opportunities in biomedicine. Finally, fluorescent-magnetic nanoparticles can serve as the building blocks for theranostics, which uses controlled medication release or special features of the nanoparticles themselves to both diagnose and cure disease. Additional advantages will become apparent as fluorescent-magnetic nanoparticle technology develops, possibly resulting in a new class of diagnostic and therapeutic materials.

ACKNOWLEDGEMENTS

This work was supported by the Science & Engineering Research Board (SERB), Department of Science and Technology, Government of India, New Delhi (File no. SRG/2021/002092). V.J.R. thanks the SERB, Government of India, New Delhi, for his research fellowship.

REFERENCES

1. Siegel, Rebecca L., Kimberly D. Miller, Hannah E. Fuchs, and Ahmedin Jemal. Cancer Statistics, 2022. CA: A Cancer Journal for Clinicians. 2022; 72 (1): 7–33.
2. Somarelli, Jason A. The Hallmarks of Cancer as Ecologically Driven Phenotypes. Frontiers in Ecology and Evolution. 2021; 9: 226.
3. Chen, Kai, and Xiaoyuan Chen. Design and Development of Molecular Imaging Probes. Current Topics in Medicinal Chemistry. 2010; 10 (12): 1227.
4. Bailey, Robert E., Andrew M. Smith, and Shuming Nie. Quantum Dots in Biology and Medicine. Physica E: Low-Dimensional Systems and Nanostructures. 2004; 25 (1): 1–12.
5. Barroso, Margarida M. Quantum Dots in Cell Biology. Journal of Histo Chemistry and Cyto Chemistry. 2011; 59 (3): 237.
6. Tan, Yang Fei, Prashant Chandrasekharan, Dipak Maity, Cai Xian Yong, Kai Hsiang Chuang, Ying Zhao, Shu Wang, Jun Ding, and Si Shen Feng. Multimodal Tumor Imaging by Iron Oxides and Quantum Dots Formulated in Poly (Lactic Acid)-d-Alpha-Tocopheryl Polyethylene Glycol 1000 Succinate Nanoparticles. Biomaterials. 2011; 32 (11): 2969–78.
7. Wang, Guannan, and Xingguang Su. The Synthesis and Bio-Applications of Magnetic and Fluorescent Bi functional Composite Nanoparticles. Analyst. 2011; 136: 9.
8. Lin, Alex W.H., Chung Yen Ang, Pranab K. Patra, Yu Han, Hongwei Gu, Jean Marie Le Breton, Jean Juraszek, et al. Seed-Mediated Synthesis, Properties and Application

of γ-Fe2O3–CdSe Magnetic Quantum Dots. Journal of Solid State Chemistry. 2011; 184 (8): 2150–58.

9. Wahajuddin, and Sumit Arora. Superparamagnetic Iron Oxide Nanoparticles: Magnetic Nanoplatforms as Drug Carriers. International Journal of Nanomedicine. 2012; 7: 3445.
10. Mahajan, Kalpesh D., Qirui Fan, Jenny Dorcena, Gang Ruan, and Jessica O. Winter. Magnetic Quantum Dots in Biotechnology–Synthesis and Applications. Biotechnology Journal. 2013; 8 (12): 1424–34.
11. You, Xiaogang, Rong He, Feng Gao, Jun Shao, Bifeng Pan, and Daxiang Cui. Hydrophilic High-Luminescent Magnetic Nanocomposites. Nanotechnology. 2007; 18(3): 035701.
12. Acharya, Amitabha. Luminescent Magnetic Quantum Dots for in Vitro/in Vivo Imaging and Applications in Therapeutics. Journal of Nanoscience and Nanotechnology. 2013; 13 (6): 3753–68.
13. Walling, Novak, and Shepard. Quantum Dots for Live Cell and In Vivo Imaging. Int. J. Mol. Sci. 2009, 10
14. Pietryga et al. Pushing the Band Gap Envelope: Mid-Infrared Emitting Colloidal PbSe Quantum Dots. J. Am. Chem. Soc. 2004, 126, 38, 11752–11753
15. Murray, Norris, and Bawendi. Synthesis and Characterization of Nearly Monodisperse CdE (E = S, Se, Te) Semiconductor Nanocrystallites. J. Am. Chem. Soc. 115, 8706–8715.
16. Bruchez M Jr, Moronne M, Gin P, Weiss S, Alivisatos AP. Semiconductor Nanocrystals as Fluorescent Biological Labels. Science. 1998; 281(5385): 2013–6.
17. Cho et al., Fluorescent, Superparamagnetic Nanospheres for Drug Storage, Targeting, and Imaging: A Multifunctional Nanocarrier System for Cancer Diagnosis and Treatment. ACS Nano 2010; 4(9):5398–404.
18. Wu X, Liu H, Liu J, et al. Immuno Fluorescent Labeling of Cancer Marker Her2 and other Cellular Targets with Semiconductor Quantum Dots. Nat Biotechnol. 2003; 21 (1):41–46.
19. Jaiswal et al., Long-Term Multiple Color Imaging of Live Cells Using Quantum Dot Bioconjugates. Nature Biotechnology. 2003; 21: 47–51.
20. Chan WC, Nie S. Quantum Dot Bioconjugates for Ultrasensitive Nonisotopic Detection. Science. 1998; 281 (5385): 2016–2018.
21. Panchuk-Voloshina et al., Alexa Dyes, a Series of New Fluorescent Dyes That Yield Exceptionally Bright, Photostable Conjugates. The Journal of Histochemistry & Cytochemistry, 1999; 47(9): 1179–1188.
22. Wegner and Hildebrandt, Quantum Dots: Bright and Versatile in Vitro and in Vivo Fluorescence Imaging Biosensors. Chemical Society Reviews, 2015, 44(14): 4792.
23. Hanahan, Douglas. Hallmarks of Cancer: New Dimensions. Cancer Discovery. 2022; 12 (1): 31–46.
24. Wang, Shizhong, Benjamin R. Jarrett, Susan M. Kauzlarich, and Angelique Y. Louie. Core/Shell Quantum Dots with High Relaxivity and Photoluminescence for Multimodality Imaging. Journal of the American Chemical Society. 2007; 129 (13): 3848–56.
25. Bhandari, Satyapriya, Rumi Khandelia, Uday Narayan Pan, and Arun Chattopadhyay. Surface Complexation-Based Biocompatible Magnetofluorescent Nanoprobe for Targeted Cellular Imaging. ACS Applied Materials and Interfaces. 2015; 7 (32): 17552–57.
26. Chen, Guoning, Qianqian Hu, Hua Shu, Lu Wang, Xia Cui, Jili Han, Kamran Bashir, Zhimin Luo, Chun Chang, and Qiang Fu. Fluorescent Biosensor Based on Magnetic

Cross-Linking Enzyme Aggregates/CdTe Quantum Dots for the Detection of H2O2-Bioprecursors. New Journal of Chemistry. 2020; 44 (41): 17984–92.

27. Ruan, Gang, Greg Vieira, Thomas Henighan, Aaron Chen, Dhananjay Thakur, R. Sooryakumar, and Jessica O. Winter. Simultaneous Magnetic Manipulation and Fluorescent Tracking of Multiple Individual Hybrid Nanostructures. Nano Letters. 2010; 10 (6): 2220–24.
28. Thakur, Dhananjay, Shuang Deng, Thierno Baldet, and Jessica O. Winter. PH Sensitive CdS–Iron Oxide Fluorescent–Magnetic Nanocomposites. Nanotechnology. 2009; 20 (48): 485601.
29. Pellegrino, Teresa, Stefan Kudera, Tim Liedl, Almudena Munoz Javier, Liberato Manna, and Wolfgang J. Parak. On the Development of Colloidal Nanoparticles towards Multifunctional Structures and Their Possible Use for Biological Applications. Small. 2005; 1 (1): 48–63.
30. Tufani, Ali, Anjum Qureshi, and Javed H. Niazi. Iron Oxide Nanoparticles Based Magnetic Luminescent Quantum Dots (MQDs) Synthesis and Biomedical/Biological Applications: A Review. Materials Science and Engineering: C. 2021; 118: 111545.
31. Zhao, Jianhong, Junwei Chen, Shengnan Ma, Qianqian Liu, Lixian Huang, Xiani Chen, Kaiyan Lou, and Wei Wang. Recent Developments in Multimodality Fluorescence Imaging Probes. Acta Pharmaceutica Sinica B. 2018; 8 (3): 320–38.
32. Quarta, Alessandra, Riccardo Di Corato, Liberato Manna, Andrea Ragusa, and Teresa Pellegrino. Fluorescent-Magnetic Hybrid Nanostructures: Preparation, Properties, and Applications in Biology. IEEE Transactions on Nanobioscience. 2007; 6 (4): 298–308.
33. Wagner, Angela M., Jennifer M. Knipe, Gorka Orive, and Nicholas A. Peppas. Quantum Dots in Biomedical Applications. Acta Biomaterialia. 2019; 94: 44.
34. Ibanez-Peral, Raquel, Peter L. Bergquist, Malcolm R. Walter, Moreland Gibbs, Ewa M. Goldys, and Belinda Ferrari. Potential Use of Quantum Dots in Flow Cytometry. International Journal of Molecular Sciences. 2008; 9 (12): 2622.
35. Lee, Jong Soo, Maryna I. Bodnarchuk, Elena V. Shevchenko, and Dmitri V. Talapin. Magnet-in-the-Semiconductor FePt-PbS and FePt-PbSe Nanostructures: Magnetic Properties, Charge Transport, and Magnetoresistance. Journal of the American Chemical Society. 2010; 132 (18): 6382–91.
36. Santra, Swadeshmukul, Heesun Yang, Paul H. Holloway, Jessie T. Stanley, and Robert A. Mericle. Synthesis of Water-Dispersible Fluorescent, Radio-Opaque, and Paramagnetic CdS:Mn/ZnS Quantum Dots: A Multifunctional Probe for Bioimaging. Journal of the American Chemical Society. 2005; 127 (6): 1656–57.
37. Shi, Donglu, Hoon Sung Cho, Yan Chen, Hong Xu, Hongchen Gu, Jie Lian, Wei Wang, et al. Fluorescent Polystyrene–Fe_3O_4 Composite Nanospheres for In Vivo Imaging and Hyperthermia. Advanced Materials. 2009; 21: 2170–73.
38. Cho, Hoon Sung, Zhongyun Dong, Giovanni M. Pauletti, Jiaming Zhang, Hong Xu, Hongchen Gu, Lumin Wang, et al. Fluorescent, Superparamagnetic Nanospheres for Drug Storage, Targeting, and Imaging: A Multifunctional Nanocarrier System for Cancer Diagnosis and Treatment. ACS Nano. 2010; 4 (9): 5398–5404.
39. Armaselu. Quantum Dots and Fluorescent and Magnetic Nanocomposites: Recent Investigations and Applications in Biology and Medicine. Nonmagnetic and Magnetic Quantum Dots. Chapter 13, 2018; Intechopen 221–235.
40. Corr, Rakovich, Gun'Ko. Multifunctional Magnetic-Fluorescent Nanocomposites for Biomedical Applications. Nanoscale Res Lett. 2008; 3(3): 87–104.
41. Koole et al., Magnetic Quantum Dots for Multimodal Imaging. Wiley Interdiscip Rev NanomedNanobiotechnol. 2009 Sep-Oct;1(5):475–91

42. Rahin Ahmed et al., Quantum Dots Incorporated Magnetic Nanoparticles for Imaging Colon Carcinoma Cells. Journal of Nanobiotechnology Volume 11, Article number: 28 (2013) https://doi.org/10.1186/1477-3155-11
43. Tan, Yang Fei, Prashant Chandrasekharan, Dipak Maity, Cai Xian Yong, Kai Hsiang Chuang, Ying Zhao, Shu Wang, Jun Ding, and Si Shen Feng. Multimodal Tumor Imaging by Iron Oxides and Quantum Dots Formulated in Poly (Lactic Acid)-d-Alpha-Tocopheryl Polyethylene Glycol 1000 Succinate Nanoparticles. Biomaterials. 2011; 32 (11): 2969–78.
44. Ma, Qiang, Yuko Nakane, Yuki Mori, Miyuki Hasegawa, Yoshichika Yoshioka, Tomonobu M. Watanabe, Kohsuke Gonda, Noriaki Ohuchi, and Takashi Jin. Multilayered, Core/Shell Nanoprobes Based on Magnetic Ferric Oxide Particles and Quantum Dots for Multimodality Imaging of Breast Cancer Tumors. Biomaterials 2012; 33: 8486–94.
45. Park et al., Micellar Hybrid Nanoparticles for Simultaneous Magneto Fluorescent Imaging and Drug Delivery. Angew Chem Int Ed Engl. 2008; 47(38): 7284–7288.

11 Carbon Quantum Dots-Based Magnetic Nanoparticles for Bioimaging

Sharuk L. Khan[1], *Falak A. Siddiqui*[1], *Md. Rageeb Md. Usman*[2], *Prashant Subhash Palghadmal*[3], *Nilesh S. Patil*[4], *Poonam Talwan*[5], *Rokeya Sultana*[6], *and Fahadul Islam*[7]

[1]Department of Pharmaceutical Chemistry, N.B.S. Institute of Pharmacy, Ausa, Maharashtra, India

[2]Department of Pharmacognosy, Smt. Sharadchandrika Suresh Patil College of Pharmacy, Chopda, Maharashtra, India

[3]SND College of Pharmacy Yeola Babhulgaon, Nashik, Maharashtra, India

[4]RD & SH National College & SWA Science College, Bandra, Maharashtra, India

[5]Himachal Institute of Pharmaceutical Education and Research, Nadaun, Himachal Pradesh, India

[6]Yenepoya Pharmacy College and Research Centre, Yenepoya (Deemed to be University), Deralakatte, Mangalore, India

[7]Department of Pharmacy, Faculty of Allied Health Sciences, Daffodil International University, Dhaka, Bangladesh

CONTENTS

DOI: 10.1201/9781003319870-11

11.1 INTRODUCTION

Minimally invasive molecular imaging has emerged as a powerful tool for the early diagnosis and fundamental research of malignant tumors [1]. This is because it allows the visualization of cellular function and molecular process in living organisms without causing any disruption to the organisms' normal functioning [2]. Because of the fast growth in nanotechnology, it is now possible to perform sensitive clinical detection of biomolecules, which has been a topic of interest to researchers for the previous few decades. As a result of the numerous qualities of nanomaterials, such as their high surface-to-volume ratio, active sites on their surfaces, and electrocatalytic activity, biosensors based on diverse nanomaterials have been widely used in biosensor production [3-5]. Noninvasive molecular imaging may be performed using a variety of imaging modalities, including magnetic resonance (MR), X-ray computed tomography (CT), ultrasound, positron emission tomography (PET), fluorescence, and single photon emission CT, among others [6-11]. Every unique imaging modality which is dependent on the imaging principle for generating the visual information, has its own merits and shortcomings, resulting in difficulty in speedy and reliable data collection and illness identification when depending on a single imaging mode for diagnosis [12-15]. Consider the MR imaging modality, which has the advantages of non-ionizing radiation, high spatial resolution, and deep tissue penetration, but also has the disadvantages of having a low sensitivity and a limited range of motion [16, 17]. Fluorescence imaging, on the other hand, has the capability of detecting single cells and achieving subcellular resolution, but it has limited spatial resolution and tissue penetration capabilities [18-21].

In the world of carbonaceous nanomaterials, CQDs are among the most recent additions. Because of their optical and biological features, CQDs have piqued the curiosity of researchers ever since they were accidentally discovered [22]. Chemistry-Quantum Dots (CQDs) are simple to produce, biocompatible, extremely water-soluble, low toxicity, exhibit sustained fluorescence, and are capable of delivering a significant electrochemical reaction [23-25]. The superior features of CQDs allow them to be used in a variety of applications such as drug administration, photoelectrochemistry, bioimaging, optoelectronics, electrochemiluminescence, biosensing, and other fields [26-29]. In addition to providing substantial carboxylic and amine groups for further functionalization with biomolecules, CQDs made from citric acid as a carbon source and ethylenediamine (EDA) as a passivating agent are water-soluble, allowing them to be used in a variety of applications [30]. As a consequence, there is significant potential to investigate CQDs for surface functionalization of Fe_3O_4 NPs, which might result in CQDs that inhibit agglomeration of Fe_3O_4 NPs by coating their surfaces. CQDs can also functionalize Fe_3O_4 NPs in two ways: first, they can reduce the likelihood of agglomeration of Fe_3O_4 NPs, resulting in a high surface-to-volume ratio for better adsorption of biomolecules; and second, they can lessen the probability of agglomeration of Fe_3O_4 NPs, leading to a low surface-to-volume ratio for superior sorption process of biomolecules. First and foremost, the presence of functional groups (carboxylic and amine) on the surface of CQDs may facilitate the biofunctionalization (proteins, antibodies) of NPs in preparation for subsequent application. This was the motivation for the current research, which involved the fabrication of Fe3O4NPs functionalized with CQDs (CQD@Fe_3O_4 NPs) utilizing wet-chemical procedures. Using these CQD@Fe_3O_4 NPs, a uniform dispersion of chitosan (CHIT) solution was achieved, resulting in a uniform thin film suitable for electrochemical biosensor application. Sarkar et al. have earlier investigated the electrochemical behavior of CQDCHIT for the detection of vitamin D2 [31]. It is mentioned here that a similar investigation was conducted with CQD@Fe3O4CHIT to assess the electrochemical response of these bioelectrodes for the detection of vitamin D2. CHIT has been extensively researched for its superior film formation, high biocompatibility, low non-toxicity, and high mechanical strength, and it is now being employed as a matrix for the dispersion of nanomaterials.

11.2 SYNTHESIS PROCEDURES OF MAGNETIC QUANTUM DOTS

An emerging area of study that has the potential to have an influence on a wide variety of technological fields is the approach of fabricating a material that can simultaneously store more than one functional component. The term "multifunctional material" refers to these kinds of materials. In particular, metal–organic framework particles (MFNPs) have come to be recognized as an emerging group of materials that are capable of being utilized in sophisticated applications [32, 33]. MNPs are a category of smart composites exhibiting superparamagnetism, a magnetic response, a large specific surface area, and a tiny particle size. MNPs have the potential to heat up when exposed to an external magnetic AC field, which opens the door to the possibility of hyperthermic properties [34-36]. In addition, external magnetic fields have the potential to transport particles to a specific location, making them a candidate for participation in site-specific delivery vectors [37-39]. There are several types of magnetic QDs which are used for multifunctional applications.

The presence of both magnetic and photoluminescent properties in a single material paves the way for the development of cutting-edge nanomaterials. MFNPs can have a variety of coatings, including silica, fluorescently tagged lipids, fluorescently labelled surface engineered MFNPs with macromolecular anchors, magnetofluorescent quantum dots, magnetofluorescent heterocrystals, and magnetic and luminescence bifunctionalized Janus particles. The MFNPs have the potential to act as a multisensory probe for use in medical diagnostics, in-vitro and in-vivo biolabelling, as well as fluorescence microscopy, MRI, and other drug delivery systems. Other potentially useful uses of MFNPs include cytometry, the tracking of cells, and magnetic separation; all of these processes may be easily controlled and monitored using confocal or fluorescence microscopy as well as MRI. Consequently, these fascinating nanostructures have garnered a lot of attention in recent years due to their attraction.

MQDs can be synthesized using a variety of methods, such as I core/shell nanostructure, in which the magnetic core material (such as Co, FeP, Fe_3O_4, FePt, or Fe_2O_3) is synthesized first, and then the QDs (such as FePt/CdSe, Co/CdSe, or heterodimer structures/dumbbell-like NPs) are coated on top of the core/shell nanostructure [40]. In other cases, they are also generated as heterodimer structures or NPs of magnetic nanomaterials and QDs in the shape of dumbbells by using

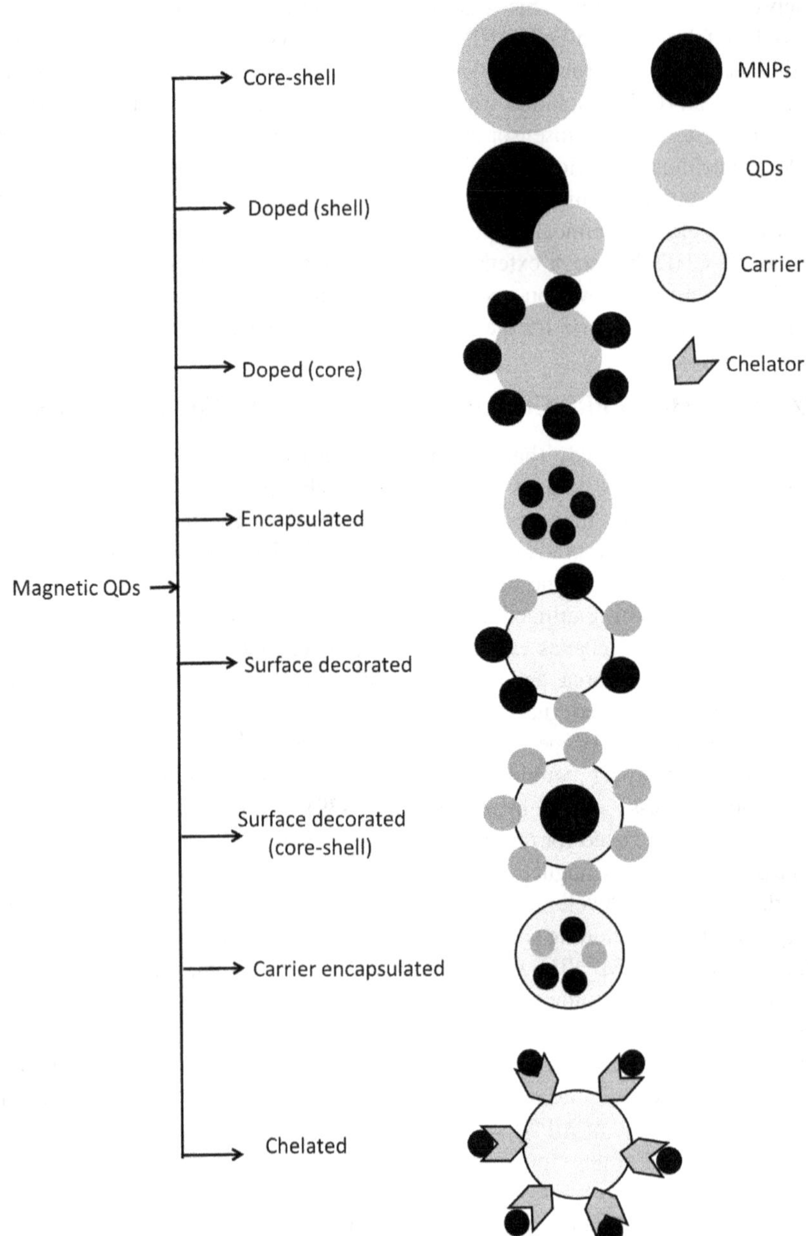

FIGURE 11.1 Different types of magnetic QDs and their probable architecture.

heterocrystalline growth, for example. FePt-CdSe, Fe_2O_3-CdSe, FeP-CdTe. Doping of quantum dots (QDs) with paramagnetic ions, such as Mn^{2+}, Gd^{3+}, Co^{2+}, and Ni^{2+}, has also been seen by many studies [41]. The encapsulation of composite nanostructures consisting of MNPs and QDs inside of the carrier material (polymer, for example PbS QDs and Fe_3O_4 NPs encapsulated in poly(lactic-co-glycolic-acid) or PLGA) has been described in other places [42]. In addition to that, silica matrix anchored QDs along the lines of Fe_3O_4/SiO_2-CdSe/ZnS MQDs were described [43]. In this field, quantum dots (QDs) with a paramagnetic coating of gadolinium (Gd)-chelates are also known to exist, and they have the requisite magnetic properties [44].

11.3 MAGNETIC QDs FOR IMAGING APPLICATIONS

An interesting new class of materials for use in bioimaging is being developed using multimodal contrast agents. These agents are based on highly luminous quantum dots (QDs) and can also contain magnetic nanoparticles (MNPs) or ions. It is possible to generate a sensitive contrast agent for two imaging techniques that are extremely strong and highly complementary by integrating two capabilities into a single nanoparticle. These imaging techniques include fluorescence imaging and magnetic resonance imaging (MRI). The use of nanosized semiconductor materials for imaging in living organisms, sometimes known as "in-vivo" imaging, is one of the uses of nanotechnology in biomedical research that is advancing at the fastest rate and offering the greatest promise. These nanosized semiconductor materials, also known as quantum dots (QDs), display a variety of outstanding optical features, which significantly boost the potential of fluorescence-based bioimaging. These QDs are also known as quantum dots. In particular, the chemistry of the surface of QDs makes it possible to include entities that increase pharmacokinetics and bioapplicability, as well as to conjugate particular targeted molecules. The field has undergone a sea change ever since the early papers that described the first biological application of QDs as fluorescent biological markers appeared in the late '90s. These findings provide evidence that QDs can be utilized for the labelling of cultured cells. Furthermore, since the publication of phospholipid micelle–coated QDs, for in-vivo imaging, a great number of other studies have made use of the exceptional fluorescent properties of QDs for the same objective. It is interesting to note that recent breakthroughs in the field of nanochemistry have made it possible to manufacture QD-based nanoparticulate materials that can also be detected using other imaging methods. These QDs are referred to as multimodal QDs. Positron emission tomography (PET) imaging has recently revealed that the pairing of quantum dots (QDs) with a radiolabel is a useful tool for studying in vivo the pharmacokinetics and biodistribution of QDs. In a further illustration, it was recently demonstrated that QDs (even without alteration) may be identified by computed tomography (CT) due to the high electron density of QD cores. In addition to this, QDs are capable of being engineered in such a way that they exhibit superparamagnetic characteristics, allowing for their detection using magnetic resonance imaging (MRI). As a diagnostic tool, magnetic resonance imaging (MRI) is distinguished by its capacity to produce three-dimensional images of opaque and soft tissue at relatively high spatial resolution and tissue contrast. As a result, MRI is the imaging method that offers the greatest degree of versatility among those that are offered in clinical settings. The magnetic resonance imaging (MRI) technique can provide metabolic and functional characteristics in addition to morphological information. Combinatory probes, that display both fluorescence and superparamagnetic characteristics, offer considerable benefits due to the complementary qualities of optical methods and MRI. Compositional interceptors display both fluorescent and superparamagnetic attributes. In contemporary memory, a number of research organizations have begun creating superparamagnetic QDs with the goal of using this application. The formation of a shell of a material with a larger bandgap around the QD core is an important stage in the process of improving the quality of the QE and the stability of the QDs. Because of the larger bandgap shell's confinement of electrons and holes within the core, surface interactions with elements like oxygen are diminished, leading to an increase in the material's resistance to photo-oxidation. In an ideal scenario, the material of the shell would have a lattice matching that of the core. This would eliminate interfacial imperfections, which can bring QE down. The enhanced stability of these so-called core/shell quantum dots not only results in a longer shelf life (measured in years when kept at room temperature), but it also results in a better bleaching behavior. When compared to photons emitted by fluorescent dye molecules, the number of photons emitted by quantum dots before they are bleached is significantly larger. There is a lot of potential for QDs to be used as labels for fluorescence imaging due to their one-of-a-kind features, as were mentioned before.

On the basis of the inner filter effect (IFE) technique of nitrogen, cobalt co-doped carbon dots (N,Co-CDs) with 2,3-diaminophenazine, Huang et al. have created a novel type of ratiometric fluorescence probe that is both extremely sensitive and well selective (Figure 11.2). This probe can be used to determine levels of cholesterol and uric acid in human blood serum (DAP) [45]. The majority of MRI contrast agents create contrast by locally reducing the amount of time needed for ^{1}H relaxation in the water around them. The method that describes a return to thermal equilibrium of protons in a magnetic field after they were initially excited by the application of a radio frequency magnetic field pulse in an MRI scanner is referred to as relaxation. Relaxation is the process that describes a return to thermal equilibrium. The relaxation may often be broken down into two primary processes: spin–lattice relaxation, also known as longitudinal relaxation (with a characteristic duration of T_1 [s] or a relaxation rate of R_1 [s^{-1}]), and spin–spin relaxation, also known as transverse relaxation (T_2, R_2). When the transverse relaxation process is sped up by magnetic field inhomogeneities, the relaxation process is referred to as T_2^*. This is because the transverse relaxation process may be accelerated. In the next sections of this chapter, we will just refer to T_2 due to the desire to keep things as simple as possible; nevertheless, it is important to note that whenever we refer to T_2, we are also referring to T_2^*. By fine-tuning the MRI sequences in the right manner, the MR scanner's sensitivity may be adjusted to accommodate variances and shifts in the relaxation times. T_1-weighted magnetic resonance (MR) sequences create pictures with areas of low T_1 that are bright (have a positive contrast), whereas T_2-weighted MR sequences produce images with areas of low T_2 that are dark (have a negative contrast) (negative contrast). T_1 contrast agents are typically non-toxic complexes formed from paramagnetic metal ions, such as Gd^{3+}, which have been coordinated to a protective chelate. The electron–proton dipolar coupling that the agent causes is the primary action that it has, and it weakens very quickly as the distance increases. This indicates that in order for relaxation to

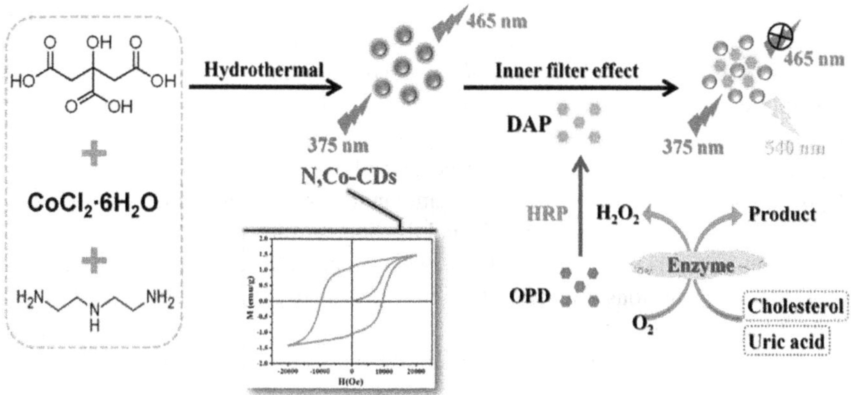

FIGURE 11.2 Schematic of N,Co-CD preparation and cholesterol and uric acid detection. Reproduced with the permission from ref. [45]. © 2019 American Chemical Society.

take place effectively, water need to be immediately coordinated to the paramagnetic ion, and ideally, it ought to go through rapid water exchange. The improvement in relaxation is due to the slowing down of the rotating motion of the complex, which may be accomplished, for example, by binding the contrast agent to a nanoparticle. Utilizing compounds that are typically referred to as superparamagnetic contrast agents is the most efficient method for lowering T_2. The majority of the time, they are magnetite (Fe_3O_4) or maghemite (γ-Fe_2O_3) nanoparticles, which are both types of iron oxide [46]. Because of the local field inhomogeneities that surround the magnetic particles, the transverse magnetization dephasing happens very quickly, which results in a fast T2 relaxation period. These nanoparticles become superparamagnetic after they are heated over the so-called blocking temperature (Tb), which occurs when the magnetocrystalline anisotropy is overcome by temperature variations. These results in the magnetic moment of the particle arbitrarily changing the direction it is pointing within itself.

Incorporating magnetic characteristics with fluorescence into a single nanoparticle can be accomplished using a direct process known as doping paramagnetic luminous ions into quantum dots (QDs). Doping semiconducting materials with paramagnetic ions is a practice that has a long history. The "trap" states that define the wavelength and lifespan of the doped QD luminescence are introduced into the system by paramagnetically charged ions that have energy levels that fall inside the QD bandgap. The luminescence may have its genesis in the paramagnetic ion itself (for instance, through an intraconfigurational $3d^n$ transition), or it may be the result of recombination to band-gap states that are associated with the paramagnetic dopant. Research into QDs doped with (paramagnetic) transition-metal ions, such as ZnS: Mn^{2+} nanocrystals, has seen a dramatic rise in luminescence quantum yield and a dramatic decrease in luminous lifetime [47]. The process of actually incorporating the dopant ions into the core of the semiconductor is a key part of the process. Providing evidence that the dopant is located inside the semiconductor particle rather than being absorbed on the particle's exterior is not a simple task. Spectroscopic techniques such as luminescence spectroscopy, electron paramagnetic resonance (EPR), and X-rays spectroscopy (e.g., extended X-ray absorption fine structure (EXAFS)) have been applied to verify the location of the dopant ions. These techniques have shown that the incorporation of the dopant depends critically on the synthesis method. Doped quantum dots may be synthesized using basically one of four different sorts of processes. The process of growing semiconductor QDs in a solution containing precursors for the semiconductor QD and the dopant ion while also incorporating a passivating ligand is the approach that is the least complicated and most straightforward. Some dopant ions are integrated into the growing semiconductor nanocrystal while it is contained within the shell of coordinating ligands. This is given that the dopant ions are chemically comparable to the semiconductor cation. The synthesis of the well-known and extensively researched ZnS: Mn^{2+} and ZnO: Co^{2+} QDs are two examples of the effective creation of doped quantum dots utilizing this approach [48]. It is possible to alter the well-known hot injection approach for the production of undoped QDs by introducing a dopant precursor in the reaction mixture. This may be done. This methodology was successful in achieving the desired results for ZnSe:Mn^{2+} [49]. On the other hand, it was shown that Mn2+ cannot be integrated into CdSe:Mn^{2+} QDs using the hot injection approach. This was considered to be connected to the crystal structure of CdSe [50]. The so-called cluster approach is the fourth way, and it was proven that it is possible to incorporate Mn^{2+} into CdSe QDs using this method. In this method, well-selected clusters of organometallic cations are used as the precursor in a process that is similar to the hot injection method (in that it involves similarly coordinating ligands and growth at elevated temperatures), but it also involves a controlled rise to the temperature of the reaction [51]. Only recently was it found that QDs that have been doped with paramagnetic ions may be used as multimodal probes in imaging, performing both magnetic and luminous analysis. In the past, the majority of research efforts were concentrated on the application of doped quantum dots (QDs) in light-emitting devices and on the utilization of the magnetic properties of QDs in the field of spintronics, with a particular emphasis on ferromagnetism in dilute magnetic semiconductor (DMS) nanoparticles. The usage of CdS:Mn/ZnS QDs for bimodal imaging was demonstrated in a work that was written by Santra et al. [52]. 3.1-nm particles were produced and coated with a thin coating of silica using the inverse micelle technique. The silica layer was then modified with APS to contain amine surface groups for bioconjugation. The particles were then used in bioconjugation. The quantum dots had a strong magnetic response and an emission that resembled a brilliant yellow Mn^{2+}.

Using a carrier material to generate a composite particle in which magnetic and semiconductor nanocrystals may be merged is yet another method for adding magnetic functions to QDs. This method involves the use of carrier material. The various NPs can either be integrated into the carrier material itself, attached to the exterior of the carrier material, or a mix of the two. For instance, the QDs are embedded within the carrier particle, while the MNPs are affixed to the outside of the carrier particle. As will be seen in the next section, silica and polymer matrices are also viable options for the role of carrier material. This can result in a construct that is typically bigger than particles of type I or type II. In the first half of this section, we will talk about composite particles that use silica as the carrier material, and in the second part, we will talk about polymer capsules that are used as the carrier material.

11.4 SUPPORTED MATRIX MAGNETIC QDs FOR IMAGING APPLICATIONS

In most cases, the high-temperature organometallic approach results in the production of QDs of a high grade. Due to the fact that they are capped with organic surfactants such as trioctylphosphine oxide TOPO and hexadecyl amine, the QDs in their natural state after synthesis are insoluble in water. Nevertheless, in order for the hydrophobic QD surface to be useful for real biological applications, it must be converted into a hydrophilic state and made susceptible to surface modification and functionalization. This condition is satisfied by the silica surface. However, due to the fact that QDs are exceedingly tiny in size, it is extremely difficult to accomplish silica coating of individual QDs using the Stober technique. solubility in water and functional NPs are essential for a wide variety of applications in the biological field. However, the production of robust functional NPs is a very difficult task because the majority of the good synthetic techniques that are now available for magnetic oxides, QDs, and noble metals yield hydrophobic NPs as a result of the hydrophobic surfactant coating. Because of this, water solubilization and functionalization are the most important challenges to address before applying them, and here is where the relevance of coating resides [53]. The coating contributes to the transformation of hydrophobic NPs into hydrophilic water-soluble particles and adds chemical activity to the particle surface, making it possible for a variety of chemicals and biomolecules to be covalently bonded. The transformation of hydrophobic nanoparticles into hydrophilic and functional NPs often involves one of two common coating techniques. The first strategy utilizes hydrophilic ligands, including thiols or other functional groups, to perform ligand exchange on the original surfactant [54]. When compared to other systems, thiol-based ligand exchange is the most popular method producing noble metal nanoparticles (NPs). This is due to the fact that thiol creates a significant chemisorption on the surface of noble metals. In addition, a number of other ways that are based on thiols have been developed in order to generate a coating that is stable. These methods include the use of ligands that include multiple thiols, thiolated dendrimers, dendrons, or the crosslinking of surface ligands [55]. The hydrophobic ZnS-capped CdSe QDs that have already been manufactured can be coated with silica using a direct one-pot reverse microemulsion approach. Igepal can be used as the nonionic surfactant in this process [56]. In the beginning, in order to create reverse micelles, the nonionic surfactant Igepal and the solvent cyclohexane were used in the synthesis process. With the use of this direct silica coating method, a broad range of hydrophobic NPs or QDs, such as CdSe and PbSe, magnetic NPs, such as Fe_2O_3, and bifunctional NPs or heterodimers, such as CdSe-ZnS/Fe_2O_3, have been able to be encased within spherical silica particles [57, 58]. Recent developments in live cell imaging have seen the development of QDs with a silica coating [57]. Silanization in reverse microemulsion led to the formation of a thin silica coating on the surface of naked CdSe QDs, which included surface NH_2 groups. The attachment of amine groups to the surface of QDs was facilitated by the addition of aminopropyl triethoxysilane APS. The formation of a reverse microemulsion was accomplished with the consecutive addition of tetramethylammonium hydroxide in 2-propanol/methanol and water. Hydrolyzing and condensing APS methoxy groups exposed surface amine groups on silanized QDs SiO2/QDs for conjugation with bioanchored membrane. The presence of superparamagnetic properties in multifunctional nanoparticles with a size of less than 15 nm is crucial for applications such as magnetic resonance imaging (MRI), magnetically guided site-specific drug administration, and alternating current magnetic field aided cancer treatment [59]. In current history, researchers have been concentrating their efforts on the production of very uniform size iron oxide nanoparticles (NPs), notably magnetite (Fe_3O_4) and maghemite (γ-Fe_2O_3). We have reported a gram-scale synthesis of almost monodispersed γ-Fe_2O_3 magnemite nanoclusters. This was accomplished with an oxidant that was less costly and did not need the use of a potentially dangerous iron pentacarbonyl FeCO5 or iron acetylacetonate precursor [60].

11.5 SUMMARY AND FUTURE PERSPECTIVES

The use of quantum dots (QDs) in biolabelling has gradually developed into a mature technique over the past decade, particularly in the field of cell-based imaging. Because cadmium is fundamentally carcinogenic, in-vivo applications continue to be concerned about the toxicity of QDs. In-vitro experiments using robust coating processes demonstrated that QDs are harmless [Citation needed]. According to the findings of current studies, the cell uptake and clearance of QDs are determined by their size, charge, and the substance used for their coating. In the field of biomedical research, multifunctional nanoparticles (NPs) that feature fluorescent, magnetic, and targeting properties are helpful. Even if there have been significant advancements in recent years, the use of multifunctional NPs in in-vivo imaging is still in its infant stages. This opens the door for the creation of more versatile systems that use fluorescence probes that are less hazardous and leave out materials that are known to cause cancer, such as cadmium and lead. When it comes to cellular absorption as well as tumor targeting, the size of the NPs plays a pivotal role. If the particles are less than 10 nanometers in size, it will be much simpler for the animal's body to eliminate them. Growing a thin hydrophilic shell around QDs or multifunctional NPs is the best way to reduce their overall size while maintaining their functionality. The reverse microemulsion approach may be used to produce a thin shell of silica or polymer, which can then be functionalized with additional biomolecules of interest. This process has the potential to be very useful.

The creation of "smart" contrast agents that are able to monitor particular cellular and molecular activities in vivo will be the obstacle that we will need to overcome in the future. This will be the task that we confront. Recent research has been directed toward the advancement of flexible multipurpose paramagnetic NPs for diagnostic imaging in a number of clinical pathologies including early cancer diagnosis and cellular trafficking in stem cell therapy and immunoregulatory initiatives. This type of imaging can be used in a variety of clinical initiatives. The precise administration of medications is a vital component of modern medical practice. Future research will look to provide solutions to early cancer diagnosis and targeted delivery of therapeutics by conjugating multiple components, such as fluorescent QDs or dyes or UCNPs, tumor-targeting groups, anticancer drugs, or siRNA, to the MPs. This will be done in an effort to improve patient outcomes.

REFERENCES

1. Wendler, T., et al., *How molecular imaging will enable robotic precision surgery.* European Journal of Nuclear Medicine and Molecular Imaging, 2021. 48(13): p. 4201–4224.
2. Wegrzyniak, O., M. Rosestedt, and O. Eriksson, *Recent progress in the molecular imaging of nonalcoholic fatty liver disease.* International Journal of Molecular Sciences, 2021. 22(14): p. 7348.
3. Das, P., et al., *Zinc and nitrogen ornamented bluish white luminescent carbon dots for engrossing bacteriostatic activity and Fenton based bio-sensor.* Materials Science and Engineering: C, 2018. 88: p. 115–129.
4. Maruthapandi, M., et al., *Microbial inhibition and biosensing with multifunctional carbon dots: Progress and perspectives.* Biotechnology Advances, 2021. 53: p. 107843.
5. Das, P., et al., *A simplistic approach to green future with eco-friendly luminescent carbon dots and their application to fluorescent nano-sensor 'turn-off'probe for selective sensing of copper ions.* Materials Science and Engineering: C, 2017. 75: p. 1456–1464.
6. Alavi, A., J.W. Kung, and H. Zhuang. *Implications of PET based molecular imaging on the current and future practice of medicine.* in *Seminars in Nuclear Medicine*. 2004. Elsevier.
7. Emelianov, S.Y., P.-C. Li, and M. O'Donnell, *Photoacoustics for molecular imaging and therapy.* Physics today, 2009. 62(8): p. 34.
8. Predina, J.D., et al., *A phase I clinical trial of targeted intraoperative molecular imaging for pulmonary adenocarcinomas.* The Annals of Thoracic Surgery, 2018. 105(3): p. 901–908.
9. Ganguly, S., I. Grinberg, and S. Margel, *Layer by layer controlled synthesis at room temperature of tri-modal (MRI, fluorescence and CT) core/shell superparamagnetic IO/human serum albumin nanoparticles for diagnostic applications.* Polymers for Advanced Technologies, 2021. 32(10): p. 3909–3921.
10. Bartling, S.H., et al., *First multimodal embolization particles visible on x-ray/computed tomography and magnetic resonance imaging.* Investigative Radiology, 2011. 46(3): p. 178–186.
11. Aviv, H., et al., *Radiopaque iodinated copolymeric nanoparticles for X-ray imaging applications.* Biomaterials, 2009. 30(29): p. 5610–5616.
12. Das, P., et al., *Tailor made magnetic nanolights: Fabrication to cancer theranostics applications.* Nanoscale Advances, 2021.
13. Skaat, H. and S. Margel, *Synthesis of fluorescent-maghemite nanoparticles as multimodal imaging agents for amyloid-β fibrils detection and removal by a magnetic field.* Biochemical and Biophysical Research Communications, 2009. 386(4): p. 645–649.
14. Skaat, H., G. Shafir, and S. Margel, *Acceleration and inhibition of amyloid-β fibril formation by peptide-conjugated fluorescent-maghemite nanoparticles.* Journal of Nanoparticle Research, 2011. 13(8): p. 3521–3534.

15. Ahmed, S.R., et al., *Strong nanozymatic activity of thiocyanate capped gold nanoparticles: an enzyme–nanozyme cascade reaction based dual mode ethanol detection in saliva.* New Journal of Chemistry, 2022. 46(3): p. 1194–1202.
16. Das, P., et al., *Converting waste Allium sativum peel to nitrogen and sulphur co-doped photoluminescence carbon dots for solar conversion, cell labeling, and photobleaching diligences: a path from discarded waste to value-added products.* Journal of Photochemistry and Photobiology B: Biology, 2019. 197: p. 111545.
17. Ahmed, S.R., et al., *Positively Charged Gold Quantum Dots: An Nanozymatic "Off-On" Sensor for Thiocyanate Detection.* Foods, 2022. 11(9): p. 1189.
18. Frangioni, J.V., *In vivo near-infrared fluorescence imaging.* Current Opinion in Chemical Biology, 2003. 7(5): p. 626–634.
19. Ntziachristos, V., C. Bremer, and R. Weissleder, *Fluorescence imaging with near-infrared light: new technological advances that enable in vivo molecular imaging.* European Radiology, 2003. 13(1): p. 195–208.
20. Leblond, F., et al., *Pre-clinical whole-body fluorescence imaging: Review of instruments, methods and applications.* Journal of Photochemistry and Photobiology B: Biology, 2010. 98(1): p. 77–94.
21. Zhou, H., et al., *Targeted fluorescent imaging of a novel FITC-labeled PSMA ligand in prostate cancer.* Amino Acids, 2022. 54(1): p. 147–155.
22. Du, J., et al., *Carbon dots for in vivo bioimaging and theranostics.* Small, 2019. 15(32): p. 1805087.
23. Ganguly, S., et al., *Microwave-synthesized polysaccharide-derived carbon dots as therapeutic cargoes and toughening agents for elastomeric gels.* ACS Applied Materials & Interfaces, 2020. 12(46): p. 51940–5195-.
24. Li, M., et al., *Review of carbon and graphene quantum dots for sensing.* ACS Sensors, 2019. 4(7): p. 1732–1748.
25. Wagner, A.M., et al., *Quantum dots in biomedical applications.* Acta Biomaterialia, 2019. 94: p. 44–63.
26. Das, P., et al., *Carbon dots for heavy-metal sensing, pH-sensitive cargo delivery, and antibacterial applications.* ACS Applied Nano Materials, 2020. 3(12): p. 11777–11790.
27. Jacak, L., P. Hawrylak, and A. Wojs, *Quantum dots.* 2013: Springer Science & Business Media.
28. Bacon, M., S.J. Bradley, and T. Nann, *Graphene quantum dots.* Particle & Particle Systems Characterization, 2014. 31(4): p. 415–428.
29. Jamieson, T., et al., *Biological applications of quantum dots.* Biomaterials, 2007. 28(31): p. 4717–4732s
30. Zhai, X., et al., *Highly luminescent carbon nanodots by microwave-assisted pyrolysis.* Chemical Communications, 2012. 48(64): p. 7955–7957.
31. Sarkar, T., H.B. Bohidar, and P.R. Solanki, *Carbon dots-modified chitosan based electrochemical biosensing platform for detection of vitamin D.* International Journal of Biological Macromolecules, 2018. 109: p. 687–697.
32. Li, X., et al., *Cyclodextrin-based metal-organic frameworks particles as efficient carriers for lansoprazole: study of morphology and chemical composition of individual particles.* International Journal of Pharmaceutics, 2017. 531(2): p. 424–432.
33. Xiao, B., Q. Yuan, and R.A. Williams, *Exceptional function of nanoporous metal organic framework particles in emulsion stabilisation.* Chemical Communications, 2013. 49(74): p. 8208–8210.
34. Ganguly, S. and S. Margel, *Design of Magnetic Hydrogels for Hyperthermia and Drug Delivery.* Polymers, 2021. 13(23): p. 4259.

35. Ganguly, S. and S. Margel, *Remotely controlled magneto-regulation of therapeutics from magnetoelastic gel matrices.* Biotechnology Advances, 2020. 44: p. 107611.
36. Chang, D., et al., *Biologically targeted magnetic hyperthermia: Potential and limitations.* Frontiers in Pharmacology, 2018. 9: p. 831.
37. Chang, M., et al., *Recent Advances in Hyperthermia Therapy-Based Synergistic Immunotherapy.* Advanced Materials, 2021. 33(4): p. 2004788.
38. Ganguly, S., et al., *Photopolymerized Thin Coating of Polypyrrole/Graphene Nanofiber/Iron Oxide onto Nonpolar Plastic for Flexible Electromagnetic Radiation Shielding, Strain Sensing, and Non-Contact Heating Applications.* Advanced Materials Interfaces, 2021. 8(23): p. 2101255.
39. Paulides, M., et al., *Recent technological advancements in radiofrequency-andmicrowave-mediated hyperthermia for enhancing drug delivery.* Advanced Drug Delivery Reviews, 2020. 163: p. 3–18.
40. Gu, H., et al., *Facile one-pot synthesis of bifunctional heterodimers of nanoparticles: a conjugate of quantum dot and magnetic nanoparticles.* Journal of the American Chemical Society, 2004. 126(18): p. 5664–5665.
41. Malik, P., et al., *II-VI core/shell quantum dots and doping with transition metal ions as a means of tuning the magnetoelectronic properties of CdS/ZnS core/shell QDs: A DFT study.* Journal of Molecular Graphics and Modelling, 2022. 111: p. 108099.
42. Ortgies, D.H., et al., *In vivo deep tissue fluorescence and magnetic imaging employing hybrid nanostructures.* ACS Applied Materials & Interfaces, 2016. 8(2): p. 1406–1414.
43. Xu, Y., et al., *Multifunctional Fe3O4 cored magnetic-quantum dot fluorescent nanocomposites for RF nanohyperthermia of cancer cells.* The Journal of Physical Chemistry C, 2010. 114(11): p. 5020–5026.
44. Mahajan, K.D., et al., *Magnetic quantum dots in biotechnology–synthesis and applications.* Biotechnology Journal, 2013. 8(12): p. 1424–1434.
45. Huang, S., et al., *Nitrogen, cobalt co-doped fluorescent magnetic carbon dots as ratiometric fluorescent probes for cholesterol and uric acid in human blood serum.* ACS Omega, 2019. 4(5): p. 9333–9342.
46. Gupta, A.K., et al., *Recent advances on surface engineering of magnetic iron oxide nanoparticles and their biomedical applications.* Nanomedicine (Lond), 2007 Feb; 2(1): 23–39.
47. Bhargava, R., et al., *Optical properties of manganese-doped nanocrystals of ZnS.* Physical Review Letters, 1994. 72(3): p. 416.
48. Yu, I., T. Isobe, and M. Senna, *Optical properties and characteristics of ZnS nano-particles with homogeneous Mn distribution.* Journal of Physics and Chemistry of Solids, 1996. 57(4): p. 373–379.
49. Norris, D.J., et al., *High-quality manganese-doped ZnSe nanocrystals.* Nano Letters, 2001. 1(1): p. 3–7.
50. Erwin, S.C., et al., *Doping semiconductor nanocrystals.* Nature, 2005. 436(7047): p. 91–94.
51. Hanif, K.M., R.W. Meulenberg, and G.F. Strouse, *Magnetic ordering in doped Cd1-x Co x Se diluted magnetic quantum dots.* Journal of the American Chemical Society, 2002. 124(38): p. 11495–11502.
52. Santra, S., et al., *Synthesis of water-dispersible fluorescent, radio-opaque, and para-magnetic CdS: Mn/ZnS quantum dots: a multifunctional probe for bioimaging.* Journal of the American Chemical Society, 2005. 127(6): p. 1656–1657.
53. Gerion, D., et al., *Synthesis and properties of biocompatible water-soluble silica-coated CdSe/ZnS semiconductor quantum dots.* The Journal of Physical Chemistry B, 2001. 105(37): p. 8861–8871.

54. Aldana, J., et al., *Size-dependent dissociation pH of thiolate ligands from cadmium chalcogenide nanocrystals.* Journal of the American Chemical Society, 2005. 127(8): p. 2496–2504.
55. Medintz, I.L., et al., *Quantum dot bioconjugates for imaging, labelling and sensing.* Nature Materials, 2005. 4(6): p. 435–446.
56. Selvan, S.T., T.T. Tan, and J.Y. Ying, *Robust, non-cytotoxic, silica-coated CdSe quantum dots with efficient photoluminescence.* Advanced Materials, 2005. 17(13): p. 1620–1625.
57. Selvan, S.T., et al., *Synthesis of silica-coated semiconductor and magnetic quantum dots and their use in the imaging of live cells.* Angewandte Chemie, 2007. 119(14): p. 2500–2504.
58. Tan, T.T., et al., *Size control, shape evolution, and silica coating of near-infrared-emitting PbSe quantum dots.* Chemistry of Materials, 2007. 19(13): p. 3112–3117.
59. Lee, Y., et al., *Large-scale synthesis of uniform and crystalline magnetite nanoparticles using reverse micelles as nanoreactors under reflux conditions.* Advanced Functional Materials, 2005. 15(3): p. 503–509.
60. Ang, C., et al., *hymiou, ST Selvan and JY Ying.* Adv. Mater, 2009. 21: p. 869.

12 Cytotoxicity of the Magnetic Quantum Dots

Sharuk L. Khan[1], *Navin Chandra Pant*[2], *Mohd Vaseem Fateh*[3], *Sarfaraz Ahmed*[4], *Mayank Yadav*[5], *Himani Bajaj*[5], *Neha Chauhan*[6], *and Disha Dutta*[7]

[1]Department of Pharmaceutical Chemistry, N.B.S. Institute of Pharmacy, Ausa, Maharashtra, India

[2]Six Sigma Institute of Technology and Science, Rudrapur, Uttarakhand, India

[3]Department of Pharmacy, Sam Higginbottom University of Agriculture, Technology and Sciences (SHUATS), Naini, Allahabad, Uttar Pradesh, India

[4]Department of Pharmaceutical Chemistry, Global Institute of Pharmaceutical Education and Research, Baksaura, Uttarakhand, India

[5]Adarsh Vijendra Institute of Pharmaceutical Sciences (AVIPS), Shobhit University, Gangoh, Saharanpur, Uttar Pradesh, India

[6]Laxminarayan Dev College of Pharmacy, Bharuch, Gujarat, India

[7]Devsthali Vidyapeeth College of Pharmacy, Rudrapur, Uttarakhand, India

CONTENTS

12.1 INTRODUCTION

Researchers and doctors are excited about the potential of nanomaterials for a wide range of biomedical research and treatment [1]. They exhibit new physicochemical features, which may be attributed to their tiny size, chemical composition, surface structure, solubility, and shape. These qualities are rapidly being employed in medicine

DOI: 10.1201/9781003319870-12

for the purposes of diagnostics, imaging, and drug administration [2]. There are a variety of applications for fluorescence tracking, including cells. Quantum dots are a form of semiconductor nanoparticle that emits fluorescent light. These nanoparticles have a diameter of between 2 and 10 nanometers and are composed of a core material that is encased within a shell of another semiconductor material [3-5]. The size of the QDs has a dependent relationship with the attributes such as optical property, absorbance, and photoluminescence [6-8]. The very word "quantum dots" alludes to the quantum confinement and optical characteristics of the material. Because of this particular characteristic, they are good candidates for studies including biological function and imaging [9, 10]. They have a significant amount of untapped potential in the areas of imaging, sensing, tracking, and monitoring in real time. Quantum dots are chosen for use in a variety of applications according to their specific properties. Typical applications of QDs have been tabulated in Figure 12.1.

Polymeric materials have discovered uses in practically every imaginable product, from those used in the home to those used in industry. In the field of biomedicine, it also has some potential applications. Biocompatibility is one of the requirements that must be met before a polymer can be used for any application in the biomedical field [11]. The term "biocompatibility" of the material is used to characterize the

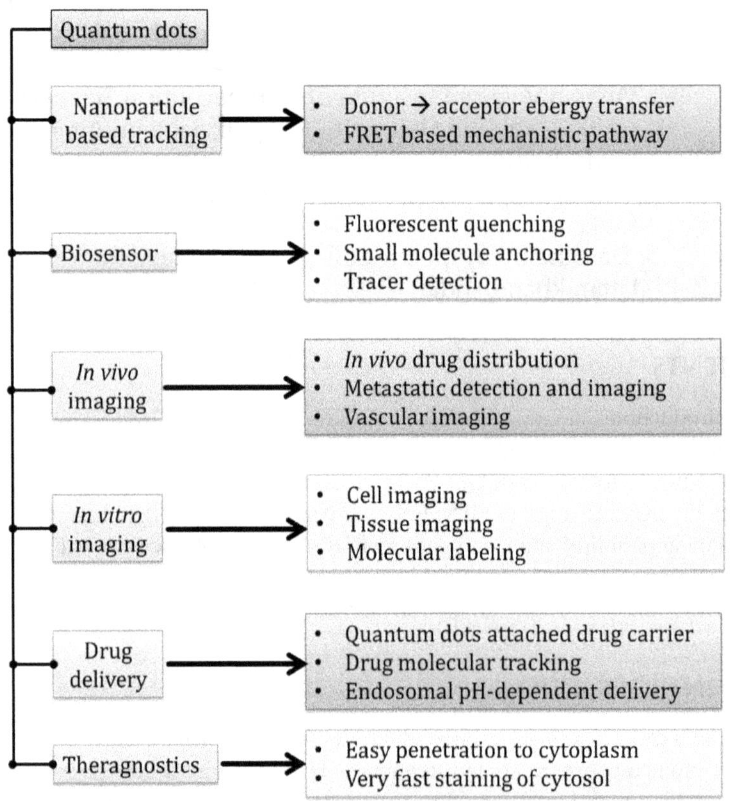

FIGURE 12.1 Different types of applications and related featured of QDs.

capacity of the material to function normally in the presence of biological fluids or the body itself [12]. Any substance will be considered biocompatible if it does not interfere with the normal functioning of the body in any way, whether it be through the development of anaphylactic reactions or even other unfavorable side effects. Natural polymers, on the whole, are biocompatible and biodegradable, and for this reason they are utilized in various biomedical applications [13-15]. However, the natural polymer has subpar qualities, such as weak mechanical and barrier properties. These properties are a drawback. Therefore, polymer-nanocomposites are used as new era materials with better properties. They also suit the bill as the material, which is biodegradable, biocompatible, and environmentally friendly polymer materials. The range of polymer nanocomposites known as polymer-quantum dot composites finds uses in a variety of fields, including drug delivery, bioimaging, tissue engineering, and biosensors, amongst others [16].

In this chapter, the authors cover the classification of polymers, as well as the methodologies and tests that can be used to determine whether or not a polymer is biocompatible and biodegradable. The chapter also examines the many procedures that researchers have implemented in order to render polymer-quantum dots biocompatible. Finally, other biocompatible polymer-quantum dot systems from the published literature also have been reviewed in detail. These systems are used in a variety of applications because of their biocompatibility.

12.2 DIFFERENT POLYMERS AND THEIR BIOCOMPATIBILITY

Polymers are lengthy chains that consist of monomer units that are repeated repeatedly. Natural polymers, synthetic polymers, and semi-synthetic polymers are the three distinct types of polymers that can be distinguished by the origin from whence they came [17-19]. The several types of polymers are subdivided in Figure 12.2.

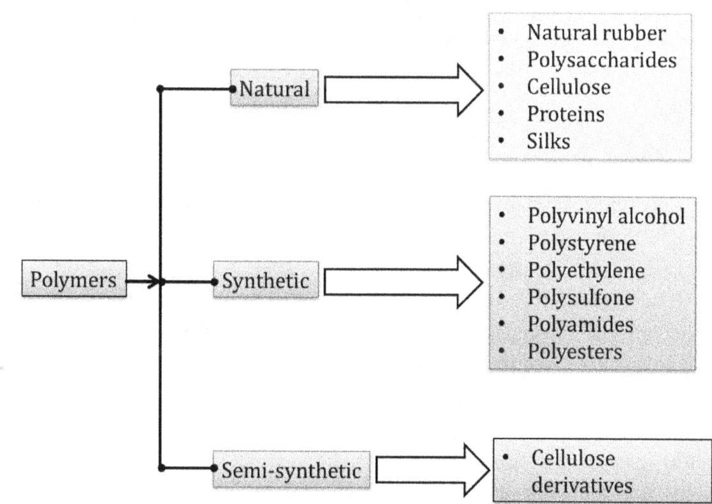

FIGURE 12.2 Different types of polymers.

The endurance of synthetic polymers to enzymatic degradation has emerged as a major cause for concern and is gaining less and less legitimacy with each passing day. On the other hand, it is possible that this is not the case with all synthetic polymers. Polyvinyl alcohol (PVA) and polysulfone are two examples of materials that have a shown track record of being biocompatible and are, as a result, utilized in various biomedical applications. The use of materials made of biopolymers is becoming increasingly important in today's medical systems. In the field of human health, the biocompatibility of biomaterials, also known as the compatibility of biomaterials with cells, plays an essential role in the development of devices as well as the drug-delivery system. The use of cutting-edge science and technology has resulted in the development of cutting-edge medical equipment as well as novel drug delivery systems that are constructed of biocompatible materials and are utilized to treat the human body. Before putting any material into use, the human body should be tested to ensure that it is biocompatible with the material and that it does not cause any harm to the body. Before the material could be utilized in such applications, it was subjected to the technique that is considered to be the industry standard. When conducting first clinical trials, international standards, including ISO 10993-1, are typically adhered to. Additionally, in-vitro and in-vivo research were conducted on the initial screening of the material before any tests were performed on animals. These studies were conducted to test for cytotoxicity, irritant, and hemocompatibility.

12.3 IN-VITRO CYTOTOXICITY OF QD-POLYMER COMPOSITES

The in-vitro cytotoxicity tests investigation is the first step in the process of determining whether or not a material is biocompatible. The effects of the experimental material on the cultivated cell are studied through in vitro testing, which evaluates the substance by observing how it directly interacts with the cultured cells. If it can be demonstrated that the substance is compatible with cells, then it is thought to be risk-free. The investigation of the harmful effects that substances have on cells is known as cytotoxicity. In the existence of a chemical, the toxic effect of the cell can be determined by the change in shape, an adjustment in the pace of cell growth, or by causing cells to die. These three outcomes are all possible outcomes. When conducting research on the cytotoxicity of a substance, in vitro testing is almost always recommended as the method of choice because it takes less time, is less expensive, is reliable, and raises fewer ethical considerations. On the other hand, the results of the in-vitro study cannot be attributed to the outcome of the experiment performed on a complicated biological system (living tissue). 1 Directly implanting the cell onto the surface of the proposed material is one method that may be utilized to determine the cytotoxicity of the material. When determining whether or not materials are biocompatible, the selection of cell lines, controls, the type of biochemical test used, and the amount of time spent in culture are all highly critical considerations. In addition, the selection of cell types has to be predicated on the particular uses of the substance that is the subject of the inquiry. For instance, keratinocytes and fibroblasts cell lines were utilized in research on the cytotoxic potential of various wound dressing materials.

12.4 IN-VIVO CYTOTOXICITY OF QDS: POLYMER NANOCOMPOSITES

To determine the level of danger posed by the substance, researchers conduct tests known as "in vivo" on living species like animals, plants, or even entire cells. In the field of medical research, in-vivo studies are extremely important, particularly for clinical trials [20]. The substance is often investigated as a transitional step between in vitro studies and clinical trials involving humans [21]. The primary kinds of in-vivo experiments were performed on animals, particularly rats or mice being the most recommended animal model to use. Therefore, there are ethical considerations to be made during the technique of biological research while doing an in-vivo study [22]. The selection of animal models occurs in a manner that is congruent with the objectives of the research. There are three possible types of animal models: homologous, isomorphic, and predictive. When it comes to the creation and use of nanomedicines, targeted drug delivery systems that integrate imaging and therapeutic modalities in a single macromolecular construct may provide certain benefits [23]. In order to incorporate the distinctive optical properties of luminescent quantum dots (QDs) into immunoliposomes

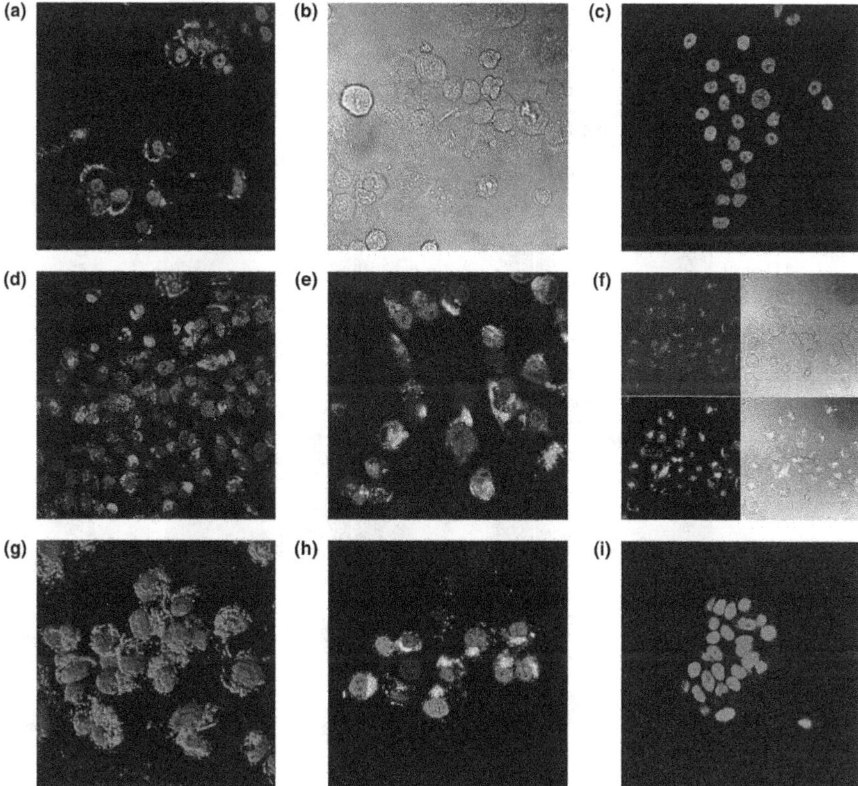

FIGURE 12.3 Internalization of QDs in different living cells. Reproduced with the permission from ref. [24]. © 2008 American Chemical Society.

for the diagnosis and treatment of cancer, Weng et al. describe the synthesis, biophysical characterization, tumor cell-selective internalization, and anticancer drug delivery of QD-conjugated immunoliposome-based nanoparticles. This was done in order to develop immunoliposomes that could be used in the diagnosis of cancer (Figure 12.3). Immunoliposomes will be able to identify and treat cancer more effectively as a result of this development (QD-ILs) [24]. Additionally, the pharmacokinetic features of QD-ILs as well as their imaging capacity in vivo were studied. Visualization of bare QDs, liposome controls, nontargeted QD-conjugated liposomes (QD-Ls), and QD-ILs was achieved by the use of freeze-fracture electron microscopy. In every respect, human illness, physiology, and therapy may be accurately modeled using homologous animal

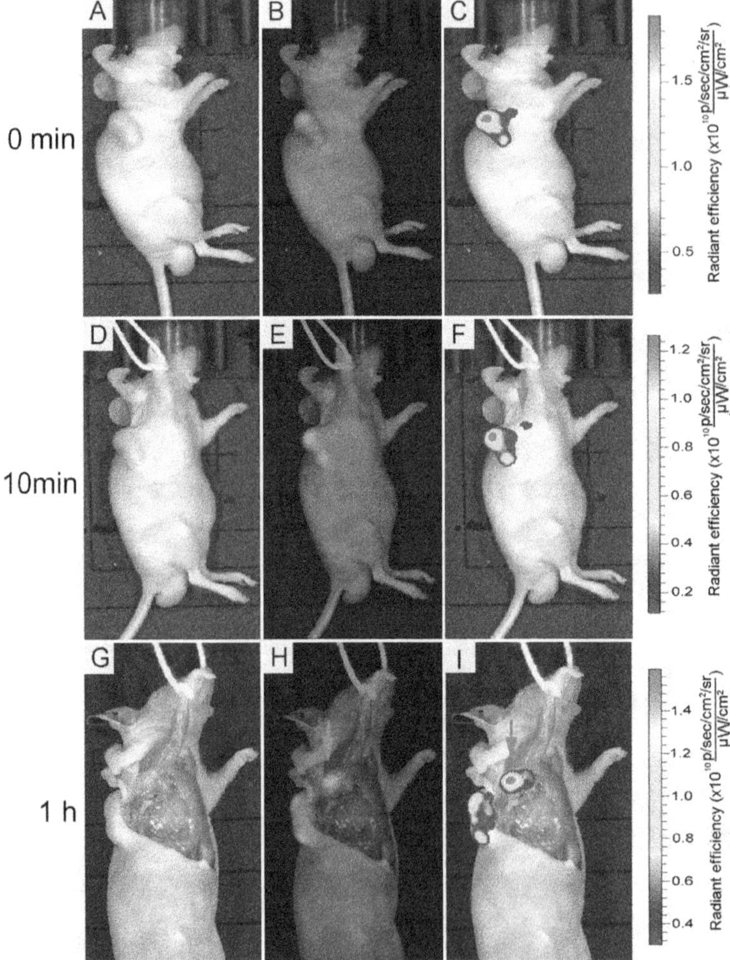

FIGURE 12.4 Images of H460 tumor-bearing mice after intratumoral injection of AIS/ZnS QDs at different time points. Reproduced with the permission from ref. [27]. © 2008 American Chemical Society.

models [25]. Isomorphic models are perfect representations of human conditions, but predictive animal models do not have anything in common with human conditions. When conducting tests in vivo, both invertebrate and vertebrate models of organisms are utilized. In most cases, researchers will employ an invertebrate called zebrafish as a model for an animal [26]. Animal models have traditionally been made using vertebrates such as baboons, macaques, cows, dogs, rabbits, guinea pigs, rats, and mice. Other examples include guinea pigs, rabbits, and rats. For instance, Sun et al. have published the results of an in-vivo investigation on mice using fluorescent AgInS/ZnS QDs (AIS/ZnS QDs) with red emission for rapid tumor drainage lymph node imaging (Figure 12.4). AIS/ZnS QDs were easily diffused in water and had hydrodynamic radii of 295 nm, which made them ideal for imaging lymph nodes. Fluorescence AIS/ZnS QDs were injected intradermally into mice. Mice were used. After 10 minutes had passed since the injections, imaging was performed on tumor-bearing mice that had been given an intratumor injection [27].

12.5 BIOCOMPATIBILITY AND CONJUGATION OF QD-POLYMER NANOCOMPOSITES

The polymer-quantum dot material possesses outstanding features, including cytocompatibility, non-toxicity, biocompatibility, and friendliness to the environment. As a result, there has been a recent surge in interest in the production of such materials, which have found widespread use in a variety of fields, such as tissue engineering, drug delivery, and bioimaging, amongst others. In the field of biomedical applications, fluorescent quantum dots, also known as QDs, are an interesting and promising material that may be used for cellular labelling and tracking, in-vivo imaging, and even as drug delivery vehicles. The fluorescence of quantum dots can be readily functionalized and adjusted because to their small size. There have been a great number of studies and papers written about how poisonous QDs are. The core composition of the QDs, the surface coating of the QDs, the hydrodynamic diameter of the QDs, and the particle surface charge are the four common parameters that are associated to the generations of toxicity that CDs cause when they interact with biological systems.

The term "bioconjugation" refers to the process of immobilizing biomolecules on the surface of quantum dots. In general, there are two different kinds of interactions that biomolecules are bonded to: covalent linking and non-covalent binding. Hydrophobic, electrostatic, and affinity interactions between biomolecules and the quantum dot surfaces are the three types of interactions that can influence non-covalent binding. Because there is no way to regulate the orientation or valency of the binding, simple non-covalent adsorption of the biomolecule to the surface is not a very attractive process. Non-covalent attachment results in relatively weak coordination binding, which is dependent on a number of parameters including ionic strength, pH, and competitive interactions with other anions. In addition, this type of binding is caused by non-covalent attachment. Working with lower concentrations of the biomolecules is made possible by the attachment of additional charged groups to the surface of the quantum dot. The procedure of connecting a multivalent polymer is not an easy one; the final geometry is determined by a number of variables, including the aggregation of nanocrystals and the cross-linking of polymers between the particles. The optical characteristics of quantum dots can be altered to the color blue or red at times due of a process called bioconjugation [28]. Conjugations of quantum dots with biopolymers are the most suitable biocompatible conjugate materials for this reason (as described earlier). Here are some examples of biocompatible polymer-QDs composites that may be employed in a variety of applications.

After alpha-lactalbumin and beta-lactoglobulin, the BSA protein is the third most prevalent whey protein in milk. However, it makes up about 10 percent of the total whey protein and is the most abundant whey protein overall. There is one free sulfydryl group among the protein's 35 Cys residues, which are connected to one another by 17 intramolecular disulphide bridges. BSA is also recognized as the most prevalent plasma protein in bovines and is made up of a single chain of 583 amino acid residues. BSA may be found in many different forms. Because of its low cost, high tolerance, and ease of production, bovine serum albumin, often known as BSA, has found widespread use in the field of drug administration. Because of the structural qualities of BSA, which contain a wide diversity of amino acid residues, it is possible to bind pharmaceuticals or bioactive chemicals that exhibit a wide variety of physicochemical properties. BSA has a great affinity for binding to QDs, and it may also conjugate with the surface of the QDs to form a protein shell that is more stable [29]. It is possible to coat the surfaces of QDs with BSA in order to make them biocompatible and to better target certain fluorescent probes in cell imaging. The BSA coating on the QDs brought about a reduction in their toxicity while also providing protection from photooxidation [30]. For instance, a one-pot synthesis of BSA-stabilized Ag_2S QDs in an aqueous solution was developed, and these particles were later bioconjugated with anti VEGF for targeted in-vivo cancer imaging [31].

In terms of its chemical composition, gelatin is made up of 18 distinct complexes of amino acids. Glycine, proline, and hydroxyproline make up the majority of the complexes' primary constituents [32]. Hydroxyproline is also present. Because of gelatin's ability to gel, it plays an important part as a binding agent in the food and pharmaceutical sectors. Additionally, it has been utilized in the stabilization of nanoparticles and quantum dots (QDs) [33]. The aqueous synthesis of gelatin stabilized CdTe/CdS/ZnS core/double-shell quantum dots with a biocompatible character, as demonstrated by Parani et al., who also demonstrated a 50% reduction in cytotoxicity. The transferrin-conjugated gelatin-stabilized CdTe/CdS/ZnS core/double-shell QDs proved to be an effective tool for fluorescence imaging [34].

Chitosan is an antibacterial, non-toxic, and biodegradable bioactive cationic polysaccharide. Chitosan is a bioactive polysaccharide that is cationic. These features are exciting for a number of different biological applications, including wound healing, as excipients for medication administration, and in tissue engineering. However, the use of pure chitosan as a wound dressing material is limited because of platelet adhesion, the formation of thrombus, and protein adsorption, all of which might boost the bacteria's ability to adhere to the wound. The fluorophore properties of the polymer chitosan QDs nanoconjugates have been utilized in a number of biological and pharmacological applications. Cytotocompatibility of the bismuth sulfide quantum dots (Bi2S3) functionalized with chitosan nanoconjugates material utilized as a fluorophore in biomedical and pharmaceutical applications. In instances when chitosan was performing effectively as a polymer ligand for nucleating Bi2S3 QDs and functioning as a stabilizer for them [35]. In a separate line of research, a biocompatible chitosan/CDs nanocomposites film was discovered. This film has the potential to be used as an efficient wound-dressing material and to monitor the pH of wounds [35]. Mathew et al. have produced a chitosan- and carbon dot-based conjugated nanocomposite that is loaded with dopamine. They have also effectively proven the dopamine drug's sustained release. The carbon dots, often known as CDs, were produced by the carbonization of chitosan [36]. Polyethylene glycol, often known as PEG, is a kind of polyether that has many uses, but medicine is one of its primary fields of use. PEG also helps enhance the biocompatibility of QDs for use in medical settings. PEG is soluble in water, ethanol, acetonitrile, benzene, and dichloromethane, but it is insoluble in diethyl ether and hexane. Hexane and dichloromethane are insoluble in PEG as well. Due to the presence of hydrophobic organic ligands on the surface of organic phase produced QDs, the use of QDs in biological applications is restricted. Coating the QDs with PEG will make them even more hydrophobic, which is a desirable attribute. With the assistance of PEG surface coating on the QDs, it is possible to add biocompatibility into the QDs for use in bioimaging applications. For the purpose of in-vivo imaging of xenograft tumors, Dai and Wang produced 6PEG-Ag2S QDs. These 6PEG-Ag2S QDs emit light in the second NIR range (1000–1350 nm). After using DHLA for the coating of Ag2S QDs, the particles were then subjected to a reaction with amino-functional six-armed PEG in order to produce 6PEG-Ag2S QDs. The 6PEG-Ag2S QDs that were obtained eventually become biocompatible [37].

Polyvinyl alcohol, sometimes known as PVA, is a kind of vinyl polymer that is solely coupled by carbon-carbon bonds. PVA is a biodegradable organic substance that is water-soluble, has high biocompatibility, is non-toxic and favorable to the environment, and does not cause cancer in animals. PVA has been employed in biomedical applications because of its biocompatibility and the ease with which it can be processed and molded. Additionally, PVA is a good host matrix for QDs because of its great properties. By using the drop-casting approach, Baoting Suo and colleagues were able to create a poly (vinyl alcohol) (PVA) nanocomposite thin film that was coupled with core/shell cadmium selenide (CdSe)/zinc sulfide (ZnS) QDs. PVA was functioning as a host material in the composite we were working with here [38]. Wang and colleagues developed a method to monitor antibiotics using a lanthanum-doped carbon quantum dot that was placed in a PVA fluorescent film [39]. PVA is an excellent option that can be utilized as a surface passivating agent to stabilize CDs and it can also be employed as a functionalization agent for CDs that are used in a variety of biomedical applications. CDs that have had their surfaces PVA-passivated in order to be employed as implant and drug delivery vehicle material have been designed. Take, for instance: It has been demonstrated by Ayaz et al. that surface passivation of CDs with PVA shows good potential to be used as anti-inflammatory implant materials. This is especially true when the material is introduced into textiles, cosmetics, and bandages to prevent the rejection of the implant by the immune system or the generation of unwanted inflammation [40].

Poly(lactic acid) (PLA) is a sustainable material that may be derived from the starch of either maize or sugar beets by the fermentation of glucose and, as a result, transformed to lactic acid. This process is known as polylactic acid fermentation. PLA is a desirable polymer material because to the one-of-a-kind features it possesses, which include biodegradability, biocompatibility, and nontoxicity. As a result, PLA has been included into the production of scaffolds for the regeneration of tissue and nerves, surgical sutures, and medication loading and release devices. Nanofibrous scaffolds of p-phenylenediamine functionalized CQDs mediated silk fibroin PLA (CQDs-SF-PLA) were manufactured by Yan et al. These scaffolds are appropriate for cardiac tissue engineering and nursing care. The incorporation of CQDs into SF-PLA scaffolds resulted in an increase in the characteristic properties of the SF-PLA scaffolds. These included an improvement in the cell proliferation, adhesion, and mRNA expression of cardiac genes (Tnnc1, Tnnt2, Cx43, and Atp2a2) in the absence of an external electrical supply [41]. An excellent biocompatible multifunctional nanocomposite of poly(L-lactide) (PLA) and polyethylene glycol (PEG)-grafted GQDs (f-GQDs) was developed by Dong et al. for the simultaneous imaging analysis of intracellular microRNAs (miRNAs) and integrated gene delivery application. As a result of the functionalization of GQDs with PEG and PLA, the nanocomposite became super physiological and maintained its photoluminescence stability across a wide pH range. This is an essential property for cell imaging (Figure 12.5). In addition, the authors demonstrate that f-GQDs successfully delivered a miRNA probe in the HeLa cell for intracellular miRNA imaging investigation [42].

Silk is obtained from the Bombyx mori silkworm and has been utilized in the textile industry for thousands of years. Silkworms are responsible for producing silk. Silk is made up of natural protein called silk fibroin (SF), which has H-chain, L-chain,

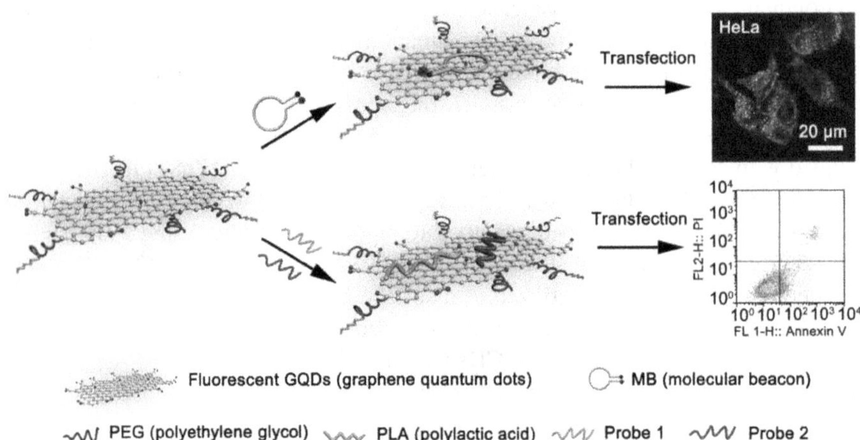

FIGURE 12.5 The f-GQDs-based cell imaging and combination particular gene-targeting drugs delivery are presented here in a schematic format. Reproduced with the permission from ref. [42]. © 2015 American Chemical Society.

and glycoprotein P25 in its structure [43]. Silk fibroin has been utilized in a variety of biomedical applications, including that requiring drug administration, tissue engineering, and implantable devices, due to its high level of biocompatibility and its low level of inflammatory reaction in vivo. The combination of nanomaterials has the potential to produce silk's characteristics [44]. Post-functionalization procedures via physical and chemical processing have been used to integrate nanoparticles into silk. Silk fibroin solution is combined with nanomaterials and processed into 2D and 3D structures using this approach. Silk fibroin solution was combined with CdTe QDs to produce the fluorescent film material assembly by chemical bonding, such as hydrogen bonds between the carboxylic group of the QDs and NH groups of silk fibroin, for instance [45]. The second one is to provide silkworms with diets that have been changed to include nanomaterials. This technology for intrinsically modifying silk gave a procedure that was beneficial to the environment. Directly feeding silkworms with graphene quantum dots (GQDs) or CdSe/ZnS core-shell quantum dots was the approach that Cheng et al. developed in order to provide a green and sustainable method for the fabrication of luminous silk from silkworms (CQDs). This produced QDs-reinforced fluorescent silkworm silk offers higher mechanical strength, good biocompatibility, toughness, and stable fluorescence compared to regular or fluorescent dye-colored silk. Silkworms are responsible for producing this silk [46].

12.6 SUMMARY

In conclusion, biocompatible polymer-quantum dot composite has recently emerged as a popular option for the material of choice in a wide range of applications. In most cases, biopolymers that are primarily sourced from natural sources, such as those based on BSA or peptides, gelatin, alginate, chitosan, or cellulose, are the ones that are utilized for this purpose. In addition, many additional man-made polymers are utilized in the production of nanocomposites including quantum dots. Many different kinds of quantum are manufactured, but recent interest has been focused on nanocomposites based on quantum dots (QD) because of their lower toxicity and quicker synthesis methods, among other advantages. However, following modification, other types of quantum dots, such as conjugated perovskite quantum dots (PQD), one of the newest additions to this family of quantum dots, which also has huge possibilities, are being employed. Techniques such as ligand exchange, encapsulation, and bioconjugation with ligands and materials such as imidazole, lipoic acid, and dihydrolipoic acid, amongst others, are among the many strategies that are utilized in the process of rendering the quantum dot biocompatible. The polymer-quantum dot material possesses outstanding features, including cytocompatibility, non-toxicity, biocompatibility, and friendliness to the environment. as a result, have seen widespread use in a variety of applications, including medication administration, bioimaging, tissue engineering, and many more. Numerous biocompatible polymer-quantum dot composites made of proteins, peptides, Geletin, chitosan, cellulose, Alginate, PLA, PEG, silk, and PVA have been employed to an extent degree in a wide variety of applications, and the number of these composites is increasing.

REFERENCES

1. Geszke-Moritz, M. and M. Moritz, *Quantum dots as versatile probes in medical sciences: synthesis, modification and properties.* Materials Science and Engineering: C, 2013. 33(3): p. 1008–1021.
2. Das, P., et al., *Tailor made magnetic nanolights: Fabrication to cancer theranostics applications.* Nanoscale Advances, 2021.
3. Ganguly, S., et al., *Advancement in science and technology of carbon dot-polymer hybrid composites: a review.* Functional Composites and Structures, 2019. 1(2): p. 022001.
4. Das, P., et al., *Graphene based emergent nanolights: a short review on the synthesis, properties and application.* Research on Chemical Intermediates, 2019. 45(7): p. 3823–3853.
5. Ahmed, S.R., et al., *Target specific aptamer-induced self-assembly of fluorescent graphene quantum dots on palladium nanoparticles for sensitive detection of tetracycline in raw milk.* Food Chemistry, 2021. 346: p. 128893.
6. Ortega, G.A., et al., *Rodlike particles of polydopamine-cdte quantum dots: An actuator as a photothermal agent and reactive oxygen species-generating nanoplatform for cancer therapy.* ACS Applied Materials & Interfaces, 2021. 13(36): p. 42357–42369.
7. Ahmed, S.R., et al., *Positively charged gold quantum dots: An nanozymatic "off-on" sensor for thiocyanate detection.* Foods, 2022. 11(9): p. 1189.
8. Li, M., et al., *Review of carbon and graphene quantum dots for sensing.* ACS sensors, 2019. 4(7): p. 1732–1748.
9. Das, P., et al., *Immobilization of heteroatom-doped carbon dots onto nonpolar plastics for antifogging, antioxidant, and food monitoring applications.* Langmuir, 2021. 37(11): p. 3508–3520.
10. Saravanan, A., et al., *Applications of N-doped carbon dots as antimicrobial agents, antibiotic carriers, and selective fluorescent probes for nitro explosives.* ACS Applied Bio Materials, 2020. 3(11): p. 8023–8031.
11. Das, P., et al., *Dual doped biocompatible multicolor luminescent carbon dots for bio labeling, UV-active marker and fluorescent polymer composite.* Luminescence, 2018. 33(6): p. 1136–1145.
12. Kavitha, T. and S. Kumar, *Turning date palm fronds into biocompatible mesoporous fluorescent carbon dots.* Scientific reports, 2018. 8(1): p. 1–10.
13. Ganguly, S., P. Das, and N.C. Das, *Characterization tools and techniques of hydrogels,* in *hydrogels based on natural polymers.* 2020, Elsevier. p. 481–517.
14. Ganguly, S., et al., *Design of psyllium-g-poly (acrylic acid-co-sodium acrylate)/cloisite 10A semi-IPN nanocomposite hydrogel and its mechanical, rheological and controlled drug release behaviour.* International Journal of Biological Macromolecules, 2018. 111: p. 983–998.
15. Mondal, D., et al., *Effect of poly (vinyl pyrrolidone) on the morphology and physical properties of poly (vinyl alcohol)/sodium montmorillonite nanocomposite films.* Progress in Natural Science: Materials International, 2013. 23(6): p. 579–587.
16. Ganguly, S., I. Grinberg, and S. Margel, *Layer by layer controlled synthesis at room temperature of tri-modal (MRI, fluorescence and CT) core/shell superparamagnetic IO/human serum albumin nanoparticles for diagnostic applications.* Polymers for Advanced Technologies, 2021. 32(10): p. 3909–3921.
17. Saha, N.R., et al., *Studies on methylcellulose/pectin/montmorillonite nanocomposite films and their application possibilities.* Carbohydrate Polymers, 2016. 136: p. 1218–1227.

18. Mondal, D., et al., *Effect of clay concentration on morphology and properties of hydroxypropylmethylcellulose films.* Carbohydrate Polymers, 2013. 96(1): p. 57–63.
19. Ganguly, S., et al., *Starch functionalized biodegradable semi-IPN as a pH-tunable controlled release platform for memantine.* International Journal of Biological Macromolecules, 2017. 95: p. 185–198.
20. Weissleder, R., *A clearer vision for in vivo imaging.* Nature Biotechnology, 2001. 19(4): p. 316–317.
21. Politis, M. and P. Piccini, *In vivo imaging of the integration and function of nigral grafts in clinical trials.* Progress in Brain Research, 2012. 200: p. 199–220.
22. Arbab, A.S., et al., *In vivo cellular imaging for translational medical research.* Current Medical Imaging, 2009. 5 (1): p. 19–38.
23. Dubertret, B., et al., *In vivo imaging of quantum dots encapsulated in phospholipid micelles.* Science, 2002. 298(5599): p. 1759–1762.
24. Weng, K.C., et al., *Targeted tumor cell internalization and imaging of multifunctional quantum dot-conjugated immunoliposomes in vitro and in vivo.* Nano Letters, 2008. 8(9): p. 2851–2857.
25. Davidson, M., J. Lindsey, and J. Davis, *Requirements and selection of an animal model.* Israel Journal of Medical Sciences, 1987. 23(6): p. 551–555.
26. Zon, L.I. and R.T. Peterson, *In vivo drug discovery in the zebrafish.* Nature Reviews Drug Discovery, 2005. 4(1): p. 35–44.
27. Sun, X., et al., *Fluorescent Ag–In–S/ZnS quantum dots for tumor drainage lymph node imaging in vivo.* ACS Applied Nano Materials, 2021. 4(2): p. 1029–1037.
28. Garg, M., et al., *Amine-functionalized graphene quantum dots for fluorescence-based immunosensing of ferritin.* ACS Applied Nano Materials, 2021. 4(7): p. 7416–7425.
29. Sahoo, S.L., et al., *Biocompatible quantum dot-antibody conjugate for cell imaging, targeting and fluorometric immunoassay: crosslinking, characterization and applications.* RSC Advances, 2019. 9(56): p. 32791–32803.
30. Zhang, B., et al., *Effective reduction of nonspecific binding by surface engineering of quantum dots with bovine serum albumin for cell-targeted imaging.* Langmuir, 2012. 28(48): p. 16605–16613.
31. Wang, Y. and X.-P. Yan, *Fabrication of vascular endothelial growth factor antibody bioconjugated ultrasmall near-infrared fluorescent Ag 2 S quantum dots for targeted cancer imaging in vivo.* Chemical Communications, 2013. 49(32): p. 3324–3326.
32. Sultana, S., M.E. Ali, and M.N.U. Ahamad, *Gelatine, collagen, and single cell proteins as a natural and newly emerging food ingredients*, in *Preparation and processing of religious and cultural foods.* 2018, Elsevier. p. 215–239.
33. Byrne, S.J., et al., *"Jelly dots": synthesis and cytotoxicity studies of CdTe quantum dot–gelatin nanocomposites.* Small, 2007. 3(7): p. 1152–1156.
34. Parani, S., K. Pandian, and O.S. Oluwafemi, *Gelatin stabilization of quantum dots for improved stability and biocompatibility.* International Journal of Biological Macromolecules, 2018. 107: p. 635–641.
35. Ramanery, F.P., et al., *Biocompatible fluorescent core-shell nanoconjugates based on chitosan/Bi2S3 quantum dots.* Nanoscale Research Letters, 2016. 11(1): p. 1–12.
36. Mathew, S.A., et al., *Luminescent chitosan/carbon dots as an effective nano-drug carrier for neurodegenerative diseases.* RSC Advances, 2020. 10(41): p. 24386–24396.
37. Hong, G., et al., *In vivo fluorescence imaging with Ag2S quantum dots in the second near-infrared region.* Angewandte Chemie International Edition, 2012. 51(39): p. 9818–9821.

38. Suo, B., et al., *Poly (vinyl alcohol) thin film filled with CdSe–ZnS quantum dots: Fabrication, characterization and optical properties.* Materials Chemistry and Physics, 2010. 119(1-2): p. 237–242.
39. Wang, M., et al., *Antibacterial fluorescent nano-sized lanthanum-doped carbon quantum dot embedded polyvinyl alcohol for accelerated wound healing.* Journal of Colloid and Interface Science, 2022. 608: p. 973–983.
40. Ayaz, F., M.O. Alas, and R. Genc, *Differential immunomodulatory effect of carbon dots influenced by the type of surface passivation agent.* Inflammation, 2020. 43(2): p. 777–783.
41. Yan, C., et al., *Photoluminescent functionalized carbon quantum dots loaded electroactive Silk fibroin/PLA nanofibrous bioactive scaffolds for cardiac tissue engineering.* Journal of Photochemistry and Photobiology B: Biology, 2020. 202: p. 111680.
42. Dong, H., et al., *Multifunctional poly (l-lactide)–polyethylene glycol-grafted graphene quantum dots for intracellular microRNA imaging and combined specific-gene-targeting agents delivery for improved therapeutics.* ACS Applied Materials & Interfaces, 2015. 7(20): p. 11015–11023.
43. Omenetto, F.G. and D.L. Kaplan, *New opportunities for an ancient material.* Science, 2010. 329(5991): p. 528–531.
44. Tsukada, M., et al., *Physical and chemical properties of tussah silk fibroin films.* Journal of Polymer Science Part B: Polymer Physics, 1994. 32(8): p. 1407–1412.
45. Lin, N., et al., *Engineering of fluorescent emission of silk fibroin composite materials by material assembly.* Small, 2015. 11(9-10): p. 1205–1214.
46. Cheng, L., et al., *Quantum dots-reinforced luminescent silkworm silk with superior mechanical properties and highly stable fluorescence.* Journal of Materials Science, 2019. 54(13): p. 9945–9957.

13 Challenges and Future Prospects of Magnetic Quantum Dots

Sayan Ganguly
Department of Chemistry and Bar-Ilan Institute for Nanotechnology and Advanced Materials, Bar-Ilan University, Ramat Gan, Israel

CONTENTS

13.1 INTRODUCTION

In recent years, magnetic nature carbon dots (MNCDs) have been investigated as a novel class of fluorescent green nanomaterials. These magnetic carbon dots have demonstrated exceptional applicability in sensing and biological applications. Research in the fields of molecular biology and medicine is benefiting from advances in the conception and production of biomaterials of the next generation, which are used in the creation of medical instruments [1]. There is a possibility that pure carbon dots (CDs) and pure magnetic nanoparticles will not offer anything useful for multimodal imaging applications [2]. Magneto fluorescent carbon dots have recently begun to play major roles in contemporary biomaterials, particularly as drug delivery, bioimaging, and magnetic resonance imaging agents in the domains of molecular biology and biomedicine [3, 4]. The magnetic nature of magneto fluorescent carbon dots is due to the presence of magnetic elements or nanocomposites [5], which is primarily responsible for the carbon dots' diverse surface chemistry. These features are largely responsible for the interesting characteristics of magneto fluorescent carbon dots. Because of their magnetic properties, magneto carbon dots have also been used

DOI: 10.1201/9781003319870-13

as probes in a variety of applications. These applications include sensing, bioimaging, cell separation, contrast agents for magnetic resonance imaging (MRI), cancer therapy, and drug administration. As a result, this study offers a considerable amount of information on the progression of magneto fluorescent carbon dots and their applications in analytical chemistry (including separation, sensing, and bioimaging), as well as in the disciplines of biomedicine. Carbon dots have been evaluated in the literature for their use in photocatalysis, optoelectronics, sensing, and bioimaging. Several reviews have thus far provided helpful explanations of carbon dot characteristics, characterization methods, and methods of synthesis. For instance, Melvin Ng and colleagues conducted a literature assessment of CD production techniques, including hydrothermal and microwave-assisted processes, and their uses for detecting heavy metal ions [6-8]. Ansari et al. outlined the synthetic techniques for the doping of heteroatoms onto CDs and their biological applications (photothermal and photodynamic treatments, biosensing, in-vitro and in-vivo research) (photothermal and photodynamic therapies, biosensing, in-vitro and in-vivo studies) [7, 9]. The uses of CDs in imaging investigation of cancer cells were summarized by Chen's team [10]. Miao et al. provided a comprehensive review of the synthesis methods, characteristics, and uses of CDs doped with heteroatomic elements [6, 11]. Detailed research on their synthesis pathways, characteristics, surface functionalization, and analytical and biological applications has been highlighted in these studies. It stands out that there is no summary of research on the analytical and medicinal uses of magnetic nanocarbon dots (MNCDs).

13.2 VERSATILITY AND APPLICATIONS

Recent developments in nanoscience and technology have opened up a plethora of new avenues for the incorporation of a variety of nanoparticles into a single hybrid nanoarchitecture. In recent years, attention has been drawn to functional materials based on QDs and MNPs due to the superiority of these materials and their promise for the largest application value [12, 13]. The structures of materials are mostly responsible for determining their characteristics, whereas the properties of materials are primarily responsible for determining their applications. Because of their unique features, MNCDs are useful for a wide variety of applications, including sensing, bioimaging, separation, and biological research. Surface functionalization of MNCDs with various ligands (such as organic molecules/inorganic composites, metal ions, polymers, and biomolecules (nucleic acids, proteins, peptides, and amino acids) is an efficient pathway for tuning the properties of MNCDs [14]. This can be done by adding ligands to the surface of the MNCDs. As a consequence of this, MNCDs have been successfully combined with a variety of sample preparation equipment and separation procedures in order to successfully separate trace level analytes from complicated samples.

Researchers have created QD-based probes conjugated with cancer-specific ligands, antibodies, or peptides for the purpose of cancer imaging and diagnosis in vitro ever since biocompatible QDs were launched for imaging of cancer cells in vitro in 1998 [15].

The advances that have been made in cancer research have come about as a result of several highly appreciated synergistic therapeutic controls. These therapeutic controls combined chemotherapy with many additional diagnostics. In this light, experts working in the field of materials engineering came to the conclusion that accurate tracking of nanoparticles through the use of a variety of diagnostic instruments will play an important role in the future generation of medical miracles. The properties of magnetism and fluorescence have been combined in order to generate a unique category of fluorescent guided magnetic nanoparticles (MNPs). The in-vivo medical therapy of iron oxides nanoparticles (IONPs) and surface carbon coated magnetic nanoparticles has attracted a lot of attention recently. These materials are able to pass through the cell membrane and respond across a broad frequency range when exposed to an alternating magnetic field. These nanoparticles allowed for non-invasive tracking, imaging, and remote control of the particles' movement. The distribution of their domain sizes has an effect on their magnetism. In a typical scenario, the magnetic nanoparticles will transform into superparamagnetic particles if their size decreases to a level below 10 nanometers, in particular. Nanoparticles that exhibit superparamagnetic behavior are known as superparamagnetic nanoparticles. These nanoparticles exhibit quick and instantaneous magnetism when exposed to an external magnetic field and fast demagnetization when the field is withdrawn. When compared to larger sized nanoparticles, the biodegradation behavior of small particles is superior. Smaller particles also have a lower environmental impact. When quantum dots, also known as QDs, and magnetic nanoparticles, also known as MNPs, come together to create a product, the final result is a miracle of pure mixing of great optical quality (from QDs) and magnetism (from MNPs). The combined features of QDs and MNP have potential applications in cancer research, particularly in the identification of tumor cells, dynamic real-time tracking, fluorescence imaging, drug administration, magnetic resonance (MR) imaging, and other areas.

13.3 MAGNETIC RESONANCE-BASED BIOIMAGING

QDs are a kind of semiconductor nanoparticle that demonstrates quantum-size effects and is smaller than the spatial extension of electrons and holes in bulk semiconductors. Some significant applications have been depicted in Figure 13.1. The 1–10 nm size regimes are where you would commonly observe these effects. With decreasing particle size comes a rise in the (kinetic) energy of the electrons and holes, which in turn causes an increase in the energy difference between the conduction band and the valence band. This is because the box that the electrons and holes are confined to is very tiny (i.e., an increase in the bandgap). This causes the absorption and emission spectra of QDs to be highly dependent on the particle size, with both moving to higher energies as the particle size decreases. The emission is constrained because it comes from the lowest exciton state, which is a pair of electrons and holes. Because of the energy breadth of the conduction and valence bands, the absorption spectrum is broad, and as a result, the QD fluorescence may be triggered over a wide spectral range.

Because of their tiny size, semiconductor quantum emitters can achieve luminescence quantum efficiencies (QE) that are far greater than those of bulk semiconductors. The quantum efficiency (QE) of a material may be calculated by dividing the number of photons it emits by the number of photons it absorbs. When compared to bulk semiconductors, the chances of an electron or hole being trapped by a defect in a

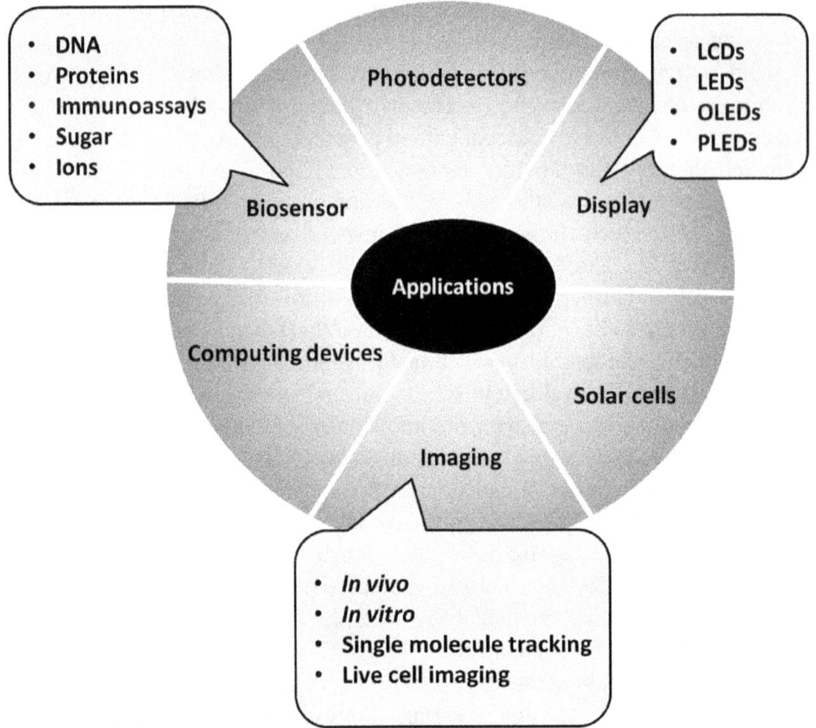

FIGURE 13.1 Different types of QD applications.

quantum dot are much lower. This is due to the fact that electrons and holes are confined within the QD, and it is possible to synthesize QDs that are defect-free. This reduces the likelihood of recombination occurring without the emission of light. When the surface of the nanocrystal is properly passivated—that is, when there are no surface quenching states—the exciton that is produced as a result of absorption is forced to recombine radiatively (by emitting a photon), which results in a high quantum efficiency. The formation of a shell of a material with a larger bandgap around the QD core is an important stage in the process of improving the quality of the QE and the stability of the QDs. Because of the larger bandgap shell's confinement of electrons and holes within the core, surface interactions with elements like oxygen are diminished, leading to an increase in the material's resistance to photo-oxidation. In an ideal scenario, the material of the shell would have a lattice matching that of the core. This would eliminate interfacial imperfections, which can bring QE down. The enhanced stability of these so-called core/shell quantum dots not only results in a longer shelf life (measured in years when kept at room temperature), but it also results in better bleaching behavior. When compared to photons emitted by fluorescent dye molecules, the number of photons emitted by quantum dots before they are bleached is significantly larger. There is a lot of potential for QDs to be used as labels for fluorescence imaging due to their one-of-a-kind features, as were mentioned before.

13.4 TODAY'S MARKET OF QDs AND ASSOCIATED COMPOUNDS

Quantum dots, often known as QDs, are a unique subset of nanomaterials that cannot be compared to any other. Nanocrystalline (nc) semiconductors (SC), metals, and magnetic materials are all capable of exhibiting exceptional quantum confinement phenomena at dimensions generally below 10 nanometers (nm). In a nutshell, when these dimensions are reached, their actual size begins to intrude onto the basic quantum confinement dimensions of circling electrons. These dimensions are specifically dictated by the atomic nucleus in which they are found. Within the range of these critical dimensions, quantum dots (QDs) display behavior that is qualitatively unique from that of their bulk form. This may be seen, for instance, in the optical, electrical, and magnetic characteristics that are qualitatively distinct.

Scientists now have the ability to accurately produce nanocrystalline materials at these key dimensions, which enables them to systematically alter the quantum confining characteristic of these materials. As a consequence of this, there is now a significant amount of desire to harness and make use of the one-of-a-kind features that QD materials show. Colloidal QD-bioconjugates are among the first wave of commercial product applications that are increasing market interest. These bioconjugates serve as a precursor to upcoming advancements in the business world. Primarily, they have swiftly created a specialized market in the communities of life sciences and biomedicine, where they give capabilities that are unrivalled in cellular imaging and therapeutic detection. A new generation of flash memory devices; nanomaterial enhancements for improving the performance of flexible organic light-emitting diodes (LEDs), as well as solid-state, white-LED lighting; and a core technology used in flexible solar panel coatings are some of the other promising prototype developments of SC QDs that are now on the horizon of commercialization. This discussion gives an up-to-date analysis of quantum dot materials, analyzing where they are today as well as where they have the potential to go in the not-too-distant future.

The sales of quantum dots by entities (organizations, solo traders, and partnerships) make up the quantum dot market. Quantum dots are nanoscale crystals of a semiconducting substance or small semiconductor particles that exhibit distinctive optical and electrical characteristics. When energy is supplied to these particles, they produce light of various colors. The colors of the light that are emitted may be altered by modifying the size, shape, and material composition of the quantum dots. Displays of many kinds, including those seen on smartphones and televisions, frequently make use of quantum dots. At a compound annual growth rate (CAGR) of 26.35%, it is anticipated that the total value of the worldwide quantum dot market would increase from $4.70 billion in 2021 to $6.03 billion in 2022. The reversal in the growth trajectory of the quantum dot market is mostly due to the fact that enterprises have stabilized their output after catering to demand that increased dramatically because of the COVID-19 pandemic in 2020. At a compound annual growth rate (CAGR) of 25.02% between now and 2026, the market for quantum dots is projected to reach $14.68 billion. The expansion of the market for quantum dots is anticipated to be driven in large part by the rising demand for smart televisions and mobile phones. A smart television is a type of internet-based television set that allows users to browse the web directly from the television's browser. The most common applications for nanoscale semiconductor particles or nanoscale crystals of a semiconducting material are found in mobile phones and televisions because of their exceptional optical and electrical qualities. According to USA Today, during the first half of 2020, there was a 50% increase in the number of sales of smart televisions with a screen size of 65 inches or greater (in units). As a result, the expansion of the market for quantum dots is being driven by the rising demand for intelligent televisions and mobile phones. Increasing investments in research and development is a crucial trend that is gaining popularity in the market for quantum dots. In today's world, a great number of businesses are investigating emerging technologies and doing research in the quantum dot sector. For example, in October 2020, Samsung Electronics, which is situated in South Korea and is a maker of electrical gadgets, produced blue Quantum Dot light-emitting diodes through its research and development centre, the Samsung Advanced Institute of Technology (SAIT) (QLEDs). These quantum dots have a maximum brightness of 88,900 nits and a QLED lifespan of 16,000 hours; their luminous efficiency has been enhanced by 20.2%.

In the global market for quantum dots in 2021, the region with the biggest share was North America. It is anticipated that Asia Pacific would have the greatest growth rate in the market for quantum dots. The study on the market for quantum dots examines the following geographical areas: Asia-Pacific, Western Europe, Eastern Europe, North America, South America, the Middle East, and Africa. The study on the market for quantum dots covers the following countries: Australia, Brazil, China, France, Germany, India, Indonesia, Japan, Russia, South Korea, the United Kingdom, and the United States of America.

13.5 BIOASSAY COMMERCIALIZATION: ISSUES AND SOLUTIONS

In contrast to the widespread adoption of glucose meters, relatively few other uses of biosensors have made it to market. The commercial viability of existing biosensors has been slowed down because of a lack of versatility and a low price, two of the most important requirements for cost-effective tests. This is compounded by a number of technological challenges that have not yet been resolved.

Because the vast majority of published works do not include an analysis of the cost-effectiveness of the proposed strategies, it is difficult to determine whether or not they are applicable for use in the real world, which in turn limits their potential for immediate or near-future application in contexts such as clinical or diagnostic settings. Bioassays that can be used commercially need to have biorecognition elements that are stable over time and have a long lifetime, but they also need to be able to deliver a result in a matter of minutes. However, only a small number of the studies that were given in this study conducted an analysis of the biorecognition element's potential to maintain its integrity and be reused. Even though they have the capability of detecting very low concentrations, assays based on nucleic acids still require time-consuming and inefficient purification procedures at the beginning of the process. On the other hand, immunodevices require a relatively lower number of steps for sample preparation, and as a result, they can be carried out in a shorter amount of time [16]. Due to the fact that research has traditionally been done using only laboratory materials, matrix interference is one of the primary problems that must be overcome in order to build tools for bioanalysis [17]. Across the entirety of this analysis, a primary focus has been placed on works that documented the usage of samples taken from the actual world in the testing of the assays, with some of those tests yielding encouraging results. Other issues that frequently arise include sensor fouling brought on by the adsorption of endogenous components in the test sample, signal drift, and the possibility of microbial contamination.

In conclusion, in order to successfully transfer technologies from research laboratories to commercial settings, successful approaches need to enable automation and integration of biosensing platforms with sampling, fluid handling, separation, and other detection principles for continuous and remote detection of multiple complex analytes. Combining chemical and biological components onto a single platform is now possible thanks to the convergence of microfluidics and biosensor/bioassay technologies, which also offers the benefits of portability, disposability, real-time detection, unprecedented accuracy, and simultaneous analysis of multiple analytes. In this regard, some published publications have taken system integration one step further by utilizing the twofold approach with magnetic beads and QDs in microfluidic devices. These devices have been described as "microfluidic devices." In 2010, Tennico et al. came up with a proposal for one of the first automated integrated systems that was based on quantum dots and magnetic beads [18]. In order to undertake on-chip washing processes, the scientists built a disposable microfluidic system that was simple, robust, and quick to use. This device was based on the utilization of two distinct DNA-thrombin aptamers. Additionally, the results of this research demonstrated that the assay was effective in removing an excess of unbound matrix and that there was no interference with the components of the serum. The assay worked with a limit of detection (LOD) of 10 ng/mL and had acceptable reproducibility and linear range (similar to those of ELISA in well format). Additional benefits were the reduced need for reagents and the capability of multiplexing with QDs.

13.6 QDs' MARKET IN MICROFLUIDICS

In 2013, Yu and his colleagues came up with an additional strategy that was based on microfluidic chips and featured a variety of magnetic bead designs that were distributed throughout various branch channels. The immunoassay was verified for the dual detection of the cancer biomarkers carcinoma embryonic antibody and AFP in serum in forty minutes, with a strong anti-interference capacity and findings that were equivalent to hospital diagnostic procedures. This was accomplished with the immunoassay [19]. The elimination of non-specific adsorption by continuous washing made it possible to reuse the chip at least 16 times before it became unusable. Both antigens had a limit of detection and linear range that was in the ng/mL range.

Computerized several QD barcode assays based on magnetism and microfluidics have also been suggested for the simultaneous detection of various harmful organisms, including HIV, hepatitis B, and syphilis. These assays have been developed to detect several pathogens at the same time. Even though detection in the nM range was possible in only 20 minutes, the authors reported a significant variation in signal intensities between different barcodes. This variation was attributed to differences in the hybridization efficiency of various DNA sequences, which was determined by the size of the sequences as well as their secondary structure [20]. And lastly, QDs adsorbed onto the surface of screen-printed electrodes have been combined into microfluidic chips for sensitive electrochemical detection, with an immunoassay employing magnetic beads as a pre-concentration platform.

Lastly, the detecting module is an essential part of any integrated system. Numerous optical detection methods need costly and cumbersome tools like microscopes and laser sources, which are sometimes impractical in the real-world circumstances where the biodevices are meant to be used [17]. In recent years, researchers have been concentrating their efforts on developing portable detecting systems that make use of optical, electrical, and magnetic sensing technologies. Along the course of this study, several studies that made use of autonomous, small, economical, and portable detection systems for detection employing the magnetic bead-QD method have been discussed [21]. In this context, our team has developed a number of robust portable optical, amperometric, and/or conductometric detection instruments that have up to six independent cells for performing simultaneous tests on a number of different types of biomarkers in sample volumes as small as 100 L. This was done in order to satisfy the current market demands for compact devices that end-users can use independently for onsite testing, as the market is in need of compact devices that can be used independently by end users [22].

13.7 ONE STAGE FURTHER: IN THE DIRECTION OF CREATIVE METHODS

In reference to the one-of-a-kind qualities of inorganic nanoparticles, such as their nanometer dimensions, tunable imaging properties, and multifunctionality, a number of recent studies have concentrated on the design of biocompatible, inorganic nanoparticle-based multimodal probes. These studies have been conducted over the course of the past few years. Nanoparticles that are primarily based on quantum dots, iron oxide, gold, and silica have been combined with other materials such as biomolecules, polymers, and radiometals to increase their functionality for molecular targeting, therapeutic delivery, and multimodal imaging. QDs are a type of nanoparticle that is characterized by its ability to form quantum dots [23]. Magnetic-fluorescent composite nanoparticles combine the fluorescent, magnetic, and physical features of the nanosize, have the ability to be steered by magnetic forces, and may offer extensive information on biodistribution when observed using a fluorescence microscope, for instance [24]. The bifunctional nanoparticles that were produced as a result have a broad variety of applications, such as the detection and bioseparation of proteins and DNA, as well as bioimaging, cell localization, and drug administration [25]. In addition, some research has demonstrated their biocompatibility in cell survival and uptake studies, which helps overcome the toxicity issues that are associated with QDs [26]. Now, researchers are working to find solutions to the difficulties inherent in the multistep synthesis and purification required to produce these multifunctional nanoparticles, as well as their instability and aggregation issues, with promising results [24]. The identification of cancer cells by DNA cyclic amplification has also been accomplished with the use of QDs that possess high electrochemiluminescent and magnetic characteristics [27].

13.8 POTENTIAL FUTURE VISTA

The images on future QD displays, whether they be TVs or monitors, will be of a high quality since they will use QD materials. Since 2015, QD technology has been used in TVs, and there have been 1.3 million shipments of these products throughout the world. It was anticipated that shipments of quantum dot televisions will rise to 18.7 million in 2018. We anticipate that by the year 2025, sixty percent of televisions and fifty one percent of monitors will use quantum dots [28]. In addition, recently developed prototypes of consumer goods, such as the iWatch and the circlet bracelet, are expected to enter the market over the next three years. There are additional items that are crucial to security, such as those that have the ability to prevent counterfeiting and have a signature that is just theirs. Because of this, they have the capacity to manipulate both the absorption and emission spectra of the QD that is injected into it. In addition, quantum dots can be utilized in applications related to defence and counter-espionage, such as the prevention of friendly fire incidents by the incorporation of quantum dots into dust that identifies and locates adversaries.

Due to the fact that cancer was until recently the leading cause of death throughout the world, early cancer detection and treatment are currently at the forefront of cancer QD research. These studies focus on therapy as well as diagnosis for a variety of cancers, including leukemia, ovarian cancer, breast cancer, prostate cancer, and pancreatic cancer.

It is anticipated that in the not-too-distant future, the combination of certain materials or processes, such as magnetic and electric material or signal application techniques, would result in potent biosensing quantum dots (QDs). Now, scientists will go on with the synthesis of QDs using an environmentally friendly method that places a premium on improved stability, biocompatibility, and distinctive optical features. Because LEDs have the ability to produce lightning, they may be utilized in the process of sending data from one computer to another. The bottleneck of data transmission technology is overcome by Li-Fi, which results in more advancements than those produced via the use of Wi-Fi technology. In contrast to Wi-Fi, which employs radio frequencies as the carrier for data transmission and networking, Li-Fi makes use of visible light, which spans a broad range of wavelengths.

13.9 LINGERING ISSUES, DIFFICULTIES, AND UNSETTLED DEBATES

The commonly observed and frequently high toxicity levels of these nanoparticles were and continue to be one of the most significant challenges faced by the researchers who are working toward the goal of developing QDs for nanomedical applications. Oxidative stress appears to be the common key factor involved in the cytotoxicity of these nanomaterials. This problem was previously thought to be primarily attributable to "classical" semiconductor nanocrystals, which were composed of naturally toxic metals such as Cd and Pb. However, a growing amount of data is bearing witness that this problem is, to a certain extent, also pertinent to many new emerging types of QDs.

It has been demonstrated that Si/SiO_2 QDs are capable of causing cytotoxicity in lung cells [29]. Stan et al. established in this work for the very first time that silicon QDs had a deleterious effect on the cellular redox balance and glutathione distribution in the human MRC-5 lung fibroblastic cell line. This was reflected in the increased amounts of reactive oxygen species and malondialdehyde, in addition to the decreased content of glutathione. Interestingly, they observed that the initial localization of glutathione in the cytoplasm has been gradually shifting to the nucleus over longer incubation intervals, suggesting a complex multi-step mechanism of this process. This same group has also shown that silicon-based QDs are able to generate inflammation in lung cells and cause an imbalance in extracellular matrix turnover [30]. This is accomplished through a differential regulation of matrix metalloproteinases and tissue inhibitor of metalloproteinase-1 protein expression. Silicon-based QDs have been shown to have these effects.

Researchers led by Qin and colleagues looked at the impact graphene QDs have on macrophages as well as the underlying chemical processes. According to the findings of their research, graphene QDs had a very slight impact on the macrophages' cell viability and cell membrane integrity, but they significantly increased the production of reactive oxygen species, apoptotic cell death, and raised expression levels of Bax, Bad, caspase 3, caspase 9, beclin 1, and LC3-I/II. This latter finding indicated autophagy as an additional nanotoxicity-related mechanism [31]. In a p38 mitogen-activated protein kinase MAPK-dependent manner, the expression of several cytokines, such as tumor necrosis factor-, interleukin-1, and IL-8, was also significantly enhanced by low concentrations of graphene QDs; however, high concentrations of these QDs elicited opposite effects on the production of the cytokines. At concentrations more than 25 g/mL, graphene QDs were shown to be harmful in an in-vivo zebrafish embryo model [32]. On a study conducted by Wang et al. in the fibroblastic cell line NIH-3T3 cells, cytotoxicity and genotoxicity of graphene QDs were investigated. The increased production of p53, Rad 51, and OGG1 proteins indicating the DNA damage caused by reactive oxygen species was documented in cells that were exposed to graphene QDs at concentrations more than 50 g/mL(-1), despite the fact that the cells did not demonstrate any substantial cytotoxicity [33]. Few experimental studies have shown the coagulation-modifying side effects of QDs, despite the critical significance of recognizing the causes of nanoparticle-associated side effects on the blood components that are likely to develop after potential systemic administration of nano-enabled imaging probes and drug nanocarriers [34].

When mice were injected with amine and carboxyl modified CdSe/ZnS QDs, they practically instantly developed pulmonary vascular thrombosis and died [34]. When it came to causing pulmonary vascular thrombosis, the carboxyl-functionalized QDs were significantly more effective than their amine counterparts. Because of this, it has been hypothesized that carboxyl QDs activate the coagulation cascade via the intrinsic mechanism. This is due to the fact that the thrombotic impact was not observed following the administration of heparin to the mice beforehand. These findings have a good deal of agreement with the publications that claim CdSe QDs can stimulate the aggregation of platelets in vitro [35]. It is certain that the dramatic impacts of various types of nanoparticles, including QDs, on the coagulation cascade will emerge as a topic that is at the forefront of the attention of research organizations all over the world [36].

As a result, it is abundantly clear that additional research needs to be carried out in order to achieve an in-depth comprehension of the particulars of the interactions that take place between biological macromolecules and QDs, which will facilitate the rational design of QDs as ideal instruments for use in nanomedical applications. In addition, a safe-by-design multi-tier strategy with systematic testing undertaken at each stage of QDs production and modification for biomedical use recently described by Movia et al. should be adhered to in order to give a realistic possibility for their future clinical translation [37]. These tests should be carried out at each stage of the creation and modification of QDs for biomedical use.

13.10 SUMMARY

Reviewing the findings of current studies on the use of QDs in nanomedicine reveals that these nanoparticles have tremendous promise for advancing our understanding of disease and developing novel approaches to diagnosis and treatment. Existing prototype diagnostic tools show promising potential for rapid translation into clinical practice. In spite of the emergence of new nanoparticles into the class of QDs with improved properties and reduced cytotoxicity, more in-depth research is needed to characterize the physicochemical and functional aspects of the QDs-based systems to guarantee their efficacy and safety in in-vivo application scenarios in humans. Investment-wise, the global QDs market is expected to grow from $316 million in 2013 to $5040 million in 2020, at a projected growth rate of nearly 30% per annum, as stated in the latest Allied Market Research Report on Quantum Dots (Global Analysis, Growth, Trends, Opportunities, Size, Share and Forecast through 2020). The QDs market in the field of biological imaging is the most developed in terms of revenue, and it is predicted to have steady expansion in the coming years. Since medical experts have been working on producing QDs-based medical devices for over two decades, this end-user category presently accounts for the largest proportion of the market. Considered as a whole, the scientific and economic progress made thus far in the study of quantum dots (QDs) bodes well for their application in nanomedicine.

It is possible that the short-term development of commercial biodevices based in QDs and magnetic beads will be constrained by the relative difficulty of sandwich strategies, their limited applicability in complex samples, their cost, and the difficulties in transferring knowledge from academic institutions to the private sector. The employment of fluorescent nanoparticles and magnetic particles together, however, has the potential to make downsizing methods, multiplex detection systems, and research based on nanomaterials easier to use in the creation of future bioanalysis instruments. It is hoped that the exploration of new methods based on multimodal nanoparticles and the refinement of existing tactics, including direct detection, would allow for simpler and sensitive biosensing for applications in the real world. These developments are now being researched.

REFERENCES

[1] Ashrafizadeh M, Mohammadinejad R, Kailasa SK, Ahmadi Z, Afshar EG, Pardakhty A. Carbon dots as versatile nanoarchitectures for the treatment of neurological disorders and their theranostic applications: a review. Advances in Colloid and Interface Science. 2020;278:102123.

[2] Yao H, Su L, Zeng M, Cao L, Zhao W, Chen C, et al. Construction of magnetic-carbon-quantum-dots-probe-labeled apoferritin nanocages for bioimaging and targeted therapy. International Journal of Nanomedicine. 2016;11:4423.

[3] Wang L, Wang Y, Hu Y, Wang G, Dong S, Hao J. Magnetic networks of carbon quantum dots and Ag particles. Journal of Colloid and Interface Science. 2019;539:203–13.

[4] Das P, Ganguly S, Saha A, Noked M, Margel S, Gedanken A. Carbon-dots-initiated photopolymerization: An in situ synthetic approach for MXene/poly (norepinephrine)/copper hybrid and its application for mitigating water pollution. ACS Applied Materials & Interfaces. 2021;13:31038–50.

[5] Perelshtein I, Perkas N, Rahimipour S, Gedanken A. Bifunctional carbon dots—magnetic and fluorescent hybrid nanoparticles for diagnostic applications. Nanomaterials. 2020;10:1384.

[6] Miao S, Liang K, Zhu J, Yang B, Zhao D, Kong B. Hetero-atom-doped carbon dots: Doping strategies, properties and applications. Nano Today. 2020;33:100879.

[7] Ansari L, Hallaj S, Hallaj T, Amjadi M. Doped-carbon dots: Recent advances in their biosensing, bioimaging and therapy applications. Colloids and Surfaces B: Biointerfaces. 2021;203:111743.

[8] Das P, Ganguly S, Mondal S, Ghorai UK, Maity PP, Choudhary S, et al. Dual doped biocompatible multicolor luminescent carbon dots for bio labeling, UV-active marker and fluorescent polymer composite. Luminescence. 2018;33:1136–45.

[9] Ganguly S, Das P, Bose M, Mondal S, Das AK, Das N. Strongly blue-luminescent N-doped carbogenic dots as a tracer metal sensing probe in aqueous medium and its potential activity towards in situ Ag-nanoparticle synthesis. Sensors and Actuators B: Chemical. 2017;252:735–46.

[10] Chen BB, Liu ML, Huang CZ. Recent advances of carbon dots in imaging-guided theranostics. TrAC Trends in Analytical Chemistry. 2021;134:116116.

[11] Das P, Ganguly S, Margel S, Gedanken A. Immobilization of heteroatom-doped carbon dots onto nonpolar plastics for antifogging, antioxidant, and food monitoring applications. Langmuir. 2021;37:3508–20.

[12] Ballou B, Lagerholm BC, Ernst LA, Bruchez MP, Waggoner AS. Noninvasive imaging of quantum dots in mice. Bioconjugate Chemistry. 2004;15:79–86.

[13] Ganguly S, Das P, Das S, Ghorai U, Bose M, Ghosh S, et al. Microwave assisted green synthesis of Zwitterionic photolumenescent N-doped carbon dots: An efficient 'on-off'chemosensor for tracer Cr (+ 6) considering the inner filter effect and nano drug-delivery vector. Colloids and Surfaces A: Physicochemical and Engineering Aspects. 2019;579:123604.

[14] Das P, Ganguly S, Margel S, Gedanken A. Tailor made magnetic nanolights: Fabrication to cancer theranostics applications. Nanoscale Advances. 2021.

[15] Gao X, Cui Y, Levenson RM, Chung LW, Nie S. In vivo cancer targeting and imaging with semiconductor quantum dots. Nature Biotechnology. 2004;22:969–76.

[16] Koedrith P, Thasiphu T, Weon J-I, Boonprasert R, Tuitemwong K, Tuitemwong P. Recent trends in rapid environmental monitoring of pathogens and toxicants: potential of nanoparticle-based biosensor and applications. The Scientific World Journal. 2015;2015.

[17] Sin ML, Mach KE, Wong PK, Liao JC. Advances and challenges in biosensor-based diagnosis of infectious diseases. Expert Review of Molecular Diagnostics. 2014;14:225–44.

[18] Tennico YH, Hutanu D, Koesdjojo MT, Bartel CM, Remcho VT. On-chip aptamer-based sandwich assay for thrombin detection employing magnetic beads and quantum dots. Analytical Chemistry. 2010;82:5591–7.

[19] Yu X, Xia H-S, Sun Z-D, Lin Y, Wang K, Yu J, et al. On-chip dual detection of cancer biomarkers directly in serum based on self-assembled magnetic bead patterns and quantum dots. Biosensors and Bioelectronics. 2013;41:129–36.

[20] Gao Y, Lam AW, Chan WC. Automating quantum dot barcode assays using microfluidics and magnetism for the development of a point-of-care device. ACS Applied Materials & Interfaces. 2013;5:2853–60.

[21] Bruno JG, Phillips T, Carrillo MP, Crowell R. Plastic-adherent DNA aptamer-magnetic bead and quantum dot sandwich assay for Campylobacter detection. Journal of Fluorescence. 2009;19:427–35.

[22] Buonasera K, Pezzotti G, Scognamiglio V, Tibuzzi A, Giardi MT. New platform of biosensors for prescreening of pesticide residues to support laboratory analyses. Journal of Agricultural and Food Chemistry. 2010;58:5982–90.

[23] Swierczewska M, Lee S, Chen X. Inorganic nanoparticles for multimodal molecular imaging. Molecular Imaging. 2011;10:7290.2011. 00001.

[24] Labiadh H, Chaabane TB, Sibille R, Balan L, Schneider R. A facile method for the preparation of bifunctional Mn: ZnS/ZnS/Fe3O4 magnetic and fluorescent nanocrystals. Beilstein Journal of Nanotechnology. 2015;6:1743–51.

[25] Heidari Majd M, Asgari D, Barar J, Valizadeh H, Kafil V, Coukos G, et al. Specific targeting of cancer cells by multifunctional mitoxantrone-conjugated magnetic nanoparticles. Journal of Drug Targeting. 2013;21:328–40.

[26] Lu Y, Zheng Y, You S, Wang F, Gao Z, Shen J, et al. Bifunctional magnetic-fluorescent nanoparticles: synthesis, characterization, and cell imaging. ACS Applied Materials & Interfaces. 2015;7:5226–32.

[27] Jie G, Zhao Y, Niu S. Amplified electrochemiluminescence detection of cancer cells using a new bifunctional quantum dot as signal probe. Biosensors and Bioelectronics. 2013;50:368–72.

[28] Mohamed WA, Abd El-Gawad H, Mekkey S, Galal H, Handal H, Mousa H, et al. Quantum dots synthetization and future prospect applications. Nanotechnology Reviews. 2021;10:1926–40.

[29] Stan MS, Memet I, Sima C, Popescu T, Teodorescu VS, Hermenean A, et al. Si/SiO2 quantum dots cause cytotoxicity in lung cells through redox homeostasis imbalance. Chemico-Biological Interactions. 2014;220:102–15.

[30] Stan MS, Sima C, Cinteza LO, Dinischiotu A. Silicon-based quantum dots induce inflammation in human lung cells and disrupt extracellular matrix homeostasis. The FEBS Journal. 2015;282:2914–29.

[31] Qin Y, Zhou Z-W, Pan S-T, He Z-X, Zhang X, Qiu J-X, et al. Graphene quantum dots induce apoptosis, autophagy, and inflammatory response via p38 mitogen-activated protein kinase and nuclear factor-κB mediated signaling pathways in activated THP-1 macrophages. Toxicology. 2015;327:62–76.

[32] Wang ZG, Rong Z, Jiang D, Jing ES, Qian X, Jing S, et al. Toxicity of graphene quantum dots in zebrafish embryo. Biomedical and Environmental Sciences. 2015;28:341–51.

[33] Wang D, Zhu L, Chen J-F, Dai L. Can graphene quantum dots cause DNA damage in cells? Nanoscale. 2015;7:9894–901.

[34] Geys J, Nemmar A, Verbeken E, Smolders E, Ratoi M, Hoylaerts MF, et al. Acute toxicity and prothrombotic effects of quantum dots: impact of surface charge. Environmental Health Perspectives. 2008;116:1607–13.
[35] Ramot Y, Steiner M, Morad V, Leibovitch S, Amouyal N, Cesta MF, et al. Pulmonary thrombosis in the mouse following intravenous administration of quantum dot-labeled mesenchymal cells. Nanotoxicology. 2010;4:98–105.
[36] Dunpall R, Nejo AA, Pullabhotla VSR, Opoku AR, Revaprasadu N, Shonhai A. An in vitro assessment of the interaction of cadmium selenide quantum dots with DNA, iron, and blood platelets. IUBMB Life. 2012;64:995–1002.
[37] Movia D, Gerard V, Maguire CM, Jain N, Bell AP, Nicolosi V, et al. A safe-by-design approach to the development of gold nanoboxes as carriers for internalization into cancer cells. Biomaterials. 2014;35:2543–57.

Index

For Product Safety Concerns and Information please contact our EU representative GPSR@taylorandfrancis.com
Taylor & Francis Verlag GmbH, Kaufingerstraße 24, 80331 München, Germany

www.ingramcontent.com/pod-product-compliance
Lightning Source LLC
Chambersburg PA
CBHW060338220326
41598CB00023B/2740

* 9 7 8 1 0 3 2 3 3 4 9 1 2 *